Solvent Recovery Handbook

Solvent Recovery Handbook

Ian Smallwood
B.A., C.Eng., F.I.Chem.E., M.Inst. Pet.
Consultant on Solvent Recovery

Edward Arnold
A division of Hodder & Stoughton
LONDON   MELBOURNE   AUCKLAND

© 1993 Ian McN. Smallwood

First published in Great Britain 1993

British Library Cataloguing in Publication Data

Smallwood, Ian McNaughton
 Solvent Recovery Handbook
 I. Title
 660.284248

 ISBN 0-340-57467-4

All right reserved. No part of this publication may be reproduced or transmitted in any form or by any means, electronically or mechanically, including photocopying, recording or any information storage or retrieval system, without either prior permission in writing from the publisher or a licence permitting restricted copying. In the United Kingdom such licences are issued by the Copyright Licensing Agency: 90 Tottenham Court Road, London W1P 9HE.

Whilst the advice and information in this book are believed to be true and accurate at the date of going to press, neither the author nor the publisher can accept any legal responsibility or liability for any errors or omissions that may be made.

Typeset in Times New Roman by Wearset, Boldon, Tyne and Wear. Printed in Great Britain for Edward Arnold, a division of Hodder and Stoughton Limited, Mill Road, Dunton Green, Sevenoaks, Kent TN13 2YA by St Edmundsbury Press Ltd, Bury St Edmunds, Suffolk.

Contents

1. **Introduction**   1

2. **Removal of Solvents from the Gas Phase**   7
   - Activated carbon   8
   - Absorption   14
   - Low-temperature condensation   17

3. **Separation of Solvents from Water**   23
   - Decanting   25
   - Solvent extraction   26
   - Membrane separation   31
   - Adsorption   33
   - Air stripping   35
   - Steam stripping   38
   - Economics of water clean-up   38

4. **Equipment for Separation by Fractional Distillation**   41
   - Heating systems for evaporation   42
   - Condensers   45
   - Fractionating columns   48
   - Storage   54

5. **Separation of Solvents from Residues**   59
   - Exotherms   59
   - Fouling of heating surfaces   60
   - Vapour pressure reduction   73
   - Odour   76

6. **Separation of Solvents**   79
   - Column testing   80
   - Relative volatility   81
   - Tray requirements   82
   - Batch vs. continuous distillation   83
   - Vacuum distillation   88
   - Steam distillation   93
   - Azeotropic distillation   94
   - Extractive distillation   97

7. Drying solvents — **105**
Fractionation — 107
Azeotropic distillation — 107
Extractive distillation — 116
Pressure distillation — 117
Adsorption — 117
Membrane separation — 120
Liquid–liquid extraction — 121
Hydration, reaction and chemisorption — 123
Salting-out — 125
Coalescing — 126
Fractional freezing — 126
Conclusion — 126

8. Options for Disposal — **129**
Liquid solvent thermal incinerators — 129
Liquid solvent to cement kilns — 131
Steam raising with waste solvents — 133
Thermal and catalytic vapour incineration — 133
Adsorption vs. incineration — 134
Biological disposal — 135

9. Good Operating Procedure — **137**
Staff — 138
Laboratory — 139
Installation design and layout — 140
Principal hazards — 141
Storage and handling of solvents — 148
Feedstock screening and acceptance — 158
Process operations — 159
Maintenance — 161
Personal protection — 165
First aid — 166
Fire emergency procedure — 167

10. Economic Aspects of Solvent Recovery — **171**
Use as a fuel — 171
Destruction — 172
Solvent recovery — 173

11. Future of Solvent Recovery — **183**
Ozone generation — 183
Ozone destruction — 185
Oil prices — 186
Environmental regulation — 186

12.	**Significance of Physical Properties of Solvents**	**189**
	Name	189
	Boiling point	189
	Freezing point	190
	Liquid expansion coefficient	190
	Specific gravity	190
	Flash point	190
	Flammable limits in air	191
	Antoine and Cox constants	192
	Odour threshold	192
	IDLH	193
	Threshold limit value (TLV)	193
	Saturated vapour concentration (SVC)	193
	Vapour density	194
	Miscibility with water	194
	Oxygen demand	194
	Autoignition temperature	194
	Hildebrand solubility parameter	194
	Dipole moment and dielectric constant	195
	Evaporation time	195
	Loss per transfer	195
	Latent heat	195
	Heat of fusion	196
	Specific heat	196
	Heat of combustion	197
	Relative volatility	197
	Azeotropes	197
	Vapour–liquid equilibrium (VLE) diagrams	197
Appendix 1.	**Physical Properties, Activity Coefficients and Recovery Notes for Selected solvents**	**201**
	Hydrocarbons:	
	n-Pentane	202
	n-Hexane	205
	n-Heptane	208
	Cyclohexane	211
	Benzene	214
	Toluene	217
	Ethylbenzene	220
	Xylenes	223
	Alcohols and glycol ethers:	
	Methanol	226
	Ethanol	230
	n-Propanol	234
	Isopropanol	240

viii   Contents

    n-Butanol 243
    Isobutanol 247
    sec-Butanol 250
    Cyclohexanol 256
    Ethylene glycol 259
    Methyl Cellosolve 262
    Ethyl Cellosolve 265
    Butyl Cellosolve 268

Chlorinated hydrocarbons:

    Methylene chloride 272
    Chloroform 276
    Carbon tetrachloride 279
    1,2-Dichloroethane 282
    1,1,1-Trichloroethane 285
    Trichloroethylene 289
    Perchloroethylene 292
    Monochlorobenzene 295

Ketones:

    Acetone 298
    Methyl ethyl ketone 302
    Methyl isobutyl ketone 308
    Cyclohexanone 311
    N-Methyl-2-pyrrolidone 313

Ethers:

    Diethyl ether 316
    Diisopropyl ether 320
    1,4-Dioxane 323
    Tetrahydrofuran 327

Esters:

    Methyl acetate 331
    Ethyl acetate 335
    n-Butyl acetate 340

Miscellaneous:

    Dimethylformamide 343
    Dimethyl sulphoxide 349
    Pyridine 352
    Acetonitrile 356
    Furfuraldehyde 360
    Water 365

Appendix 2.	Calculation of Vapour–Liquid Equilibrium	368
Appendix 3.	Conversion Factors for SI Units to Those Used Customarily in Solvent Recovery	372
Bibliography		375
Index		379

1 Introduction

From the production of life-saving drugs to typists' correction fluid, solvents play a crucial role in modern industrial society. However, they share one thing in common—all the world's production of solvents ends up by being destroyed or dispersed into the biosphere. There is a negligible accumulation of solvents in long-term artefacts so the annual production of the solvent industry corresponds closely to the discharge.

Solvents are the source of about 40% of the volatile organic compounds (VOC) entering the atmosphere. Their contribution to VOC is similar in magnitude to all the VOC arising from the fuelling and use of motor vehicles. Since the latter source is being substantially reduced by catalytic converters and other major modifications in the use and distribution of motor fuel, it is not surprising that increasing pressure will be brought to bear on solvent users to cut the harm done to our environment by their discharges.

There are several ways of diminishing the quantity of harmful organic solvents escaping or being disposed of deliberately into the air.

1 Redesigning products or processes to eliminate the use of organic solvents may be possible. For example, great changes have taken place and are continuing in the surface coating industry, which is currently the largest industrial consumer of solvents.

   The reformulation of products used domestically is the only realistic way of dealing with such solvent emissions, since the recapture of a myriad of small discharges is impractical. The annual domestic consumption of solvent per capita in industrial countries through the use of paints, adhesives, polishes, pesticides, dry cleaning and other household products is of the order of 4 kg. Although in some instances the use of a more effective solvent can reduce the solvent content of a product, the use of water suspensions or emulsions is the only complete answer to domestic discharges.

2 For locations where large amounts of solvents are used, recapture and recycling is a valid cure to the problem. Existing plants can have equipment retrofitted, although this is seldom as effective as designing solvent handling systems from scratch with, for example, pressure storage, interlinked vents and dedicated delivery vehicles necessary for very volatile solvents.

3 Where a solvent, or a mixture of solvents, provides a medium in a process scheme, the choice of solvent can have a very significant impact on the amount of recycling possible. Often, consideration of the recycling process is left too late in the process design to influence the choice of solvent from the recovery viewpoint.

2 Introduction

4 Solvents vary in the amount of harm they do to the environment. Some CFCs, for instance, display great stability and are decomposed so slowly that they reach the upper reaches of the atmosphere where they destroy the ozone layer. Other solvents are very reactive at low levels and produce ozone where it can harm people and plants directly.

Avoiding the more harmful of solvents in making formulations is clearly desirable and the Montreal Convention and the Los Angeles Rule 66, which dates from the very early 1970s, plus a lot of more recent legislation, help to enforce this approach.

5 The use of solvents of low volatility and high water solubility, which can be very easily removed from contaminated air by water scrubbing, is an attractive option. If they are too dilute to recover, they can be destroyed biologically without significant evaporation into the atmosphere.

6 When all else fails, the incineration of used solvents, if properly operated, is the one totally effective way of eliminating their escape into the environment. While it may be objected to on the grounds that this produces 'greenhouse gases,' the amount of solvents incinerated is small in comparison with the quantity of fossil fuels burnt for their heat content.

Up to about 1955, industrial solvents were considered as being wholly beneficial, apart from a few toxicity problems that were mostly due to their employment in work places with almost no ventilation. It must now be realised that their production and use should be entered into with great caution. It seems very probable that restrictions on the control of solvents, which apart from CFCs have been local or, at most, nationwide, will become international as our understanding of their drawbacks becomes more widespread.

The decision to change a solvent may be approached by considering what harm the solvent to be replaced is doing or what more desirable properties some other solvent may have while still attaining the desired result. The improved properties may be for the user's health (e.g. replacement of the suspected carcinogenic ethyl Cellosolve by propylene Cellosolve), for photochemical activity (e.g. ethylbenzene for xylenes) or for ease of recovery (e.g. dimethylacetamide for dimethylformamide). Solvent cost may also be important, but when consumption of new solvent is reduced by recycling, cost is a less important factor than heretofore.

A great increase in the number of solvents available in bulk took place in the three decades between 1920 and 1950. Although this will not recur, there is no reason to suppose that a few new solvents, much more fully tested for harmful properties than some of their predecessors, will not come on to the market. Many of the solvents we now use originated as by-products from, for instance, coal gas manufacture and the destructive distillation of wood in the 19th century. It is questionable whether they would be acceptable if they were offered as new products today.

Great changes have been made. The use of carbon disulphide, carbon tetrachloride and benzene, all once common solvents, has been greatly reduced and it is doubtful if anyone would consider including them in a new formulation. Methanol, while under consideration for use as a fuel on a large scale, is rarely found as a solvent. n-Hexane is treated with great care. Safer replacements have been found for all of them and the trend will certainly continue.

One major reason that is likely to lead to changes of solvent in the future is the need to make recovery easier. There are four reasons why solvents can need recovery because they are unusable in their present state:

1 Mixture with air
This usually occurs because the solvent has been used to dissolve a resin or polymer which will be laid down by evaporating the solvent. Recovery from air can pose problems because the solvent may react on a carbon bed adsorber or be hard to recover from the steam used to desorb it.

Replacement solvents for the duty will therefore have similar values of Hildebrand solubility coefficient and of evaporation rate. The former can be achieved by blending two or more solvents together, provided that when evaporation takes place the solute is adequately soluble in the last one to evaporate. To achieve this, an azeotrope may prove very useful. Particularly in the surface coating industry, where dipping or spraying may be involved, viscosity will also be an important factor in any solvent change.

2 Mixture with water
Whether it arises in the solvent-based process or in some part of the recapture of the solvent, it is very common to find that the solvent is contaminated with water. Removal of water is a simple matter in many cases but in others it is so difficult that restoration to a usable purity may prove to be uneconomic.

It should always be borne in mind that the water removed in the course of solvent recovery is likely to have to be discharged as an effluent and its quality is also important.

3 Mixture with a solute
A desired product is often removed by filtration from a reaction mixture. The function of the solvent in this case is to dissolve selectively the impurities (unreacted raw materials and the outcome of unwanted side reactions) in a low-viscosity liquid phase while having a very low solvent power for the product.

The choice of solvent is often small in such a case, but significant improvements in the solvent's chemical stability can sometimes be found by moving up or down a homologous series without sacrificing the selectivity of the solvent system.

A less sophisticated source of contamination by a solute occurs in plant cleaning, where solvent power for any contaminant is of primary importance but where water miscibility, so that cleaning and drying take place in a single operation, is also an important property. Low toxicity is also desirable if draining or blowing out the cleaned equipment is also involved. In this case there is seldom a unique solvent that will fulfil the requirements, and ease of recovery can be an important factor in the choice.

4 Mixtures with other solvents
A multi-stage process such as found typically in the fine chemical and pharmaceutical industries can involve the addition of reagents dissolved in solvents and solvents that are essential to the yields or even the very existence of the desired reaction. No general rule can be laid down for the choice of solvent, but consideration should be given to the problems of solvent recovery at a stage at which process modification is still possible (e.g. before FDA approval).

To achieve the aim of preventing loss of solvents to the biosphere, it is necessary to recapture them after use and then to recover or destroy them in an environmentally

4 Introduction

acceptable way. It is the objective of this book to consider the ways of processing solvents once they have been recaptured.

Processing has to be aimed at making a usable product at an economic price. The alternative to reuse is destruction so the processing will be 'subsidised' by the cost of destruction.

Probably the most desirable product of solvent recovery is one that can be used in place of purchased new solvent in the process where it was used in the first place. This does not necessarily mean that the recovered solvent meets the same specification as virgin material. The specification of new solvent has usually been drawn up by a committee formed of representatives of both users and producers, who know what the potential impurities are in a product made by an established process route. The specification has to satisfy all potential users, who are, of course, usually customers. For any given user some specifications are immaterial—low water content for a firm making aqueous emulsions, water-white colour for a manufacturer of black and brown shoe polish, permanganate time for methanol to be used to clear methane hydrate blockages, etc.

Hence the solvent recoverer may well not have to restore the solvent to the same specifications as the virgin material. On the other hand, the used solvent for recovery has passed through a process that was not considered by those who drew up the virgin specification and knew what impurities might be present. A set of new specifications will be required to control the concentration of contaminants that will be harmful to the specific process to which the solvent will be returned.

It is the drawing up of these new specifications that the recoverer, whether he be in-house or not, has a vital role to play. Specifications should always be challenged. The cost, and even the practicability, of meeting a specification that is unnecessarily tight can be very large. All too often the specification asked for by the user is drawn up, in the absence of real knowledge of its importance to the process, by consulting the manufacturer's virgin specification. It will be seen that the cost of reaching high purities by fractional distillation rises very steeply in many cases as the degree of purity increases. This is because the activity coefficients of impurities in mixtures tend to increase very fast as their concentrations approach zero. Even when it appears from an initial inspection that the appropriate relative volatility is comfortably high for a separation, this is often no longer true if levels of impurity below, say, 0.5% are called for.

Not only does working to an unnecessarily high specification add a lot to fuel costs, but also the capacity of a given fractionating column may be reduced several-fold in striving to attain a higher purity than planned for when it was designed.

In making a case on specification matters, the solvent recoverer needs to be able to predict, possibly before samples are available for test, the cost of recovery of a solvent to any required standard, since it is only by so doing that the true economics of, say, reducing water content may be calculated for the whole circuit of production and recovery. This is now possible in most cases. The properties of most binary solvent mixtures are known or can be estimated with reasonable accuracy. More complex mixtures often resolve themselves into binaries in the crucial areas and, for many ternaries, the information is in the literature. It is therefore possible for the solvent recoverer to play a part in the decision-making process rather than be presented with a solvent mixture that is impossible to recover but cannot be altered.

It is a matter of fact that there are few solvents with solvent properties so unique that they cannot be replaced at an early stage in a product development process. It is also true that the properties which the recoverer relies upon for making separations are not those which the solvent user requires. Cooperation at this early stage is important if the cost of industry's efforts to reduce solvent pollution of the environment is to be minimized.

2 Removal of solvents from the gas phase

The technology for removing volatile liquids from gases has its origins in the operations leading to the production of gas from coal. Removal of naphthalene, which tended to block gas distribution pipes in cold weather, and carbon disulphide, which caused corrosion of equipment when burnt, were both desirable in providing customers with a reliable product. Inevitably, in removing these undesirable components of the raw gas, benzene and other aromatic compounds had to be taken out. Both scrubbing with creosote oil and gas oil and adsorption on activated carbon were used on a large scale for these purposes and helped to provide some of the earliest organic solvents.

It was therefore a natural step to employ these techniques when the use of solvents on a large scale made the recapture of solvents from process effluent air attractive economically. Our present concern with the quality of air is, of course, a much later development but carbon bed adsorption and air scrubbing are still two of the most frequently used methods of removing solvents from air. To them, we can now add the low-temperature condensation of solvents from air owing to the demand for liquid oxygen and therefore the availability of very large amounts of liquid nitrogen.

To put the requirements of solvent removal from air into perspective, it is useful to compare the purity levels that are required for a variety of purposes. For this comparison (Table 2.1), all the concentrations have been reduced to parts per million (ppm) on a weight basis.

Table 2.1 Vapour concentrations (*see* Figure 2.1)

	Acetone	Ethyl acetate	Toluene
Odour threshold	100	1	0.17
TLV–TWA	1000	400	100
IDLH	20 000	10 000	2000
Atmospheric discharge[a]	62	41	26
Air ex drier[b]	7000	1920	3000
LEL	26 000	22 000	12 700
Saturated vapour at 21 °C	250 000	100 000	31 000

[a] TA Luft limit.
[b] Typical value usually set to be safely below the LEL.

To give satisfactory air pollution as far as ozone is concerned, photochemical oxidants which include most solvents should not exceed about 0.044 ppm in the atmosphere. To

8 Removal of solvents from the gas phase

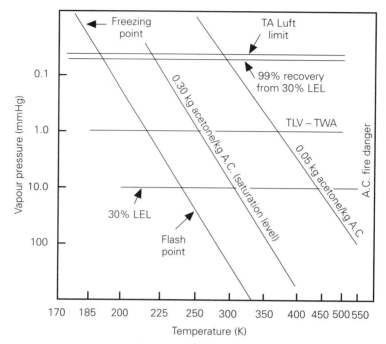

Fig. 2.1 Limits for acetone recovery by activated carbon and low temperature

achieve such levels, limits of total discharge rather than concentration limits may be necessary in areas of high industrial density. Such levels are listed in Table 2.7.

Ideal conditions for solvent recovery from a process could be achieved by recirculating a gas through the area where evaporation of the solvent takes place and then through a condenser which would knock out most of the solvent in the carrier gas and leave it lean enough to return to the evaporation zone. No discharge to atmosphere could take place and the gas could be inert and oxygen free so that there would be no explosion hazards. No moisture would be introduced into the system so that the solvent would remain dry, facilitating reuse without further processing.

Such an ideal model ignores the difficulty of getting the material being dried in and out of the evaporation zone without admitting new gas. It also ignores the very low temperatures required to condense volatile solvents at low vapour pressure. It does, however, provide a standard against which practical operations can be compared.

Since some new gas must enter any real system, it is necessary to have a way of purging from the system a similar quantity of gas which must not contain a higher solvent concentration than economics and regulations permit. The well tried way of cleaning up such a discharge is by using activated carbon adsorption.

Activated carbon

A typical activated carbon (A.C.) system (Figure 2.2) consists of two beds packed with A.C. and a valve arrangement to direct the flows. The stream of solvent-laden air is directed through the first bed until it is exhausted, or for a predetermined time, at which

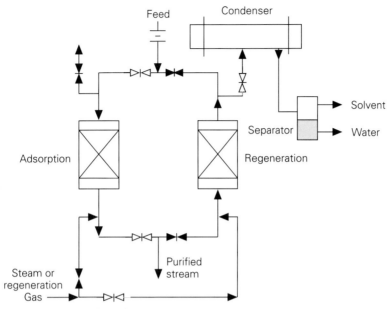

Fig. 2.2 Typical two-bed A.C. adsorption system

point it is switched to the second bed. The spent bed is then regenerated, usually with low-pressure steam, and the steam–solvent mixture is condensed. The regenerated bed is then cooled by blowing with atmospheric air before being put back on-stream.

It should be noted that regeneration of gas adsorption A.C. is very different from liquid-phase adsorption A.C. The granular material used in gas-phase operations has a very long life provided that it is protected from contamination through the use of air filtration.

The flammable solvent concentration in air arising from an evaporation process is usually limited to a maximum of 25–35% of its LEL to avoid explosion hazards. Chlorinated solvents can, of course, be safely handled at a higher limit. On the other hand, if the incoming air is primarily used to provide an acceptable working environment, the concentration for all solvents may well be below the TLV.

These concentrations are generally above what may be discharged straight to the atmosphere without treatment and are within the operating capability of A.C.

The limit to which solvents can be removed from air depends upon the design and operation of an A.C. plant. If necessary, 99% of the solvent entering the A.C. bed can be adsorbed. This would not be normal practice for economic solvent recapture, although it may be necessary to meet discharge regulations. Although twin-bed A.C. plants are normal for vapour recovery, as distinct from liquid adsorption operations, there are cases where space and economic considerations call for three-bed units where the second on-stream bed performs a polishing role. Such an arrangement can result in a 99.7% recovery efficiency. Typical operating results are given in Table 2.2.

10 *Removal of solvents from the gas phase*

Table 2.2 Typical operating results for A.C. plant operating to give 20 ppm effluent air

	Inlet concentration (ppm)	Cyclic adsorption (wt% of bed)	Steam (kg/kg)
Methylene dichloride	10000	17	1.4
Acetone	10000	21	1.4
Tetrahydrofuran	5000	9	2.3
n-Hexane	5000	8	3.5
Ethyl acetate	5000	13	2.1
Trichloroethylene	5000	20	1.8
n-Heptane	5000	6	4.3
Toluene	4000	9	3.5
Methyl isobutyl ketone	2000	9	3.5

1 The molecular weight of the solvent

All solvents with a molecular weight higher than that of air can be adsorbed and the higher the molecular weight the more readily the adsorption occurs. If two or more solvents are present, the one with the lower volatility will be adsorbed more readily and will tend to displace the lighter solvent as the bed becomes more saturated.

2 Temperature of the solvent-laden air

The equilibrium partial pressure of the solvent adsorbed on the A.C. is a function of the bed temperature and, particularly at the tail end of the bed in contact with the least rich air, the bed should be cool (*see* Figure 2.1). The temperature will be determined by the inlet temperature and the amount of solvent in the incoming air which will give out its heat of adsorption (*see* paragraphs 7 and 8 below).

3 Bed size

If a bed is fed very slowly with solvent-carrying air, it is possible for the A.C. to hold about 30% of its dry weight of solvents. In practice, although the 'front' of the bed which is in contact with air rich in solvent may reach that level (*see* Figure 2.1), the back of the bed, in contact with air fit to be discharged to the atmosphere, will have a much lower concentration in the A.C.

A.C. has a bulk density of 500–1000 kg/m^3 and for a low molecular weight solvent the average pick up will be about 5%. A typical operating cycle will occupy 3 h, with half the time spent on regeneration and the other half on adsorption. This calls for a bed size of about 3750 kg to handle each 1000 Te of solvent per year on an 8000 h/yr basis with a twin-bed unit.

A.C., being relatively light, is liable to fluidize if air is passed upwards.

4 Treatment of desorbate

Desorption and A.C. regeneration are usually carried out with low-pressure steam (5 psig). The desorbed solvent and steam are condensed in a conventional water- or air-cooled heat exchanger, after which separation by decanting may be possible if the solvent involved is not water miscible. In the case of alcohols, esters and ketones a wet solvent mixture will need to be treated downstream of the condenser or to be stored for subsequent recovery. The solvent content of the liquid from the condenser falls sharply as the steaming of the bed progresses and, if more than one solvent has been adsorbed in the earlier half of the cycle, the composition of the desorbate will vary. Owing to its changing nature, the stream does not lend itself to continuous refining without buffer storage to eliminate these fluctuations. Despite this, ethyl acetate, which is unstable in aqueous solution, will usually have to be processed continuously after condensation to minimize hydrolysis.

5 Inhibitors

Many solvents contain small concentrations of inhibitors and their fate in the evaporation, adsorption, desorption and water contacting that all form part of the recapture of solvent on A.C. adsorbers should be borne in mind. Reinhibiting immediately after water removal is required in many cases.

6 Hot gas regeneration

Hot gas can be used for regeneration although, because there is usually water adsorbed on the A.C. bed, this will not guarantee the condensate being low enough in water to be reusable without drying. It also leaves the problem of what to do with the hot gas after it has passed through the condenser and dropped most but not all of its solvent load. The degree of desorption using hot gas is not as complete as when using steam:

	Temperature (°C)	Percentage desorbed
Indirect heating	100	15
Reduced pressure	20	25
Hot gas recirculation	130	45
Direct steam	100	98

Incomplete desorption is a problem when a plant is running on campaigns handling a variety of solvents.

7 Water in bed

After steam regeneration, the bed is hot and wet and must be cooled by blowing with air. This will also remove much of the water present. Some of the water should remain on the bed, where it will in due course be displaced by the more strongly adsorbed solvent. This helps to keep the bed temperature low during the adsorption part of the cycle since the heat of desorption of the water is supplied by the solvent's heat of adsorption.

8 Bed heating and ketones

When solvents are adsorbed on A.C. they release heat (Table 2.3). Part of this is latent heat given up in the change from vapour to the liquid state. The remainder is the net heat

Table 2.3 Heat of adsorption on A.C.

	Total heat of adsorption (kcal/mol)	Latent heat (kcal/mol)
Toluene	12.2	9.2
Ethanol	12.5	9.4
n-Hexane	14.2	7.5
Acetone	11.0	7.1
n-Propanol	14.5	9.8
Methyl ethyl ketone	12.0	7.6
Methyl ethyl ketone with water	25.0	
Cyclohexanone	35.0	9.0

12 Removal of solvents from the gas phase

of adsorption on to the A.C., which should, in the absence of any other reaction, be of the order of 5 kcal/mol. During the adsorption part of the operating cycle, this heat tends to accumulate in the bed and to warm up the effluent air. Ketones tend to undergo reaction on the A.C. in the presence of water which releases a lot more heat as well as destroying the adsorbed solvent.

There is actually a danger that in the case of the higher ketones the A.C. can form hot spots and reach a temperature of about 370 °C, at which the A.C. will ignite.

9 Materials of construction

Mild steel is a satisfactory material for construction of A.C. beds handling hydrocarbons. However, stainless steel should be used for those parts of the A.C. unit in contact with ketones and esters because of their instability. Loss of inhibitor can make chlorinated hydrocarbons in contact with water somewhat unstable and some parts of the plant may require non-metallic linings.

Rekusorb process

A modification to the basic A.C. steam-regenerated operation is one which uses hot gas to regenerate in a way that meets the problems mentioned in paragraph 5 above. Known as the Rekusorb process, its adsorption step is conventional. Desorption, however, begins with a dry nitrogen purge until the level of oxygen in the desorption loop is too low to be an explosion hazard. The gas now in the desorption loop is heated and circulated (Figure 2.3).

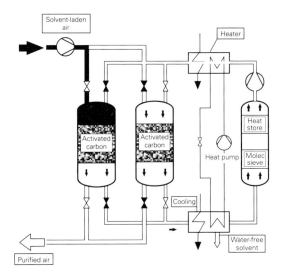

Fig. 2.3 Rekusorb process

In addition to the solvent adsorbed on the A.C., there is also moisture given up by the solvent-laden air. Owing to its volatility and low molecular weight, this is not strongly adsorbed and is desorbed preferentially. The desorption loop includes a molecular sieve drier with sieves able to take up water but not solvent (*see* Table 7.10).

Once the water is desorbed and held in the molecular sieves, the hot, dry, nitrogen-rich loop gas progressively desorbs the solvent. The rich gas passes to a cooler and condenser (or a washing tower using chilled solvent) where most of its solvent load is condensed. Heat removed in condensing is transferred to the gas heater by a heat pump and the hot gas is returned round the loop to the A.C. bed again. Once the bed is fully desorbed, the gas heater is stopped and the circulating gas starts to cool the bed. The heat picked up from the bed at this stage is used to regenerate the molecular sieves, the moisture from which is returned to the reactivated bed along with any residual solvent in the loop gas before it is discharged. Heat that is not needed to regenerate the molecular sieves is held in a heat store ready for the next regeneration cycle.

The good heat economy of this system makes it economical to regenerate the A.C. beds more frequently than with the conventional system and therefore keeps the recovery unit much more compact. However, its major advantage is that the solvent product is free from gross quantities of water and in most cases the solvent is fit for reuse without further processing.

Heat removal

Reference to Table 2.1 shows that in the evaporation zone, if the solvent-laden air is allowed to approach a saturation close to the vapour equilibrium, it would carry many times more solvent than allowed by the safety requirement, which calls for operation at 25% or so of the LEL. Even if, as is the case with non-flammable chlorinated solvents, the safety limit is not applicable, operation at such high concentrations would cause problems of bed overheating.

A.C. in a packed bed has a very low heat conductivity and the air flowing through the bed carries away much of the heat of adsorption. A tenfold reduction in the air flow would therefore be unacceptable. A solution to this problem without increasing the amount of air being discharged to the atmosphere is shown in Figure 2.4. The adsorption

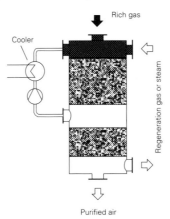

Fig. 2.4 Modified Rekusorb process

bed is split, with the part closer to the incoming air being cooled by a recycle stream. The lower bed is fed with air carrying only a small amount of solvent and so can be reduced to a very low solvent concentration in equipment of modest size.

14 Removal of solvents from the gas phase

Although A.C. adsorption is very effective for removing solvents from air, it has two major operational disadvantages. First, it is a batch process that is extremely difficult to turn into a continuous process. It is difficult to use the large amounts of low-grade heat which therefore are rejected to cooling water and the equipment is large for the quantitiy of solvent handled. Second, it is very inefficient in its use of heat, particularly when recovering high molecular weight solvents. Reference to Table 2.2 shows that the latent heat in steam used to recapture 1 kg of toluene is about 20 times the latent heat of the toluene, about 10 times the heat requirement for absorption, and subsequent desorption, in a scrubbing liquid. It is, however, very much more effective at recapturing low concentrations of volatile solvents.

Absorption

Absorption is a continuous operation and needs comparatively little plot area compared with a conventional A.C. system. It also has the advantages common to continuous plants in the way of control and the steady requirement of utilities. It lacks, however, the reserve of capacity inherent in an A.C. bed which, even when close to breakthrough, can absorb large amounts of solvent if a surge of solvent in air reaches it. This is likely to happen from time to time if a batch drier is upstream of the air cleaning equipment, which must be designed to cope with such a peak.

The problems of heat removal inherent in a fixed bed do not arise with absorption. If an air stream very rich in solvent has to be handled, inter-stage cooling can be fitted on intermediate trays in the absorber column. The restriction of the solvent concentration for safety reasons need not be applied, although flame traps may be fitted in the air ducting. If the pressure drop can be kept low enough, it is possible to position the ventilation fan downstream of the absorber where flammable vapour concentrations should never occur (Figure 2.5).

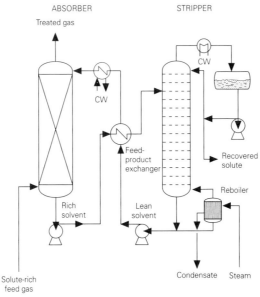

Fig. 2.5 Typical absorber

Absorption depends for its effect on the vapour pressure of the solvent to be recaptured over the absorbent liquor. In the absorption stage, it is desirable to have a high mole fraction in the liquor for a low partial pressure, i.e. a high value of x/p, where

$$\frac{x}{p} = (\gamma P)^{-1}$$

and
x = mole fraction of solvent in absorbent fluid;
p = partial vapour pressure of the solvent;
P = vapour pressure of the pure solvent at the operating temperature;
γ = activity coefficient of the solvent in the absorbent.

A high value of P corresponds to a highly volatile solvent and indicates that the absorption process is better suited to solvents with a relatively low volatility.

The value of γ is determined by the choice of absorbent and by the concentration of solvent in the absorbent. The latter is usually low and the values of γ^∞ are a good guide in comparing absorbents. As reference to Table 3.8 will show, the values of $\gamma^\infty P$ for water as the absorbent vary over a range of at least seven orders of magnitude. Values of $\gamma^\infty P$ below 500 are worthy of further consideration for water scrubbing recovery. Comparison of water with monoethylene glycol (MEG), however, (Table 2.4) shows that purely on

Table 2.4 Comparison of $\gamma^\infty P$ in water and MEG as scrubbing liquors. Lower values are better

Vapour	MEG	Water
Tetrahydrofuran	3.63	31.15
n-Butanol	6.60	52.3
Methanol	1.07	2.2

the grounds of the value of x/p there are possibly better choices for cases where water seems a favoured choice. For two solutes that have very high values of $\gamma^\infty P$ in Table 3.8 there can, as Table 2.5 shows, be a wide range of performance in other solvents.

Table 2.5 Comparison of $\gamma^\infty P$ for scrubbing benzene and n-hexane out of air

	n-Hexane	Benzene
N-Methylpyrrolidone	14.2	1.1
Dimethyl sulphoxide	64.5	3.33
Dimethylformamide	17.0	1.4
MEG	430.4	33.9
n-Hexadecane	0.9	1.1
Decahydronaphthalene	1.3	1.5
Water	489 000	1730

There is comparatively little published information on the activity coefficients of volatile solvents in liquids which have high enough boiling points to be considered as absorbents. Nevertheless, the experimental technique of using potential absorbents as the stationary phase in gas–liquid chromatographic columns and eluting the solvent through them is simple and quick.

16 *Removal of solvents from the gas phase*

In addition to a suitable value of γ, the scrubbing liquid should be chosen with attention to the following properties:

1 Vapour pressure

While the solvent-laden air will be discharged with a much reduced solvent content, it will be in equilibrium with almost pure scrubbing liquid. This presents no problem if the scrubbing liquid is water since make-up is easy and there is no restriction to discharging water vapour. If, however, an organic liquid has been selected, its vapour pressure must be low enough not to provide an unacceptable level of pollution itself. This effectively rules out most potential scrubbing liquids with boiling points below 190 °C for scrubbing operations being carried out at temperatures over 20 °C.

2 Stability

A highly volatile solvent cannot easily be condensed under vacuum so the recovery column in an absorption process usually works at atmospheric pressure. Although in a continuous column the hold-up should not be large and, hence, the residence time at the boiling point of the absorbing liquid not great, an ideal liquid should remain circulating in the recovery system for many weeks and therefore be very stable. Of the potential absorbents listed in Table 2.5, dimethylformamide, dimethyl sulphoxide and hexadecane would probably be rejected on stability grounds.

3 Azeotropes

The absorbent should ideally not form an azeotrope with the solvent being recaptured. This is because downstream processing of the recovered solvent will be hampered. Fortunately, the less volatile solvents for which water is often chosen as the scrubbing liquid (e.g. dimethylformamide) do not form aqueous azeotropes.

4 Viscosity

The scrubbing column should be operated at as low a temperature as possible. This is because values of P are approximately halved for every 17 °C fall in temperature. In trying to get the highest possible x/p value this is a modest effect compared with the range of γ, but nonetheless is not to be ignored.

Many of the potential scrubbing liquids become viscous at low temperatures and do not spread well on the column packings which are generally used for absorption. Plate columns can be used but they have a higher pressure drop for the same duty, involving more fan power to move the solvent-laden air through the system.

5 Regeneration

The best clean-up of the solvent-laden air that absorption can achieve is for the air to leave the absorption column in equilibrium with the regenerated absorption liquid. This means that the stripping column must remove the solvent to a very low level if some form of back-up (e.g. a small A.C. unit) does not have to be fitted to prepare the air for final discharge. The possibility of returning the air to the evaporation stage avoids this problem and is theoretically very attractive. The high value of x/p that aided absorption is a handicap to regeneration.

The absorption column handles large amounts of comparatively lean gas and needs to have a large diameter, short column and low pressure drop. In contrast, the stripper has a large liquid load and a comparatively small amount of vapour (the recaptured solvent), tending to lead to a tall column with a small diameter.

Since the stripping column acts through fractional distillation, there is no reason why, by using a modest amount of reflux to fractionate the high boiling absorbent liquid out of the recaptured solvent, it cannot produce a solvent ready for use in many cases.

With good heat exchange between the stripper bottoms and the solvent-rich stripper feed, the heat requirement for absorption is likely to be less than 0.5 kg of steam per kg of recovered solvent. This will depend on the latent heat of the solvent and the amount of reflux required on the stripper. Reference to Table 2.2 shows that conventional A.C. adsorption needs considerably more energy than this.

Low-temperature condensation

Adsorption and absorption systems bear little relationship to the 'ideal' model of a solvent vapour recapture process. They do, however, achieve the low concentrations in effluent air that are required. Comparatively recently a new process relying upon low temperatures has become available.

Table 2.6 shows the very low temperatures that must be reached to condense solvent from a gas stream sufficiently well to make it fit to work in or, assuming that TA Luft or some similar regulations become the international standard, to discharge to the atmosphere. In many cases the low temperature required is below a pure solvent's freezing point.

Table 2.6 Equilibrium temperature of pure solvents required to attain air purity standards

Solvent	TA Luft		TLV–TWA		Freezing point (°C)
	Category[a]	Limit (ppm)	Equilibrium temperature (°C)	Limit (ppm)	Equilibrium temperature (°C)	
n-Pentane	3	50	−103	600	−84	−130
n-Hexane	3	42	−85	100	−78	−95
n-Heptane	3	36	−67	400	−46	−91
Benzene	C	1.5	−97	10	−86	+5.5
Toluene	2	26	−64	100	−53	−95
Xylenes[b]	2	23	−49	100	−36	−100
Ethylbenzene	2	23	−50	100	−36	−95
Cyclohexane	3	43	−78	300	−62	+6.6
Methanol	3	112	−67	200	−62	−98
Ethanol	3	78	−53	1000	−32	−112
n-Propanol	3	60	−43	200	−33	−127
Isopropanol	3	60	−56	400	−39	−86
n-Butanol	3	49	−29	50	−28	−80
Isobutanol	3	49	−41	50	−41	−108
sec-Butanol	3	49	−39	100	−33	−115
Cyclohexanol	3	36	−24	50	−20	+24
Ethylene glycol	3	58	+11	100	+15	−11
Methyl Cellosolve	2	32	−47	5	−62	−85
Ethyl Cellosolve	2	27	−40	100	−28	−69
Butyl Cellosolve	2	20	−24	50	−15	−75
Methylene dichloride	3	42	−99	100	−92	−97

Table 2.6 Continued

Solvent	TA Luft Category[a]	TA Luft Limit (ppm)	TA Luft Equilibrium temperature (°C)	TLV–TWA Limit (ppm)	TLV–TWA Equilibrium temperature (°C)	Freezing point (°C)
Chloroform	1	4	−103	10	−97	−63
Carbon tetrachloride	1	3	−102	10	−95	−23
1,2-Ethylene dichloride	1	5	−88	10	−84	−81
Trichloroethylene	2	18	−70	100	−59	−73
Perchloroethylene	2	14	−31	100	−13	−19
Monochlorobenzene	2	21	−56	75	−45	−45
Acetone	3	62	−86	1000	−62	−95
Methyl ethyl ketone	3	50	−73	20	−62	−95
Methyl isobutyl ketone	3	36	−50	50	−48	−86
N-Methylpyrrolidone	3	36	−50	50	−48	−85
Diethyl ether	3	49	−102	400	−86	−106
Diisopropyl ether	3	35	−83	250	−68	−68
Tetrahydrofuran	2	33	−86	200	−72	−65
Dioxane	1	5	−83	25	−71	+10
Methyl acetate	2	32	−86	200	−71	−99
Ethyl acetate	3	41	−73	400	−54	−82
Butyl acetate	3	31	−46	150	−32	−76
Dimethylformamide	2	32	−34	10	−44	−58
Dimethyl sulphoxide	2	31	−11	—	—	+18.5
Dimethylacetamide	2	28	−21	10	−30	−20
Pyridine	1	6	−68	5	−69	−42
Acetonitrile	2	—	—	40	−83	−41
Furfural	1	5	−46	2	−53	−39

[a] See Table 2.7.
[b] Xylenes comprise a mixture of *ortho*, *meta* and *para* isomers with ethylbenzene. Frequently the *para* content is low because it is removed at the refinery to make terephthalic acid. If the *para* isomer concentration is high, it will begin to freeze at about −20 °C.

Table 2.7 TA Luft solvent discharge limits

Category	Concentration limit (mg/m^3)	Quantity limit (kg/h)
C[a]	5	0.025
1	20	0.10
2	100	2.00
3	150	3.00

[a] C = carcinogen.

However, many of the solvent systems used in coating technology are not pure and have very much lower freezing points than their pure components. Indeed, if it is decided to employ low-temperature condensation at an early stage in the process development it may be worth considering the choice of a mixed solvent because of its low freezing point.

Aliphatic hydrocarbon solvents are seldom pure, single chemicals, but rather a mixture of normal- and iso-alkanes with some naphthenes, lying within a boiling range of 5–15 °C. As a result, they tend to have very low freezing points and are unlikely to cause any problems in solidifying during recapture by cooling to a low temperature.

The presence of water causes problems with low-temperature operations since the very

Low-temperature condensation 19

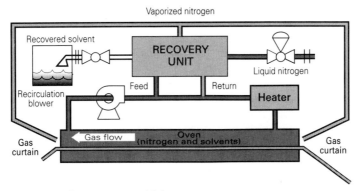

Fig. 2.6 Low-temperature solvent recovery with inert gas

cold surfaces used tend to become coated with ice and lose their effectiveness. This can be overcome by having switch condensers with one on-line while the other is heated to melt off the ice.

Airco process (Figures 2.6 and 2.7)

This is a method introduced fairly recently that suits continuous operation particularly well, such as is common in paper, metal coil and fabric coating. Ideally it should be part of the original equipment since it needs to exclude air (as a source of oxygen) from the evaporation zone and this is a function not easily retrofitted to existing plant. It is very compact so that space near the evaporation zone, often very limited, is kept to a minimum. A measure of the problem is that a skid-mounted module with a plot area of 3 m by 2 m and an overall height of 3.75 m has a solvent capacity of 450 l/h (Figure 2.8).

In this method, inert gas (nitrogen with less than 7% oxygen) is circulated between the evaporation zone and multi-stage condensation unit. Because the gas is inert, the restriction which calls for solvent concentration never to exceed a fraction of the LEL is not applicable. The circulating gas can pick up as much solvent per pass as the limits set by product quality allow. A concentration of ten times the LEL is typical of what can be

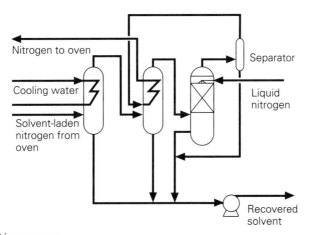

Fig. 2.7 Schematic Airco process

20 *Removal of solvents from the gas phase*

Fig. 2.8 Modular Airco unit: skid-mounted condensation-based recovery unit manufactured under licence from the BOC Group (courtesy of Biwater Process Plant Ltd and Airco Industrial Gases)

achieved leaving the evaporation zone. Thus, for a given amount of solvent evaporated, a 30-fold reduction in gas to be handled in the evaporation zone is theoretically possible.

The first stage of removing solvent from the rich gas is straightforward cooling and condensation using cooling water. In the case of a fairly high-boiling solvent such as xylene or cyclohexanone, the rich gas may leave the evaporation zone at 80 or 90 °C with about 12% of solvent in it; most of the solvent load will be removed in cooling to 20 °C. For low-boiling solvents this stage of condensation will be much less effective. Cooling water is much the cheapest medium for removing heat so as much cooling as practicable should take place at this stage. This means that the circulating gas should be loaded with as much solvent as is practically possible.

In the second condensation stage, very cold nitrogen gas from the third-stage condenser is used to cool the partially depleted circulating gas counter current (*see* Figure 2.5).

The third stage of condensation is by heat transfer between already very depleted circulating gas and liquid nitrogen in a unit that vaporizes the latter. This gas forms the gas curtains that stop air leaking into the evaporation zone through the inlet and exit openings of the material being dried.

The gas used for the curtains is the only gas that is discharged to the atmosphere and, provided that the curtains work effectively, very little of the solvent-rich gas in the evaporation zone mixes with them. Therefore, the amount of liquid nitrogen evaporated in the third stage is determined primarily by the requirements of the gas curtains. If the solvent is not very volatile there is little solvent still in the circulating gas at this point and to return it to the evaporation zone rather than condensing it does little harm to the efficiency of the system.

Thanks to the gas curtains, very little air from outside, with a normal water content of about 0.3%, leaks into the circulating gas. However, at the low condensation temperatures any water will join the solvent stream and, if it is miscible, will build up there. Although the quantity involved per circuit of the system should not amount to more than about 0.1% in the solvent, it will eventually reach an unacceptable level and the recovered solvent will need to be dried. It is possible that the material being dried may also contribute a small amount of water to any build up in the solvent stream.

For solvents not miscible with water, such as hydrocarbons, the danger of a build-up of ice exists and may justify swing condensers at the third condensation stage.

Modifications to Airco process

There is no reason, however, why liquid nitrogen must be the source of the curtain gas. It can be generated on-site using package membrane separation or adsorption plants. Since very pure nitrogen is not needed for the curtain units of this sort, which have high capacities, nitrogen containing 2–3% oxygen is suitable. The gas is produced at ambient temperature and so does not have any role to play in the condensation stages.

Similarly, the coldness arising from the latent heat of evaporation of the liquid nitrogen and from the sensible heat to raise the gas to near ambient temperature can be replaced by refrigeration operating at the required temperature.

The criteria by which these alternatives should be judged are purely economic. On a site close to a bulk oxygen plant, where liquid nitrogen is a large-volume by-product, it is relatively cheap to truck in 14 000 m^3 tanker loads of nitrogen and the capital cost of the installation is very small. Nitrogen may be required on the site at the standard purity of 99.995% or the very low temperature of −196 °C may be used, and this cannot easily be obtained by standard refrigeration units.

3 Separation of solvents from water

Consideration of how aqueous effluent contaminated with solvent may be disposed of should have a prominent place in deciding the solvents to be used in any new process. This is particularly so when biological treatment may be involved since long residence times and, therefore, large site areas may be required. Some solvents (e.g. dimethyl sulphoxide) can give rise to unacceptable odour nuisances when disposed of biologically and others may have high BODs and long lives even in the most active conditions. Hence the removal of most of the solvent from aqueous wastes for recovery may be economic despite the possibility that the recovery cost may be more than the price of new solvent.

This is becoming truer since the cheapest way of removing many low-boiling solvents from waste water has been by air stripping or evaporation from effluent ponds or interceptor surfaces. Such avoidable contributions to volatile organic compounds (VOC) will become increasingly unacceptable as standards for air quality are raised. This also applies to marine dumping since volatile solvents are mostly evaporated before degradation takes place.

The future choice will lie between recovery and destruction of solvents and not merely the transfer of pollution from water to the atmosphere. If destruction is to be chosen then incineration, with or without heat recovery, is an alternative to biodegradation. The low calorific value of dilute aqueous effluents leads to high fuel charges and also haulage costs if the incineration is not carried out on the site of production of effluent. Partial recovery to make a concentrated solution of solvent with a high calorific value and a reduced bulk suitable for haulage to an incinerator is an option well worth considering for waste generated some distance from the incineration point.

The choice of processes leading to possible recovery of solvents from dilute solutions are the following:

- Decanting
- Solvent extraction
- Membrane separation
- Adsorption
- Air stripping
- Steam stripping

24 Separation of solvents from water

The approach to cleaning up water effluent is very different to the drying of solvents, although water cleaning will often yield solvents to be dried before reuse and the economics of the two processes involved will be interlinked. It is not the intention here to describe the various methods for dealing with effluent streams except where they impact upon the recovery of the solvents removed from the effluents.

In general, the standards of purity set for water are much stricter than those required for recovered solvents (Table 3.1). Except in cases where water or some other impurity

Table 3.1 Typical toxic pollutant effluent standards for direct discharge after biological treatment

Solvent	Concentration (ppm) One day	Monthly	Solubility of solvent in water (ppm)
Benzene	136	37	1800
Toluene	80	26	520
Ethylbenzene	108	32	200
Methylene dichloride	89	40	1820
Chloroform	46	21	790
1,1,1-Trichloroethane	54	21	1300
Trichloroethylene	54	21	1100
Perchloroethylene	56	22	150
Monochlorobenzene	28	15	490

actually reacts with the reagents in a synthesis, impurity levels in recovered solvents are in the region 0.1–1.0% (1000–10000 ppm). The standards for water purity can be set for several reasons, namely to avoid:

- toxicity to human beings when the water is discharged in such a way that it can be mixed with potable water;
- toxicity to the fauna and flora of the body of water in which it is discharged; this effect may be direct or brought about by the exhaustion of dissolved oxygen vital to life in the watercourse;
- toxicity to people working in the enclosed environment of a sewer in which vapours from the effluent may collect.

It is impossible to set a level of purity applicable to all discharges when the variety of sizes, disposal destinations and regulatory authorities is so great. The examples quoted in Table 3.1 indicate some of the standards that are required.

Table 3.2 clearly shows that however attractive it may seem to be to treat chemical

Table 3.2 Safe limits of discharge of volatile materials to sewers (ppm)

Solvent	Level of aquatic toxicity to aqueous life (ppm)	TLV (ppm)	Aqueous concentration to yield TLV (ppm)
Ethanol	250	1000	8550
Acetone	14250	1000	1030
Isopropanol	1100	400	1100
n-Butanol	500	200	1890
Toluene	1180	100	2
Pyridine	1350	5	32

effluent in a mixture with large volumes of other domestic and industrial wastes, its safe transmission to a sewage plant cannot be assumed to be straightforward. This is particularly so if a solvent that is both toxic and immiscible with water (e.g. toluene and benzene) reaches the sewer and can contaminate huge quantities of aqueous sewage to a dangerous concentration.

Decanting

Many solvents are only sparingly soluble in water, although none is completely immiscible. It is therefore important, if contamination of water is to be minimized, that uncontaminated water is not exposed to such solvents. Even when water is already 'waste' water it is undesirable to saturate it unnecessarily with a further contaminant. A phase separation of the organic from the water phase should take place as near to their source as possible.

This 'point-source' approach to the problem, which should be contrasted with the 'end-of-the-pipe' alternative in which effluent from the whole process, or even the complete site, is collected and mixed for treatment before discharge, is applicable when small-scale equipment can be used. Small package decanter units with capacities from 300 l/h of aqueous effluent up to units at least 40 times larger are commercially available.

Gravity separators can be designed to handle solvents denser or less dense than water provided that there is a density difference between the phases of about 0.03. This will depend on droplet size and viscosity. It is preferable that effluent streams to be separated by decanting should not be pumped to the decanter, since small globules of the dispersed phase settle more slowly than large ones.

If pumping is unavoidable, positive displacement pumps do less harm than centrifugal types and the throttling of flows, leading to the generation of turbulence, is to be avoided.

It is reasonable to aim to separate by unassisted gravity settling globules of about 15 μm (0.015 cm) diameter. These have a rate of rise or fall in fresh water of about $1.4(\rho-1)$ cm/s, where ρ is the density of the dispersed phase in g/cm^3. A positive figure indicates downward movement. Since the settling speed is fairly slow it is important to have:

- little vertical flow in the settler;
- a short vertical distance for the globules to move before they meet a surface on which they can coalesce;
- adequate residence time for the globule to reach such a surface.

These criteria can best be met on a small scale in a horizontal cylinder of high length to diameter ratio. The feed should enter the cylinder at a low velocity to avoid creating turbulence which could break up the settling pattern and close to the line of the interface between the two phases.

The droplet settling speed quoted above is applicable to a continuous phase of water at 20 °C. The speed is inversely proportional to the viscosity of this phase and there may be circumstances when it is better to carry out the separation at a higher than ambient temperature if the increased solvency of the solvent in water does not outweigh the advantage of a faster settling.

The throughput capacity of the separator chamber can also be increased by fitting a tilted plate pack to provide a metal surface upon which coalescence can take place after a very short vertical path. This, or a coalescer pad of wire in the separating vessel (Figure 3.1), can be retrofitted if droplet sizes are found to be smaller than foreseen and

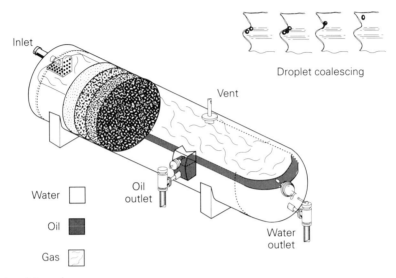

Fig. 3.1 Natco plate coalescer

therefore the performance of a simple empty vessel is found to be inadequate. Alternatively, performance-enhancing devices such as these would be fitted routinely if the residence time for the larger phase were longer than about 10 min.

For larger flows that may arise from the contaminated drainage of plants and tank storage areas, long, shallow, rectangular basins fitted with tilted plate packs are suitable. Horizontal velocities of 1 m/min are typical for such separators, with a depth to width ratio of 0.4 and maximum depths of not more than 2 m.

Although the above techniques can handle very small droplets given a long enough residence time in the separator, they are not effective against true emulsions. If these are subjected to a high-voltage electric field they can in most cases be made to coalesce into droplets that will separate under the influence of gravity.

If the density of the solvent droplets is very close to that of the aqueous phase, the action of gravity irrespective of droplet size may not be sufficient to give good separation and the volume of the decanting vessel may become inconveniently large. Centrifuges, which may occupy very little space, can enhance the effect of density difference very greatly, giving $10000\,g$ on standard machines although they cannot separate true emulsions.

Solvent extraction

Decantation alone is likely to be a sufficient method for cleaning up effluents contaminated with hydrocarbons with water solubilities of less than 0.2% and will, by removing the majority of chlorinated hydrocarbons and other sparingly water-soluble solvents at point-source, minimize their spread throughout the effluent system. However, decantation does nothing to remove materials in solution. Indeed, water-miscible solvents will help to take into solution otherwise immiscible components.

A measure of the hydrophobic nature of individual solvents is given by their log P_{ow} values, where

$$P_{ow} = \frac{\text{concentration of solvent in } n\text{-octanol}}{\text{concentration of solvent in water}}$$

A high value for P (e.g. $\log P_{ow} > 1.5$) indicates a solvent that will only be sparingly soluble in water. Similarly, a negative value of $\log P_{ow}$ indicates a solvent that is very hydrophilic and would be extremely difficult to extract from water using a third solvent. In between these two groups are a substantial number of common solvents that could be extracted from their aqueous solutions to a level that would allow discharge to biological treatment on site or into municipal sewers.

In passing, it should be noted that the very large numbers of published values for P by Pomona College were originally used as a guide to the biological effect of a compound. A high value of P, corresponding to a low concentration in water, matches a low biological effect because the solvent cannot easily invade living organisms. As will be observed in Table 3.3, the solvents that are particularly hazardous to handle because they easily pass through the skin (e.g. DMSO) have very low values of P.

Table 3.3 Log P_{ow} of solvents based on n-octanol

Solvent	Log P_{ow}[a]	Solvent	Log P_{ow}[a]
Cyclohexane	4.15	Diethyl ether	0.77
n-Hexane	3.80	Isobutanol	0.76
n-Pentane	3.23	Ethyl acetate	0.73
n-Octanol	3.15	Pyridine	0.64
Xylenes	3.00	sec-Butanol	0.61
Monochlorobenzene	2.84	Tetrahydrofuran	0.46
Carbon tetrachloride	2.83	Methyl ethyl ketone	0.29
Ethylbenzene	2.76	n-Propanol	0.25
Toluene	2.69	Furfural	0.23
Perchloroethylene	2.60	Methyl acetate	0.18
1,1,1-Trichloroethane	2.49	Isopropanol	0.05
Trichloroethylene	2.29	Acetonitrile	−0.22
Benzene	2.13	Acetone	−0.24
Diisopropyl ether	2.00	Dioxane	−0.27
Chloroform	1.97	Ethanol	−0.31
1,2-Ethylene dichloride	1.48	Ethyl Cellosolve	−0.54
Methylene dichloride	1.25	Methanol	−0.64
Cyclohexanol	1.23	Methyl Cellosolve	−0.77
n-Butanol	0.88	Dimethylacetamide	−1.01
Butyl Cellosolve	0.83	Dimethyl sulphoxide	−1.35
Cyclohexanone	0.81	Monoethylene glycol	−1.93

[a] Log $P = 4.5 - 0.75 \log S$, where S is the solubility of the solvent in water in ppm, is a reasonable correlation of the above for log $P > 0$.

When considering the use of an extraction solvent for cleaning up solvent contaminated water, the following characteristics are desirable:

1. low solubility in water (high P);
2. good solubility for the solvent to be extracted;
3. ease of separation of the extract from the extraction solvent; since distillation is the most likely method of separation, an absence of azeotropes and a much higher volatility for the extract;
4. chemical stability;
5. low BOD so that the water will be easy to dispose of;
6. safe handling properties, e.g. high flash point, high TLV;
7. high density difference from 1.0 to allow easy phase separation;
8. ready availability and low cost.

An illustration of the use of solvent extraction for cleaning up contaminated water occurs in the recovery of ethyl acetate vapour from air with an activated carbon bed. When the bed is steamed for regeneration the recovered distillate has the approximate composition

Ethyl acetate	8%
Ethyl alcohol	1%
Acetic acid	0.5%
Water	90.5%

Not only must the ethyl acetate be recovered from water but also the hydrolysis products, that have been formed during the heating of the ethyl acetate in the presence of a large excess of water, must be removed before the solvent is fit for reuse. Although this can be done by fractionation it involves separating a two-phase ternary azeotrope and the unstable nature of ethyl acetate is also a problem since the fractionation must be done at a low pressure.

The partition coefficients (hydrocarbon phase/water phase) between a C_{10} n-/isoalkane mixture are

Ethyl acetate	4.0
Ethyl alcohol	0.04
Acetic acid	<0.02

Hence by contacting the water phase with such a hydrocarbon it is possible to leave almost all the unwanted acetic acid and most of the alcohol in the water for disposal while extracting the majority of the ethyl acetate into the hydrocarbon phase. P for decane has not been published but, by extrapolating from the information in Table 3.3 it would seem very likely that it would have a value of $\log P$ of about 4 and be very hydrophobic. The solubility of water in decane and its homologues is given in Table 3.4.

The presence of the ethyl acetate in the extract phase increases the ability of the hydrocarbon to dissolve water but still leaves the ethyl acetate fairly dry and therefore reusable once it has been stripped from the hydrocarbon layer (Table 3.5).

Solvent extraction 29

Table 3.4 Saturated solubility in water in *n*-alkanes (% w/w)

Alkane	25 °C	40 °C
n-Octane	0.013	0.025
n-Nonane	0.008	0.017
n-Decane	0.007	0.014
n-Undecane	0.007	0.013

Table 3.5 Effect of ethyl acetate on solubility of water in *n*-decane (% w/w at 25 °C)

Ethyl acetate content of hydrocarbon phase	Water content of hydrocarbon phase	Water content of recovered ethyl acetate
0	0.007	—
6.5	0.008	0.11
9.0	0.011	0.12
10.0	0.014	0.14

The resulting process for the recovery of ethyl acetate is shown in Figure 3.2. Most low molecular weight solvents are more stable than ethyl acetate and the removal of impurities may therefore be of minor importance in a binary water–solvent mixture. However, the selective removal of the solvent from a ternary solvent–methanol–water mixture by this method can be attractive.

Checking the numbered criteria laid down above for the choice of a suitable solvent for extraction from water of dilute concentrations of organic solvents, it is clear that the *n*-alkanes have most of the desirable properties:

1 Solubility in water—*see* Table 3.4.

2 Partition coefficient vs. water—*see* Table 3.3.

3, 4 The stability of *n*-alkanes at their atmospheric boiling point is only moderately good and worsens as the molecular weight increases. If a lower alkane (e.g. octane) can be used without creating fractionation problems at the stripping stage, the thermal stability will be adequate. If a higher molecular weight hydrocarbon must

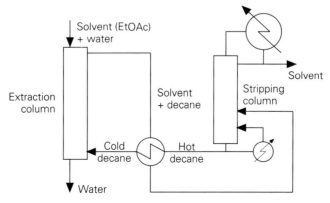

Fig. 3.2 Removal of ethyl acetate from water using *n*-decane

30 Separation of solvents from water

be used it will probably be necessary to carry out the stripping stage under reduced pressure. This may cause problems with condensation of the extract.

5   The n-alkanes of C_8 and above are readily biodegraded and their solubility in water is so low that at point source they can be easily decanted from waste water.

6   C_8 alkanes have flash points within the ambient temperature range but, as Equation 12.1 shows, any hydrocarbon with an atmospheric boiling point above 140 °C has a flash point above 30 °C (86 °F). The toxicity of alkanes is relatively low as the high P_{ow} values would lead one to expect.

7   The specific gravity of n-octane is 0.703 and that of n-undecane is 0.741, so this homologous series has very good properties as far as phase separation is concerned.

8   The individual n-alkanes in the C_8–C_{11} range are commercially available at purities of 95% or more. The impurities present are mostly isoalkanes of the same molecular weight or n-alkanes one carbon number different. For material that will be recycled many times with small losses both to water and by evaporation, their cost is low.

Compared with steam stripping, the thermal efficiency of solvent extraction is particularly noteworthy when the solvent to be recovered from water forms a homogeneous azeotrope with a substantial water content. Pyridine, which has an azeotrope that contains 40% w/w water, has a latent heat of 106 kcal/kg. Assuming perfect heat exchange with a flowsheet similar to that detailed in Figure 3.3, this should

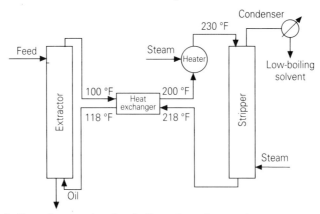

Fig. 3.3 Use of high-boiling solvent to clean low-boiling solvent from waste water

be the heat needed to remove pyridine from water. However, if, instead of a liquid–liquid extraction, either fractionation or steam stripping is used (Figure 3.4) the water–pyridine azeotrope with a latent heat of 464 kcal/kg of pyridine content has to be evaporated. This requires four times as much heat before the production of dry pyridine from its azeotrope is considered.

It is not important to operate the process with a minimum circulation of the extraction solvent (E.S.) because it does not need to be evaporated at any stage. The loss of E.S. is, of course, a function of the volume of the effluent water (which will always be saturated

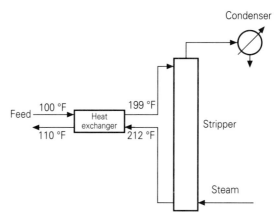

Fig. 3.4 Steam-stripping system for waste water clean-up

with E.S. on discharge) and not of the quantity of E.S. circulated. As Table 3.5 showed, a low usage of E.S., corresponding to a high concentration of ethyl acetate in the rich extract, led to a higher water content in the final recovered product. This may not always be true since different recovered solvents will alter the solubility of water in the E.S. to different extents. It is, however, worth the simple experimental work required to investigate this parameter for any proposed application.

Aqueous streams containing appreciable concentrations of high-boiling organic contaminants present problems when using solvent extraction as a clean-up technique. Once the solvent content of the aqueous phase has been removed, contaminants which are insoluble in water will either build up in the E.S. or fall out of solution in the contacting equipment.

In the former case the E.S. may have to be flashed over from time to time. This may need special equipment such as a wiped-film evaporator working under vacuum and may produce a residue that is difficult to handle. There is an alternative which may prove more economic, especially if the E.S. is a comparatively inexpensive hydrocarbon fraction with a high flash point. The E.S. containing the organic residue can be burnt as a fuel and replaced with new material.

Fouling and blockage of the contacting equipment may be avoided or mitigated by design. It should never be forgotten that deposits that may appear trivial in the laboratory may represent major problems at the plant scale.

Membrane separation

The principles of using pervaporation for removing water from solvent are covered in Chapter 7 and involve the use of a hydrophilic membrane. The removal of solvents from water acts in an identical way but with a membrane that rejects water but is lyophilic.

Such membranes are harmed by exposure to very high concentrations of solvent such as would be present in the solvent phase of a two-phase mixture. Decanting of the solvent phase is essential if dealing with a stream that is more than saturated with solvent. The permeation membrane is likely to be operated at temperatures above ambient so that there is little danger of a separate solvent phase once decantation has taken place. When pervaporation is used to clean up end-of-pipe effluents, there is a possibility of contamination of the effluent with oil emulsions and such material fouls the membrane surface, severely reducing its capacity to pass solvents.

The ability of the membrane to concentrate solvents in the permeate varies. Sparingly soluble, volatile solvents such as chlorinated hydrocarbons, benzene and heptane are concentrated up to 100-fold and can be made fit for reuse without any additional treatment other than phase separation. More importantly, in the clean up of contaminated water, the water stream leaving the plant can be reduced to a solvent content of 10 ppm or even less, at which it may be possible to discharge it or polish it at low cost with activated carbon.

The solvents with values of $\log P$ between 1.0 and 0.6 concentrate less well, typically about 40-fold. A single stage of pervaporation (Figure 3.5) will not produce an effluent fit

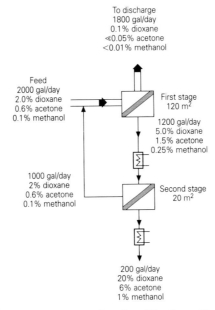

Fig. 3.5 Schematic design of two-stage pervaporation plant (Membrane Technology and Research Inc.)

for discharge and simultaneously a permeate that is near to being fit for reuse. However, this point might be approached by using two stages of pervaporation with the aqueous effluent from the second solvent-enriching stage being returned to the feed of the first stage.

Fully water-soluble solvents can only be concentrated about five-fold and treatment of effluents containing them represents primarily a volume-reduction operation. Pervaporation is a relatively new technology and membranes with improved properties are being developed by many teams, both commercial and academic. It offers the advantage of compact skid-mounted units requiring no utilities apart from electricity and cooling water. This makes it very attractive for sites where tightening restrictions on water quality of discharges require remedial action or where ground water treatment must be undertaken without a sophisticated industrial support structure.

Adsorption

Activated carbon (A.C.) is very widely used, often in a final polishing step, to reach the high purities demanded of effluents for discharge. It is a flexible technique capable of being applied to one-off situations such as spillages or changeable effluents arising from batch processes, neither of which can be satisfactorily dealt with by biodegrading.

Treatment on a fairly small scale can be carried out batchwise using powdered A.C. stirred in contact with the effluent which is removed by filtration when spent. Used A.C. of this sort is seldom regenerated on site and usually has to be disposed of by dumping along with its associated filter aids. It may be noted in passing that this technique is also used as a final stage in solvent recovery as a means of removing unacceptable colour.

A more economical use of A.C. is continuous percolation through granular beds, since it can be regenerated and, by using a series of columns, it can be ensured that the column in contact with the richest effluent is fully saturated before it is regenerated. Regeneration also avoids the need to dispose of waste sludges. A continuous percolation does involve long contact times since the adsorption is controlled by the rate of diffusion into the pores. The larger the particle size of the granular A.C., the longer the diffusion takes while the smaller the particle size the greater the pressure drop through the beds. A compromise usually leads to residence times of about 2–4 h.

The minimum usage of A.C. which can economically justify on-site regeneration is 0.3 Te/day. If the usage is smaller than this, spent A.C. can be returned to the manufacturers for regeneration although this will result in any recoverable solvent being incinerated. Since the process of regeneration is treated as an incineration operation, the equipment needed for meeting environmental regulations is considerable and a small-scale on-site unit should not be chosen without careful consideration. The solvents arising from on-site regeneration, which involves an initial desorption stage before treatment at 850 °C, are likely to be small in quantity and to have a minor effect on the overall economics.

In the sort of dilute solutions found in effluents, the take-up of solvents from water can be quantified by P_{ac}, where

$$P_{ac} = \frac{\text{concentration of solvent in A.C.}}{\text{concentration of solvent in water}}$$

the concentrations being expressed in mg of solvent per kg of carbon and in ppm in the effluent. It is possible from a laboratory batch experiment to calculate P_{ac} for any single solvent, although this will be affected by temperature, pH and salt content of the aqueous solution. Since solvents compete for positions on the adsorbent, care must be taken in extrapolating the results of single solvent isotherms to multi-component mixtures. Values are given in Table 3.6.

As Table 3.6 shows, A.C. is most effective at removing high-boiling non-polar solvents from water and has a range of effectiveness of about 10 000 with solvents of similar volatility. Mixtures of solvents in the wide ranges that are found in contaminated ground water are thus removed to widely varying extents. It is also noticeable in practice that whereas regenerated A.C. maintains its overall adsorption capacity, it is poor at adsorbing low-boiling solvents such as trichloroethylene.

The difference in effectiveness is demonstrated by two solvent-containing effluents

Table 3.6 Values of log P_{ac} for aqueous solutions of 0.1% w/w of solvent or a saturated solution if the solvent has a lower saturation. Activated carbon applied to 0.5% w/w

Solvent	Log P_{ac}	Solvent	Log P_{ac}
Perchloroethylene	5.3	Butyl Cellosolve	2.4
Trichloroethylene	5.0	n-Butanol	2.36
Monochlorobenzene	4.9	Ethyl acetate	2.31
Carbon tetrachloride	4.3	Pyridine	2.26
Xylene	4.3	Methyl ethyl ketone	2.25
1,1,1-Trichloroethane	4.0	Isobutanol	2.16
1,2-Ethylene dichloride	3.8	Ethyl Cellosolve	1.95
Benzene	3.6	Methyl acetate	1.85
Chloroform	3.6	Acetone	1.74
Ethylbenzene	3.1	n-Propanol	1.67
Butyl acetate	3.04	Methyl Cellosolve	1.50
Methyl isobutyl ketone	3.05	Isopropanol	1.46
Diisopropyl ether	2.9	Ethanol	1.35
Isopropyl acetate	2.63	Monoethylene glycol	1.16
Cyclohexanone	2.61	Methanol	0.86

that were treated with marginally insufficient A.C. (Table 3.7). In each case the more easily adsorbed solvents have been almost totally removed while a substantial amount of the more difficult ones have been left in the water phase.

Table 3.7 Treatment of solvent from contaminated pond water

	Inflow concentration (ppm)	Outflow concentration (ppm)
Solvent A containing		
Methylene dichloride	5.4	0.3
Chloroform	0.3	0.01
Trichloroethylene	0.5	N.D.
Solvent B containing		
Methylene dichloride	0.92	N.D.
Acetone	0.45	0.14
Methyl ethyl ketone	0.32	0.002
Toluene	0.32	N.D.

With the process of decanting there is much to be gained from treating effluent with a high concentration of solvent at the point source rather than using A.C. as an end-of-pipe method of clean-up, as the following example shows:

	Point source	End-of-pipe
Daily flow (l)	100 000	1 000 000
Solvent flow (kg)	200	200
Solvent concentration (ppm)		
In	2000	200
Out	20	20
P_{ac}	5000	5000
Consumption of A.C. (Te/day)	1.0	9.0

This assumes that the spent A.C. reaches only 50% of its equilibrium concentration. Not only is the amount of A.C. used much higher when treating dilute solutions, but the volume of the percolation towers and the inventory of A.C. must be larger to give the appropriate residence time.

Another advantage of using A.C. on a closely specifiable stream rather than the mixed effluent arising on a diverse site is that A.C. is liable to absorb inorganic salts that are not removed during regeneration. As a result, the capacity of the A.C. deteriorates over a series of regenerations as the active sites become blocked.

Air stripping

Many organic solvents can be removed from waste water by air stripping, to a level at which the water is fit to discharge. This applies particularly to solvents that have a low solubility in water or a high volatility with respect to water. Indeed, in extreme cases, a comparatively short residence in a shallow lagoon can result in the evaporation of a large proportion of the solvent present. Many biological treatment plants rely on the evaporation of volatile solvents for an appreciable part of their effect.

Two physical laws are available to express the vapour pressures of solvents in water in dilute solutions, Raoult's law and Henry's law. Operating in the very dilute solutions common in waste water treatment, it does not matter which law is used to obtain the system's properties. Unfortunately, the experimental work reported in the technical literature for Henry's law is expressed using a wide variety of units and because the Henry's law constant, H, is variable with temperature this law is less convenient to use. For this reason, the data tabulated here are suitable for use using Raoult's law.

The deviations from ideal behaviour according to Raoult's law are expressed by activity coefficients according to the equation

$p = x\gamma P$

where

- p = the vapour pressure of the dissolved solvent expressed here in mmHg;
- x = the mole fraction in the liquid phase of the dissolved solvent;
- P = the equilibrium vapour pressure of the pure solvent (in mmHg) that is dissolved in the water at the temperature of the operation; this can be obtained from Antoine or Cox equations;
- γ = the activity coefficient of the dissolved solvent in water.

It will be clear from the above that γ is a dimensionless number not affected by the units used. It is not completely constant with respect to temperature but, in the temperature range commonly found in air stripping, its variation can be ignored as being negligible within the engineering safety factors used in design.

For the very low concentrations of dissolved solvent commonly found in air stripping, the value of γ can be treated as γ^∞. Values of γ^∞ are available for a large number of single solvents in water (Table 3.8). They can be obtained from vapour–liquid equilibrium data and, for solvents such as hydrocarbons and chlorinated hydrocarbons that are very sparingly soluble in water, from solubility data.

36 *Separation of solvents from water*

Table 3.8 Correction factors for vapour pressure of solvents over dilute aqueous solutions

Solvent	γ^∞	P (mmHg at 25 °C)	$\gamma^\infty P$ [a]
n-Pentane	109 000	485.1	53 000 000
n-Hexane	489 000	150.1	73 000 000
Benzene	1730	95.1	164 000
Toluene	3390	28.4	96 000
Xylenes	29 733	8.8	261 000
Ethylbenzene	29 500	9.5	280 000
Cyclohexane	77 564	97.6	7 570 000
Methylene dichloride	312	448.4	140 000
Chloroform	907	197.6	178 400
Carbon tetrachloride	10 684	78.1	834 000
1,2-Ethylene dichloride	550	79.0	43 500
1,1,1-Trichloroethane	5825	124.7	726 000
Trichloroethylene	145 500	69.2	1 007 000
Perchloroethylene	3400	18	61 000
Methanol	2.2	127.0	279
Ethanol	5.9	59.0	348
n-Propanol	15.5	19.8	307
Isopropanol	11.8		
n-Butanol	52.3	6.5	340
Isobutanol	40.7	12.4	505
sec-Butanol	35.2	18.2	641
Cyclohexanol	15.8	1.5	24
Monoethylene glycol	0.27	0.17	0.05
Ethyl Cellosolve	6.7	5.7	38
Butyl Cellosolve	201	1.1	223
Acetone	9.9	230.9	2290
Methyl ethyl ketone	29.2	90.4	2640
Methyl isobutyl ketone		19.4	
n-Methylpyrrolidone	6.7	2.2	15
Cyclohexanone	74.1	4.6	341
Diethyl ether	86.6	534.2	46 200
Diisopropyl ether	4.7	149.7	703
Tetrahydrofuran	31.2	162.2	5060
Dioxane	7.6	37.4	284
Methyl acetate	23.6	216.2	5100
Ethyl acetate	45.8	94.6	4330
Butyl acetate	1016	11.3	11 500
Dimethylformamide	2.3	3.8	9
Dimethyl sulphoxide	0.23	0.6	1.4
Dimethylacetamide	1.6	1.2	1.9
Pyridine	30.9	20.1	621
Acetonitrile	9.9	91.1	901
Furfural	73.3	1.8	128

[a] Expressed to three significant figures.

The information available in the literature based on Henry's law can be applied to Raoult's law, but it is first necessary to compare the two laws.

Henry's law states that

$$P = Hc$$

where P is the partial vapour pressure of the dissolved solvent expressed in a variety of units that include atmospheres, Pascals and mmHg, and c is the concentration of dissolved solvent in water, which also can be expressed in a number of different units such as % w/w, mol per 100 l (=g-mol per 100 l), mol/m^3 and mole fraction.

If c is expressed in mole fraction, thus becoming equal to x from Raoult's law, then

$$H = \gamma P$$

and H like P is therefore a function of temperature. It is possible to calculate values of H from the literature, given the value of P by calculation at 25 °C from Antoine's equation. Corrections must be made for units used to express H, and H should be reported at 25 °C.

Values of γ^∞ can range from less than unity for solvents that are very hydrophilic to 100 000 or more for solvents that are almost completely immiscible with water. The value of x is never greater than unity and therefore the very high values of γ are only applicable at very high dilution (e.g. 1 ppm). Even so, the partial vapour pressure of the solvent to be removed from the water can be very much greater than for an ideal solution.

Solvent removed from water by air stripping is recovered by adsorption on activated carbon as the air leaves the stripper. Alternative methods of removing solvent from air are described in Chapter 2. If air stripping is used to clean water for discharge, the air leaving the process usually has a fairly low solvent concentration in comparison with other processes which give rise to solvent-rich air. This is especially true of any batch process for air stripping which is likely to aim to reduce the solvent content of the water to less than 100 ppm (w/w) and, in many cases, down to 20 ppm.

If a typical solvent is assumed to have a molecular weight of 80, the mole fraction (x) of 20 ppm of solvent in water is 4.5×10^{-6}. To reduce the solvent content of waste water by 1 ppm from 21 to 20 ppm at a $\gamma^\infty P$ of 50 000 needs 1 m^3 of air for every cubic metre of water. Thus the effluent air contains 1 mg of solvent per cubic metre. This is two orders of magnitude less than the normal concentration in the effluent in a carbon bed adsorber.

A continuous process in which the contaminated water flowing to the air stripper may contain 1000 ppm of dissolved solvent still needs a solvent of $\gamma^\infty P > 250 000$ to begin to make solvent recovery from stripping air a profitable recovery proposition. It may, of course, be necessary for achieving regulatory approval whatever the value of the recovered solvent.

The above survey of air stripping has been based on a simple binary mixture of fresh water and a single solvent. Solvents in low concentrations have no effect on each other as far as air stripping is concerned and can be treated individually in calculating their rate of stripping. The addition of concentrations of alcohols of the order of 5% w/w does have a significant impact since it changes the solubility of, say, hydrocarbons in water and therefore the $x\gamma P$ value of the hydrocarbon. A reduction of 10–15% in P would be typical for a 5% addition of alcohols.

The presence of inorganic salts has the opposite result and is very much more marked. Thus the values of γ^∞ for benzene and toluene increase up to eightfold in a sodium chloride solution of ionic strength 5 when compared with pure water. Similar results, although smaller in magnitude, occur for solvents that are more water soluble.

Steam stripping

The disadvantage of air stripping as a means of solvent recovery has been shown to be the low concentration of solvent in the effluent air, which poses a problem in recapturing the solvent. Steam stripping, although requiring a more elaborate plant for stripping the solvent from waste water, needs very much simpler equipment for trapping the stripped solvent.

The steam costs are modest provided that good heat exchange can be maintained between the hot stripped water being discharged (Figure 3.4) and the feed to the stripper. Effluent water is, however, liable to pick up impurities and there should be provision for ample heat exchange capacity and cleaning of both sides of the heat exchanger.

The combination of effluent clean-up and solvent distillation should be considered in the design of a stream stripper. For water-miscible solvents that do not form water azeotropes, such as methanol and acetone, the conversion of the stripping column into a fractionating column presents few problems (Figure 3.4). Similarly, the solvents that are sparingly water miscible can be passed through a decanter and the water phase returned to the stripper feed.

A combination stripper and distillation unit would be favoured when very consistent flows of effluent water both in quality and quantity need to be processed. This particularly applies to the more complex problems imposed by solvents that form single-phase azeotropes with water.

Steam stripping is not suitable for the water-miscible, high-boiling solvents listed in Table 3.8. These have lower values of $\gamma^{\infty} P$ than the vapour pressure of water at 25 °C, which is 23.3 mmHg. In addition to these, cyclohexanol and butyl Cellosolve require a lot of stripping stages and may be better removed from water by extraction.

A steam stripper in use for effluent clean-up operates well above the temperature at which scale is deposited by hard water. This is likely to take place at the hotter end of the feed heat exchanger and close to the point where the feed enters the column. It may be necessary to install facilities either for clearing scale or for by-passing blockages at these points.

If steam is injected directly into the bottom of the stripper column, it may bring with it chemicals added to the boiler to guard against corrosion of the steam system (e.g. cyclohexylamine). There is therefore a danger that with direct steam injection an impurity can reach the solvent circuit and it may be necessary to use a heat exchanger to prevent this.

Economics of water clean-up

Three factors contribute to the economics of removing solvent from waste water.

The water itself may have a positive value that can vary widely, depending on how plentiful it is and how pure the cleaned up effluent needs to be for use as a substitute for purchased water. If the recovered water is to be used as cooling tower make-up, its passage through the cooling tower may form part of its treatment. On the other hand, the presence of dissolved chloride salts may prevent water that has been thoroughly cleaned of its organic impurities from being used industrially.

It may not be the most economical option to clean up water to a standard at which reuse or even discharge to a water course is permissible. In, or close to, large centres of population very large quantities of non-industrial waste water are treated extremely economically. Here the dilution of industrial effluent for biodegradation may be the most economic route to take. As Table 3.2 demonstrates, the use of sewers to transport solvent-laden water may present problems and the use of municipal sewage treatment works will inevitably attract a charge. It will, however, avoid the use of valuable space on a factory site and is the ultimate in end-of-pipe treatment.

The solvents to be removed from the waste water may represent an asset or liability. Except in the case of the steam stripping of methanol and acetone, it is unlikely that the solvents arising from water clean-up will be fit for reuse. Further refining is usually necessary unless the treatment is close to the point source and therefore as free as possible from adventitious contamination. In the worst case, such as the cleaning of ground water contaminated with a variety of solvents (*see* Table 3.7), it may be necessary to dispose of the removed solvents by land-filling of the spent activated carbon or by incineration of the solvents.

Some relatively cheap solvents such as hydrocarbons and chlorinated solvents form such dilute aqueous solutions that, unless they can be recovered by decantation, their positive value, even if fit for immediate use, is trivial.

The one clear exception is methylene chloride, which is soluble in water to the extent of about 1.3% w/w and therefore when removed from saturated water contributes about $6 per cubic metre of water to the cost of extraction. The methylene chloride will be water saturated and may require dehydration before reuse.

Other chlorinated solvents have lower water solubilities and trichloroethylene, yielding about $0.8 per cubic metre of water and perchloroethylene about $0.1 on the same basis, are more typical of the credit to be expected. The likelihood is that chlorinated solvents recaptured from dilute aqueous solutions may need reinhibiting in addition to dehydrating.

Benzene is the most water-soluble hydrocarbon and for this reason the most attractive financially to remove from water. If fit for reuse, it will yield about $1 per cubic metre of water. Benzene is generally used only when extremely pure and therefore extra costs will probably be incurred in working it up for reuse. With the possible exception of toluene, no other hydrocarbon solvent has a high enough solubility in water to make a significant positive contribution to water clean-up.

Organic solvents that are soluble in water can have large values when stripped out, but because of subsequent purification costs and the large range of possible concentrations in the waste water, no helpful indication of the possible economics can be made. It will be clear when considering the costs of stripping that it is possible for the value of the recovered solvent to pay for the removal of pollution from the effluent.

Only the broadest approximation of costs for water clean-up can be made. The usual basis is cost per cubic metre of waste water treated rather than per kilogram of solvent removed. This approach tends to favour the end-of-pipe method but the point-source method is the better for total annual cost and value of solvent recaptured.

It is clear that air stripping is the cheapest technique with costs, depending on the concentration of solvent left in the water, of $0.1–0.3 per cubic metre of water treated. The capital cost is low but there is no possibility of credit for recaptured solvent and the air contamination may, in many cases, be unacceptable.

Supplementing air stripping with an A.C. unit for removing solvent from the effluent air results in an increase in price of about $0.4 per cubic metre but a credit for recovered solvent may offset that.

The use of disposable powdered A.C. to remove involatile solvents (and other high-boiling organic contaminants) from the air-stripped water is likely to raise the water to reusable quality but yields no further recovered solvent. In addition, cost is incurred for disposal of spent carbon. Costs will be affected by the value of pollutant removed by the A.C. but a further outlay of $0.4–0.5 per cubic metre would be realistic. Thus the cost for a combination of air stripping, liquid-phase polishing with A.C. and recapture of solvent from the air with A.C. will total $1.0–1.2 cubic metre less any credit for solvent and water.

Pervaporation costs more than any of these techniques at about $2 per cubic metre before allowing for solvent credits, but it is a comparatively new method. It seems likely that with improvements in membrane materials its costs will come down, whereas air stripping and A.C. treatment are by comparison well tried and mature.

Steam stripping is also long established and its cost is very dependent on the relative volatility of the solvent being stripped from the water. In favourable circumstances, when P is very large, figures below $1 per cubic metre before solvent credit may be achieved, but for methanol $3–4 would be more likely.

Solvent extraction, since it involves a stripping stage, albeit under very favourable conditions, is likely to cost between the best and worst steam-stripping figures.

4 Equipment for separation by fractional distillation

The engineer designing and building equipment to restore contaminated solvent to a reusable condition has the full range of unit operations at his disposal. However, it is most likely that he will choose distillation, which exploits differences in volatility, as the most effective and flexible technique for his purposes.

Solvent recovery by distillation can have three different objectives, any or all of which can be present in an operation:

i   separation of the solvent from heavy residues, polymers or inorganic salts (Chapter 5);
ii separation of solvent mixtures into individual components (Chapter 6);
iii separation of water from organic solvents (Chapter 7).

The equipment to achieve the desired aims will consist of:

- heating system to evaporate the solvent;
- condenser;
- fractionating column—this will always be needed for (ii) and usually for (iii), but it is often possible to carry out (i) without a column;
- storage both as part of the plant as a still kettle and to hold residue, products and feed.

For small- and medium-scale operations and if the equipment is not run on a 24 h/day basis, operations will usually be batchwise.

For large solvent recovery streams, or for streams where the plant inventory must be kept to a minimum, continuous distillation (and fractionation) is often preferred to batch operation. The essential plant components as listed above are similar whether for continuous or batch distillation, but for the former the reliance on instrumentation is very much greater and individual plant items (e.g. pumps) need to be very reliable.

In specifying the equipment or checking whether a given unit can do a specific task satisfactorily, the first consideration should be whether its materials of construction are suitable. While a literature search will often provide information on the performance of metals in contact with pure solvents, the solvents in a recovery unit are seldom pure and corrosion tests should be done routinely during laboratory evaluation of a process. Test coupons of metals should include a weld which should be stressed (e.g. sharply bent). Coupons should be in the liquid and in the vapour. Even if no weight loss (indication of general corrosion) is observed, careful examination near or at the weld may reveal pitting or cracking which can result in rapid plant failure. Such corrosion is typical of the attack of hydrochloric acid on stainless steel. Effects on other materials of construction, e.g. packing, gasket, hose and valve seats, should not be overlooked.

If general corrosion is found it may be at an allowable rate. Particularly if the plant is made of heavy gauge mild steel and/or the process is not going to be very prolonged a rate of up to 0.05 in/yr (1.25 mm/yr) might be acceptable.

Usually corrosion attack is much faster on heating surfaces and on stressed components (e.g. screw threads, expanded tube ends) than on the main body of the metal and stainless steel can be justified in heating tubes with mild steel elsewhere.

A combination of erosion and corrosion, such as can be found in the wetted parts of pumps, can cause damage and justifies the use of exotic alloys when much of the rest of the equipment will only be lightly affected.

Dirty solvents that may deposit tars in stagnant corners of the plant can be harmful to alloys such as stainless steel that depend on oxygen to repair a protective oxide coating, and plant should be designed to eliminate such vulnerable places.

Heating systems for evaporation

Electricity

In choosing the source of heat, safety must play a very important role. Hot oil and steam generated by conventional methods demand a flame fed by a substantial air flow. This represents a constant source of ignition and therefore its use requires a site sufficiently large to separate the flame from the largest credible emission of flammable vapour. If such a site is not available then the use of electricity must be considered, despite its very high cost, as a direct source of heat or as a means of raising steam or heating oil.

Direct electric heating must be used with extreme care since an electrically heated surface can reach any temperature short of its melting point as it tries to dissipate the energy put into it. Thus very high spot temperatures will be generated if the transfer of heat is hindered by fouling.

Steam

If heat is required at temperatures below 180 °C (equivalent to steam at about 130 psig), its many other uses on a site (tank heating, steam ejectors, vapour freeing of tanks, steam distillation, etc.) make it the obvious choice.

Since the most common application of steam involves using its latent heat in a heat exchanger, steam jacket or coils, it is common practice to return the hot condensate to the boiler via a hot well. This creates the possibility that flammable solvents can be brought into the boiler area. Because the hot well may be at a temperature of 80–90 °C, many comparatively high boiling solvents can reach it above their flash point.

Contamination of steam condensate can occur by leaks in heat exchangers. When steam is shut off at the end of a batch or campaign, a vacuum forms in the steam space. This vacuum can suck solvent through a leak from the process side of the exchanger. When next steam is turned on, solvent is pushed through the steam trap to the hot well.

How likely this is to happen depends on the corrosiveness of the materials being distilled for the materials of construction of the heat exchanger. If the risk of a leak cannot be regarded as negligible, the hot well should be located in the process area rather than the conventional position close to the boiler.

There are a number of ways of transferring the heat from steam into the solvent that has to be vaporized, as follows.

Direct steam injection

This is the simplest method of injecting heat into the system but is only suitable if at least one of the following conditions are met:

- the solvent to be boiled has a boiling point below 100 °C (e.g. acetone, methanol);
- it is acceptable to recover the solvent as its water azeotrope (e.g. possibly ethanol or isopropanol);
- the solvent to be recovered is not miscible with water (e.g. hexane, methylene dichloride or toluene);
- the mixture from which the solvent is to be recovered already contains a substantial amount of water.

One of the major attractions of direct steam injection is that there are no heat transfer surfaces that may become fouled. However, it usually results in an increase in the process effluent and sometimes in residues that are very hard to handle and dispose of. It can also, provided the rest of the system can accommodate an increased throughput, be enlarged in size very easily.

Because the temperature at atmospheric pressure cannot exceed 100 °C, there is little risk of baking peroxide-containing residues to their decomposition point. For solvents boiling well below 100 °C the thermal efficiency of direct steam injection is at least as good as for other methods of evaporation.

The danger of solvent being sucked back from the still into the boiler in the event of an emergency shut-down must be guarded against. Since many boilers have volatile amines and other chemical additives in them, it is important to ensure that these are not unacceptable in the solvent product.

In cases where inorganic halides are present in the feed to be vaporized, exotic materials may be needed to avoid stress corrosion in heat exchangers. Direct steam injection, by eliminating heat exchanger tubes, avoids this problem.

Shell and tube heat exchangers

Used solvent is liable to foul heat exchanger surfaces and so will almost always be on the tube side of a shell and tube heat exchanger with steam on the shell side. While it is possible to use a natural-circulation external calandria if the solvent to be evaporated is clean, forced circulation is more reliable if the solvent contains residue, despite the fact that it may be a difficult pump duty as regards both cavitation and seal maintenance.

At the bottom of a continuous column or near the end of a batch distillation, even a forced circulation system may not keep the heat-exchange surfaces clean if solvent flashes off the residue in the exchanger. Flashing can be minimized by keeping a back-pressure on the circulating residue until it has left the exchanger.

U-tube reboiler

These tend to foul easily and are hard to clean so they are seldom the best choice in general-purpose solvent recovery plant. If they have to be used, the tube spacing should be generous to make pressure jetting easy.

Internal coils

Coils with steam inside them can be installed inside a batch distillation kettle. In principle, however, this is similar to a U-tube reboiler with a very large shell. Even when a very clean service can be guaranteed, coils suffer from the disadvantage that if the residue is small at the end of a batch, the coils may be partially uncovered. Since temperature difference between steam and still contents will be falling as the batch proceeds, it becomes even more difficult to maintain heat flux if the heat transfer area is also reduced.

External jacket

Small batch distillation kettles can be jacketed, but this becomes less suitable as the size of the kettle increases since the heat transfer area per unit volume decreases with increasing size. A jacket is even more vulnerable to being out of contact with the still charge as the volume in the kettle decreases.

If the heat-transfer surface becomes severely fouled it is necessary to enter a jacketed vessel in order to clean it while coils can be withdrawn and a heat exchanger can be replaced without vessel entry. For all these reasons, an external jacket is seldom the best heat-exchange method in solvent recovery.

Scraped-surface and thin-film evaporators

These are suitable for continuous and batch operations and are the best equipment for mixtures with difficult residues and for temperature-sensitive materials that polymerize or crack when exposed to heat for long periods. They are, however, high in capital cost and need good quality maintenance. The high heat-transfer coefficients attainable with these evaporators can reduce their comparative cost when exotic metals have to be used to protect evaporators from corrosion.

Not only is the low residence time of the solvent to be evaporated an advantage because it reduces the risk of exothermic reactions, it also reduces the inventory of material involved in an exotherm compared with all other evaporating equipment, except direct steam. In general-purpose solvent recovery where exotherms are the most difficult hazards to avoid, such a reduction is a significant advantage.

Hot oil

If temperatures above 180 °C are needed to distil high-boiling liquids, hot oil with a maximum temperature of about 310 °C and capable of transferring useful amounts of heat at 270–280 °C has the advantage over steam of requiring only modest pressures. Because hot oil is always under positive pressure, the risk of contamination with the liquid being processed is very small.

A considerable number of solvents, including many of the glycol ethers, have autoignition temperatures between 200 and 300 °C. When handling such solvents, great care must be given to lagging in any place where leaks or spills might come in contact with hot oil pipes, owing to the hazard that such pipes present. Care must also be taken to cover heating surfaces with liquid before hot oil is circulated through heat exchangers. Whereas it is fairly easy to meter and control the heat supplied to a process using steam, it is much more difficult to do so when using hot oil, particularly if the overall system is a complex one including more than one heat-consuming unit. For this reason, the chance of detecting an exothermic reaction at an early stage when using hot oil as the heating medium is much less than when using steam.

Heat-transfer coefficients on the hot oil side of evaporators tend to be much lower than those for condensing steam, and this can halve the overall heat-transfer coefficient with a resultant cost penalty if the materials of construction are exotic.

The choice of heat-transfer equipment for solvent recovery is similar for steam and hot oil systems, with the exception that direct injection is impractical.

Condensers

A reliable supply of cooling medium is the most important utility for the safe operation of a distillation unit. For a very small unit, where utility cost may be negligible, mains water provides an almost totally reliable means of cooling but the cost, both in supply and disposal, is large unless the water can be used for another purpose after passing through the condenser. One possible use on the solvent recovery unit itself is for the dilution of any effluent that would be otherwise unacceptable for disposal to the sewer (e.g. because of a low flash point).

For larger units the choice of cooling medium lies between:

i water from lake, canal or river on a once-through basis;

ii water from cooling tower;

iii air.

All of these depend on electricity and may be cut off without warning if the electrically driven prime movers stop.

i Water from a natural source is very liable to be contaminated with leaves, plastic bags and other materials which can choke suction filters quickly and, especially in the case of plastic packaging material, completely.

46 *Equipment for separation by fractional distillation*

ii Cooling tower fans tend to be difficult items to maintain if they are specified to be driven by flame-proof electric motors. Since leakage of flammable solvent, via a condenser leak, into the cooling tower and its associated pond is possible, the tower should be located in the process area. However, flame-proof motors are vulnerable in very wet conditions.

iii Air tends to be less cold than cooling tower water and, because of comparatively low air-side heat-transfer coefficients, needs finned tubes on the air side. In areas where the ambient air is dirty or liable to contain (seasonally) large number of insects, the fins are difficult to keep clean.

One advantage of air-cooled condensers is that, in the event of a power failure which stops the cooling air fans, natural convection of air through the banks of hot finned tubes can provide up to 25% of the full condensing capacity, although this depends on the operating temperature. It is preferable, however, to use water as the cooling medium because of the possibility that a general-purpose plant might have to condense a product that solidifies when cold (e.g. cyclohexane, *tert*-butanol, naphthalene). Blockages from such a cause can easily be dealt with on a water-cooled condenser but are much more difficult to handle and can do much more mechanical damage on an air-cooled unit since a blocked tube remains cold and acts as a stay tube under stress from the expansion of all the other tubes in the bundle. Leaks are much more difficult to spot on an air-cooled bundle because the leaking solvent is carried away in a large airstream.

The other source of heat in distillation operations arises from the materials being processed. These may undergo an exothermic reaction with a rate of output of heat much greater than that allowed for from the heating medium. There is no realistic way of designing a general-purpose distillation plant to cope with an 'unknown' exotherm either by containing it or by venting to a safe place. It is therefore most important to test in the laboratory the materials to be processed for signs of exothermal activity under the temperature conditions proposed and, if an exotherm is found, to run the heating medium at least 20 °C below the exotherm initiation temperature. This can be achieved by controlling the steam supply pressure to the plant or governing the input hot oil temperature with a limitation not under the control of the process operator. It is not a sufficient safeguard to control the laid-down operating conditions to a set temperature since high surface temperatures of heating coils or the loss of vacuum may lead to an exotherm triggering despite the fact that bulk measured temperatures are 'safe'.

Another way to minimize the hazard from an exotherm is to run the plant with a minimum inventory (e.g. continuous rather than batch distillation). As an operational procedure rather than design matter, if batch distillation is unavoidable it is advisable to avoid recharging the kettle on top of residues from a previous batch.

If, despite all the safety precautions that have been taken, the condenser is overloaded, facilities must be provided to release the resultant pressure safely. For normal operation a vent will be needed to release the air, or inert gas, that will fill any distillation plant at start-up. This vent must be placed so that air is not trapped in the condenser with a consequent loss of heat transfer area for condensing solvent vapour. At start-up the vent may discharge a solvent–air mixture and, if this mixture is flammable, discharge should be through a gauze or flame trap. It atmospheric pressure operation only is intended and no blockage can occur between the potential source of overpressure (the boiling feedstock in the kettle) and the vent discharge, no pressure-relief valve is necessary.

More often than not, freedom from blockage cannot be assured and a pressure-relief device must be fitted. This can be a safety valve and/or bursting disc fitted on the kettle or column bottom since the column or liquid disentrainer may become blocked.

A safety valve has the advantage that it will close when the pressure is back to normal, allowing plant operation to continue. However, many feedstocks for solvent recovery contain polymerizable or subliming material that may prevent the safety valve from opening when it should.

Once burst, a bursting disc needs replacement, which may lead to a considerable loss of production time. However, a combination of a bursting disc on the process side of a safety valve keeps the latter clean until it has to operate. The safety valve can then be relied on until there is an appropriate opportunity to replace the burst disc. For such a service, a bursting disc must have an indicator to show when it needs replacement and must withstand both full vacuum and pressures up to the appropriate plant safety limit.

The choice of which system to use depends on the quality of feedstock being processed and the value of lost production time.

The discharge of vents and pressure relief pipes should be to areas safe from both the fire and toxic hazards. It should not be within a building. Since the vent of the feedstock storage tank must discharge a similar vapour in a safe place, the feedstock tank is often a suitable catchpot for the disengagement of any liquid droplets that may be carried by vent discharges, provided that sufficient ullage in the tank is maintained at all times.

In cases in which, often at start-up, hard-to-condense vapours which are also potentially toxic or environmentally unacceptable have to vented, consideration should be given to scrubbing them. If a solvent recovery unit needs vacuum-making equipment, a liquid ring vacuum pump can often also be used as a vapour scrubber in addition to its main role.

While normal vents from a distillation unit may be routed via the feedstock tank, a dedicated dump tank should be provided if there is a serious risk of an exotherm being discharged through the safety valve.

Even when a solvent recovery unit is being designed for a stream that is believed to be fully specified in quantity and quality for both feed and product, it is wise to build in spare condenser capacity. It has been shown that additional evaporation can easily be obtained with direct steam heating and extra column capacity can often be obtained with minor investment. However, additional condensation is often hard to achieve and proves the bottleneck to expansion of throughput. For a general-purpose plant that may be called upon to handle solvents from pentane to N-methylpyrrolidone (NMP), the capacity of the condenser is the most difficult feature to choose.

A dangerous circumstance that can arise in distillation is that heat continues to be supplied to a plant after the cooling medium has been interrupted. Typically this can happen when steam heat is used and the electricity fails. In such a circumstance, steam from the boiler system may flow for many minutes although gradually decreasing in pressure. Air-operated control valves set to close on compressed air failure will only shut when the air receiver is empty. Pump-circulated cooling water will stop at once. To protect against considerable quantities of vapour, possibly both toxic and flammable, being discharged to atmosphere, a safety interlock between the heating medium flow and the still pressure on vent temperature is a vital part of the plant's control instrumentation.

Fractionating columns

There will be such a large difference in volatility between a tarry residue and the solvent holding it in solution that a single separation stage (represented by the act of evaporation) may be enough to separate the solvent from its residue. Such a flash distillation does not need a fractionating column between the evaporator and the condenser, although it is often necessary to prevent droplets of residue from being carried over from the evaporator by introducing a disentrainer in the vapour stream. This is usually a pad of wire gauze upon which the droplets impinge and then coalesce. However, for separating a water-miscible solvent from water or one solvent from another using a difference in volatility, a fractionating column will be needed.

Fractionation takes place by contacting an upward flow of vapour with a downward flow of liquid over as large and turbulent vapour/liquid interfacial area as possible. The surface area is created either by bubbling the vapour through the liquid on distillation trays or by spreading the liquid very thinly over column packing in the vapour stream.

Both methods have their advantages and disadvantages in solvent recovery service and although for a column dedicated to a known stream it is usually clear which is the better, the choice for a general-purpose unit is inevitably a compromise.

The criteria for judging the right column internals for a given duty are:

- Column diameter
- Pressure drop
- Fouling
- Foam formation
- Side streams
- Feed points
- Turndown
- Wetting
- Efficiency
- Retrofitting
- Liquid hold-up
- Robustness

Column diameter

The trays in a fractionating column are almost always installed in a fabricated shell. This involves a man working inside the column and the minimum diameter in which this can safely and satisfactorily be done is 750 mm. This size corresponds to a boil-up at atmospheric pressure of about 100 kmol/h and proportionately less at reduced pressure. It is possible to design trays with less capacity if a tray column is vitally necessary for a special duty, but for small units packed columns are usually used.

There is no effective minimum diameter for packed columns, but the size of random packing elements should normally be less than one tenth of the column diameter.

Pressure drop

For a multi-purpose column it will be a requirement that reduced pressure operation is possible. The pressure drop generated by the column internals is the biggest of any part of the system. It varies greatly depending on the material being processed, the rate of boil-up and the absolute pressure of the system. However, as a very rough guide which is sufficient at this stage of considering the plant design, the pressure drops per effective transfer stage (ETS) are as follows:

- trays   3–5 mmHg;
- random packing   1–3 mmHg;
- ordered packing   0.5–1 mmHg.

A general-purpose solvent recovery column will typically have 20–30 ETS so that its pressure drop will be in the range 20–150 mmHg.

Fouling

The active surface area of a distillation tray is an area of great turbulence and accumulations of solids or tars are unlikely to settle there and block liquid or vapour flow. The downcomers are much more liable to become blocked because flow in them is much slower, but handholes can be fitted to permit cleaning without the need for entering the column. It is also possible to fit liquid by-passes around blocked trays if this facility is included in the original design.

Neither random nor structured packing can be cleaned from outside the column once a complete blockage has occurred, so if a partial blockage is suspected prompt action should be taken. Even then stagnant areas in the packed bed may be very difficult to reach with wash solvent. If very bad fouling with polymer or tar takes place, there is a real danger that structured packing may be impossible to remove since, once installed, it fits tightly in the column shell. In such a case a diameter of 750 mm would prove inadequate and enough room to work with pneumatic tools is likely to be needed. As far as random packing is concerned, to empty the column one relies on it pouring from the manhole or being sucked out with a large air-lift. Agglomerated lumps make removal of the packing difficult. In both cases, if removal of the packing from the column is necessary, replacement may be required if the packing is beyond refurbishment.

For these reasons, trays are superior to both types of packing for processing dirty feeds. No column internals are wholly satisfactory and a preliminary evaporation (so that the potential fouling material is never in contact with them) is the best way to avoid fouling problems.

Foam formation

The action by which a distillation tray works, of bubbling vapour through liquid, is one that encourages the formation of foam. A feed with a strong foaming propensity can easily fill a tray column with a stable foam, thereby making it inoperative. This is a difficult problem to diagnose.

Packing, on the other hand, does not encourage foam formation. Since the majority of laboratory columns are packed it is easy not to notice this characteristic of a solvent feedstock and it is important, if the plant unit is a tray column, that the laboratory tests should include a tray distillation.

Anti-foam agents, if they do not result in an unacceptable contamination in the recovered solvent, can cure this problem. It should be remembered that trays are spaced at 300–600 mm apart and a foam height of half the tray spacing is not seriously harmful. Foam height is not affected by column diameter so it is possible to use foam heights measured in the laboratory for extrapolation to plant-scale operation.

Side streams

For both continuous and batch fractionation it may be necessary to take a liquid or vapour side stream from a column.

Liquid

Provided that the requirement is known at the design stage, a liquid side stream could be taken from any tray of a tray column. It might be sensible to consider installing this facility for every fourth tray on a continuous column. The operational justification on a batch plant for more than a single side-stream product is difficult to see.

For a packed column the liquid is only collected in such a way that it can be taken as a side stream at the redistributors, which are normally installed every 4–6 m in the column. Since they occupy column height that might otherwise be filled with packing, excess side stream draws would have an adverse effect on column performance.

As a very rough estimate, the tray column could thus have a draw at every three theoretical stages (assuming an efficiency of 75%) against the packed column's ten theoretical stages.

Vapour

A vapour side stream from a tray column demands a larger tray spacing than normal, but the major problem posed by a multiplicity of off-takes is the associated large diameter pipework involved in vapour handling. There is seldom a justification for having a vapour side stream above the feed in a continuous column or anywhere in a batch column.

The position of a vapour side stream in a packed column is governed by the same considerations as for a liquid side stream.

Feed points

Liquid

Liquid feed can be put into any downcomer of a tray column and feed points can therefore be installed after the column has been designed and erected. One precaution that needs to be taken in their use is that the feed may be raised to a comparatively high temperature in a confined space. If the feed contains hard water, the hardness may be laid down as scale, eventually blocking the downcomer. If the feed contains an inorganic solute it also may be deposited if the water in the feed flashes off.

In a packed column, the feed can be put on to any redistributor, i.e. every 4–6 m (Figure 4.1). Modern redistributors have a large number of very small holes which are

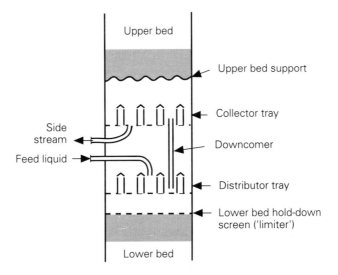

Fig. 4.1 Side stream product and feed arrangement for packed column

deisgned to handle clean distillate liquids. The feed should be filtered in the feed line to prevent pieces of rust, scale, etc., from blocking some of these holes and spoiling the distribution of reflux and feed. This is a very important factor in the performance of a packed column. In addition, the possibility of scale being formed on the redistributor should be guarded against since it would have the same effect.

Vapour

If the nature of the feedstock is such that preliminary evaporation is necessary, it is thermally efficient to feed the resulting vapour stream directly to the column in a continuous operation. In a packed column the same restrictions as above apply, i.e. the feed must enter the column between the beds of packing.

A tray column's vapour feed will need an additional space between trays to accommodate the vapour main and so unrestricted flexibility cannot easily be provided. Nevertheless, a choice of two or three vapour feed inlets could be made available as long as the tray spacing is designed appropriately.

Turndown

A multi-purpose column intended for atmospheric pressure and vacuum operation, and also possibly for extractive and azeotropic distillation, must work adequately well over a very large range of vapour and liquid loads. Even when the mode of operation is fixed, the range of molal latent heats in a solvent mixture, and therefore the vapour velocities in the top and bottom of the column, can be from 6000 to 10450 cal/mol.

From their ways of creating an interface for mass transfer between vapour and liquid phases, it is clear that there is a danger that packing may operate unsatisfactorily if there is insufficient liquid to wet its high surface area. Provided that there is a high enough vapour velocity to stop liquid leaking down the vapour holes, shortage of liquid is not a problem with trays. Hence, too much liquid, as might easily arise when trying to use a standard column for extractive distillation, is liable to overload the downcomers, which are not overlarge since this is not a part of the column area which contributes to mass transfer activity.

Maximum vapour rates on packed and tray columns are similar, being about 1.4–1.9 m/s based on the empty column area. As indicated under Pressure drop above, the pressure drops over trays will be a good deal greater so that the absolute pressure at the column top is likely to be lower in vacuum operation and the mass flow will be correspondingly less. Vendors of column internals, most of whom are able to supply both trays and column packing systems, can provide information on the details of their products in respect of their liquid and vapour performance, but a turndown of 3:1 on packed columns and 5:1 on trays should be attainable.

Wetting

Just as the problem of foaming was restricted in practice to trays, that of wetting is one that only afflicts random and structured packing.

In a dedicated plant, the surface of packing can be conditioned so that it will be wetted by the product it has to handle. A packing that must handle aqueous material, for instance, should be free from grease or oil and may with advantage be slightly etched with acid. As is likely to happen on a multi-purpose plant, the result of processing solvents of different surface tensions can be that the packing surface resists wetting and offers a very much reduced area for mass transfer. A reduction of as much as 90% is possible. This is a condition that can be cured by, for instance, washing with a powerful degreaser, but it is difficult to diagnose.

Efficiency

For a variety of reasons the height of a fractionating column should be as low as possible. In the design of a new plant, the cost of the column shell and internals is significant particularly if, for reasons of corrosion resistance, they must be made of expensive alloys. The feed, reflux, vapour, vacuum, cooling water, etc., lines tend to be more expensive as the column height grows and the pumping heads also increase. In many locations the distillation column will be the tallest structure on the site and, as an obvious sign of industrial activity, may not be welcome.

For the above reasons, the lower the height of the ETS the better. Typically a distillation tray will give between 0.65 and 0.75 of an ETS and at a normal tray spacing will compare with random and structured packing thus:

- valve tray   1.5 ETS/m
- random packing   1.7 ETS/m
- structured packing   2.4 ETS/m

Retrofitting

The retrofitting an existing plant is probably determined by the need to install either a given number of stages (in which case a packing that yields more stages may avoid the need to refit the whole column) or to obtain the best out of the available height. It should be noted that it is difficult to retrofit structured packing in a tray column shell. The support rings needed for trays have to be ground off smooth to allow for the tight fit of structured packing. If random packing is put into a tray shell the support rings need to be removed since they represent an appreciable reduction in diameter, and thus operating capacity, but it is not vital to grind the interior of the shell smooth.

Liquid hold-up

Tray columns in operation hold about a 40 mm depth of liquid on each tray and thus have a higher hold-up than packed columns. This is not very important in continuous distillation, where ideally the composition of any part of the system remains the same during a campaign once equilibrium has been established.

In a batch distillation the compositions in the system vary as the more volatile components are preferentially removed. A high inventory in the column tends to make a sharp separation more difficult to achieve.

Robustness

A multi-purpose solvent recovery plant tends, because of the variety of operating conditions it has to cover, to be submitted to unusual mechanical loads. These can arise from column flooding, sudden losses of vacuum and higher liquid densities (e.g. perchloroethylene) than are usual.

If random packing is used, bed limiters should always be fitted to prevent the packing beds being fluidized and rings being washed into distributors or even condensers and product systems.

Structured packing is not so vulnerable but standard trays, which are typically only 0.25 mm thick, can easily be bent and hence pulled off their support rings by the atypical strains put on them such as the 'bumping' that occurs in flooding.

Storage

Vessel design

The design of the still kettle must allow it to take full vacuum since it is possible, if the system vent is blocked or closed, for almost all the vapour within the solvent recovery system to be condensed once the heating medium is turned off. Since all air will have been displaced from the system in the early stages of a batch, a vacuum will form progressively as the vapour condenses.

A vessel designed to withstand full vacuum is likely to withstand the pressure at which one would wish to operate a solvent recovery unit, although this should not be taken for granted. The vessel must be designed for the relief valve pressure.

In designing a still for solvent recovery, consideration must be given to its size. The contents of a still will be a quantity of boiling solvent and, the larger the quantity, the greater is its potential for danger in the event of an accident. In addition, long process times create a greater risk of undesired reactions (e.g. polymerization, decomposition) taking place.

On the other hand, a larger batch capacity requires fewer charging and discharging operations, which are the most hazardous and labour intensive parts of the batch cycle. It also reduces the number of running samples to be tested and the number of tank changes to be made. A typical balance for round-the-clock working would be a 24 h batch cycle, whereas for non-continuous operation a batch completed in each working day, thus minimizing cooling and reheating of the still contents, is a reasonable design basis.

For a general-purpose commercial recovery unit a still kettle that can accept a charge of a full tanker load adds to the flexibility of operation. Contaminated solvents often arise in such quantities or result from mistakes in loading tank wagons and for such one-off situations it is frequently inconvenient to have to allocate a feedstock tank rather than charge the still directly from the vehicle.

Since vapour rises from the liquid surface in a still, consideration needs to be given to the still being a horizontal or a vertical cylinder. It is easier to fit heating coils inside the former and, provided a disengagement velocity of 1 m/s can be designed for, a horizontal cylinder is the conventional choice.

In the case of a fairly high-boiling solvent, such as xylene, the expansion of liquid between ambient temperature and boiling point is about 10% and, as soon as it begins to boil and bubbles of vapour lower the bulk density of the still contents, a further 3% increase in volume should be allowed for. Since the contents of a full-length external sight glass will remain cool while the batch is heating to boiling point, the expansion will not show in a sight glass but must be taken into account when fixing the size of a batch and the area over which vapour disengagement takes place. If direct steam injection is used as the method of heating, the volume of condensed steam to bring the batch to its boiling point must also be taken into account.

A further problem that may cause overfilling is that residue may not be discharged completely from a previous batch. When handling feedstocks with a large viscous residue (especially if the residue has to be cooled before discharge) it is possible that all the residue may not flow to the still outlet but the plant operator will be deceived into thinking that the still is empty. Any calculations then made on the available still volume may be wrong.

It is important that the still is not overcharged for two reasons:

- the plant safety valve should be fitted to the still since this is certain to be upstream of any blockages or points of high pressure drop in the system; if the safety valve on an overfilled still lifts it will discharge hot liquid, thus creating a different hazard to that posed by a vapour discharge;
- if there is a fractionating column between the still and condenser its internals (either trays or packing) can be damaged by vapour bubbling through the part of the column filled with liquid.

As an important safety measure, therefore, the still should be fitted with either:

- a float switch or other automatic control mechanism linked to the charging pump or to an automatic valve on the charging line to cut off the flow of feedstock when the appropriate ullage is reached;
- or an overflow line from the chosen ullage level back to the feedstock tank (of a larger diameter than the feed line).

Sight glasses and manual dipping are useful ancillary aids but are not in themselves sufficient safeguards.

Of all the vessels on a solvent recovery site, the batch still kettle is the one most likely to need entering for cleaning. It is, therefore, advisable to design it with this in mind. A large manhole at one end of a cylindrical vessel, set at a moderate fall towards the manhole, will allow liquid to be sucked out before entry. Through this manhole sludge or solids may have to be shovelled or raked so easy access at the outside must be provided. The cylinder itself should have a minimum diameter of 6 ft (1.8 m) to make manual work easy for a man equipped with life line and air line or breathing set.

Even when the still has been steamed out to a level below 10% of LEL, it is possible that when sludge is disturbed local pockets of solvent will be released, and good ventilation forcing air into the end or top of the still furthest from the manhole is desirable. Steam heating coils set in the bottom of the vessel hinder cleaning and, in circumstances where footholds are usually very slippery, are treacherous to stand on.

Residue tankage

This tank is likely to receive residues close to their boiling points. It is possible that a hydrocarbon residue may follow an aqueous one with the risk of a foam-over if the lower (water) phase boils, so it is important that dipping and draining facilities are adequate for checking on this risk.

The tank is likely to need a heating jacket or heating coils since residues may soldify when they cool. It is much more difficult to melt material that has solidified than to keep it mobile.

The residue tank may also need entry for cleaning and a large manhole with good external access is vital.

Although the flash point of a residue sample may be high, it is almost certain (because it will, on initial discharge, be near its boiling point), that the contents of the residue tank will be above its flash point and electrical equipment, bunding and regulations in the area surrounding the tank should be as for a low flash point product.

The residue may have a smell that is considered unpleasant both within the site and in the neighbourhood. Transfer by vacuum from the still kettle to residue tank and scrubbing the extracted air in a liquid ring pump can be a solution to this problem. In this case the residue tank must be able to withstand full vacuum and should be a horizontal cylinder to reduce the static head involved. Normally the residue tank should be constructed of the same material as the kettle.

Product tankage

A high proportion of the product tanks in a general-purpose solvent recovery plant should be made of stainless steel. SS304 will usually be good enough since the requirement is largely to keep the product water-white and to allow easy cleaning on product change.

If the production equipment is operated batchwise it is useful to have a facility for mixing the tank contents, preferably by circulating them with a pump so that a true sample may be taken. Mixing by rousing solvents with inert gas leads to vapour losses and possible neighbourhood odours. Air as a rousing medium or for blowing pipelines clear has these disadvantages, in addition to the risk of generating electrostatic charges in the vapour space above the liquid.

If inert gas is available on site, gas blanketing of solvents stored within their explosive range is usually justifiable. Solvents such as toluene, heptane, isopropanol and ethyl acetate fall into this classification. Air with its oxygen content reduced to below 10% will not support combustion but it is normal practice to err on the safe side and use 2–3% oxygen for tank blanketing and clearing pipelines. Solvents which are particularly prone to form peroxides, e.g. tetrahydrofuran, must be protected by nitrogen if they have to be stored uninhibited.

A few solvents (e.g. benzene, cyclohexane and *tert*-butanol) have melting points high enough to require heated storage and traced pipelines. Because of the vigorous convection currents that bottom heating generates they will not require blending facilities but will require carefully heated vents since even at modest tank temperatures enough solvent may sublime to block standard p–v (pressure–vacuum) valves.

When serving a processing unit, product storage fulfills two functions. Running tanks will hold the product from a batch or, on continuous distillation units, usually a day's operation while it is tested by quality control before being released for distribution or, possibly, rejected and returned for reprocessing. It is, therefore, necessary to have two running tanks for a product made on a continuous plant unless the runs are very short. A batch plant may have three or four running tanks in all, provided they can be cleaned easily so that they can be used for a variety of products. Vertical cylindrical tanks with bottoms sloping to a drainable sump will usually be satisfactory for this service. Their size should match that of the batch still kettle for ease both of production and for reprocessing when that is necessary.

Stock tanks, from which material for sale or reuse will be supplied, should be chosen with a view to the size of vehicle loads, length of campaigns and, for a commercial recovery plant, the operational pattern on a given solvent. Residence time in stock tanks may be long and the hygroscopic nature of many solvents may call for gas blanketing or breathers protected with silica gel or some other air desiccant. Blending facilities, so that parcels of materials from running tanks are mixed homogeneously into the stock, are desirable and may allow slightly off-specification product to be blended off, rather than reprocessed.

It is seldom that the capital cost of product storage will exceed the value of its contents and in planning storage facilities the working capital commitment that they effectively represent should not be overlooked.

The range of density of solvents likely to be processed on a commercial solvent recovery plant is very wide (pentane 0.63, perchloroethylene 1.62) and tank foundations, depth sensors and pump motor horse powers should all be considered carefully in this context.

Feedstock storage

A commercial solvent recoverer will receive feedstock by road or rail in drums or bulk at ambient temperature and it will normally not be severely corrosive to mild steel. Hence feedstock storage, where colour pick-up is unimportant, can be constructed of mild steel.

If the distillation equipment is operated batchwise, the minimum size of tank should be for a single batch. If a continuous distillation plant has to be served a minimum of 2 days feedstock should be held. In both cases a further minimum size equal to a road tanker load plus 5% ullage should be set.

The commercial recoverer is often paid to receive waste solvents and so, far from needing to provide working capital to finance his stock of raw material, the bigger the stock, the greater is the financial benefit. Operationally a large 'fly wheel' in the system helps to smooth out variations in quality in addition to giving customers for recovered solvent confidence in being able to obtain continuing supplies.

Such typical products of a recoverer as windshield wash (isopropanol) and gas hydrate solvent (methanol) are seasonal in use but feedstocks may have to be accumulated in the off-season to maintain the service to their generators.

This is a very different situation to that of in-house recycling. Here the stocks of used solvent should be kept to a reasonable minimum consistent with smooth plant operation. As typical targets of 90% recycling are attained it becomes very difficult to use up excess stock accumulating in the system, which thus consumes both working capital and storage capacity.

For smaller scale operations, a vessel fitted with vacuum capability for consolidating drums into a still charge without tying up the still for that purpose should be considered.

All feedstock tanks should have means of dipping (either by tape or dipstick) both to gauge their contents and to detect water layers lying either above or below the solvent layer. A drain valve to remove bottom water layer should be fitted and, because it is vulnerable to water freezing in it, this valve should be cast steel, not cast iron.

For removing water floating on a chlorinated hydrcarbon bottom layer, two or three drain valves at easily accessible positions should be provided. The valves should be fitted close to the tank sides to avoid the danger of the drain lines freezing up. Product tanks should be fitted with pressure-vacuum valves and with self-closing dip and sample hatches.

5 Separation of solvents from residues

Solvents for recovery are frequently contaminated with residues that have a negligible vapour pressure and are waste materials for disposal. The solvents can arise in different ways:

- Mother liquors, from which a desired product has been removed by filtration or decanting, will be saturated with product at the process temperature but are also likely to contain undesired by-products.
- Washing equipment, e.g. ball mills in paint manufacture, will produce solvents containing both resins and inorganic pigments. The used solvent will be far from saturation in the former but will contain a suspension of the latter.
- A solvent–water mixture may hold inorganic salts in solution. If the solvent is less volatile than water (e.g. DMF), the salts may come out of solution as the water is removed from the mixture.

The recovery of solvents involved in such mixtures poses four problems: exotherms, fouling of heating surfaces, vapour pressure reduction and odour.

Exotherms

There have been a considerable number of accidents during solvent recovery operations due to the triggering of exothermic reactions that have run away, causing structural damage to the recovery equipment.

When dealing with any feedstock containing unknown compounds, it is good practice to test it in the laboratory for the presence of an exotherm under the most severe conditions to which it is likely to be subjected using an accelerated rate calorimeter. Both temperature and exposure time may be important contributory factors and a test at the highest temperature that the site's heating medium (usually steam or hot oil) can offer, for the time of a batch process, should form the initial trial. If an exotherm should be found under these conditions the temperature should be reduced until no exotherm is detected and a safe operating temperature limit 20 °C less than the lowest exotherm temperature should be set for plant operation.

Very many chemical reactions have activation energies in the range 20–30 kcal/mol. This means that, in the temperature band of 100–180 °C commonly met in solvent recovery, the rate of reaction doubles or trebles for each 10 °C increase in temperature. A 20 °C margin is thus equivalent to a safety factor of 400% or more.

The practice of charging a batch on top of the residues from previous batches is potentially hazardous because the residence time of the residues may become very long and difficult to ascertain. If there is any possibility that an exotherm may occur, the residue should be completely discharged after each batch.

The damage that a runaway reaction may cause is due to the energy that is released and the inability of the equipment to remove it fast enough. It is obvious, therefore, that the likely damage will be reduced if the inventory of material in the plant is kept as small as possible. Large batch stills are not a good choice for the processing of unstable materials and the very low hold-up in the highest temperature zone of thin-film or wiped-film evaporators makes them especially suitable for such duties.

Achieving the required separation of solvent from residue at the safe operating temperature is likely to involve the use of reduced pressure, particularly towards the end of a batch when the mole fraction of volatile solvent becomes low and that of the involatile residue becomes high. Because this situation is present all the time in a continuous operation, it is likely to be under vacuum. This presents no insuperable problem for handling solvents with high boiling points since it is still possible to condense their vapours with cooling water or ambient air with an adequate temperature difference in the condenser.

For volatile solvents with boiling points below 60 °C at atmospheric pressure, vacuum operation is not a practicable proposition.

Fouling of heating surfaces

As the solvent is removed, the solution becomes supersaturated and polymers or salts begin to be deposited. The most concentrated solution tends to be immediately adjacent to the heating surface at which vapour is being generated and there is therefore a likelihood that solid will build up on the heating surface, spoiling its heat transfer.

This problem can be avoided by several methods which depend for their success on the nature of the solute. The methods can be classified as:

- eliminate the evaporator heat transfer surface;
- do not allow evaporation at the heat transfer surface;
- mechanically clean the heat transfer surface;
- flux the solute.

All the methods can be applied to continuous or batch plants and, particularly for the latter, skid-mounted package units are available in some cases (Figures 5.1 and 5.2). Although the principles of operation may seem simple, the handling of non-Newtonian tars and polymers can prove difficult and the know-how of plant manufacturers in this field is valuable.

Fouling of heating surfaces 61

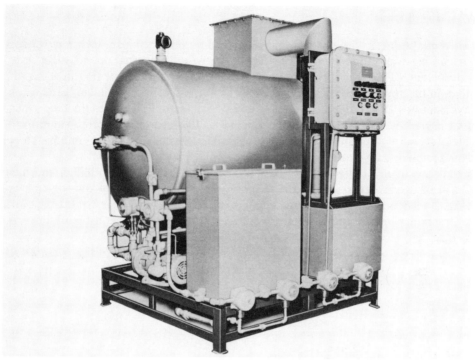

Fig. 5.1 Automatic batch steam distillation unit (Interdyne)

Fig. 5.2 Sussmeyer solvent recovery unit

Steam distillation

All solvents boiling below 100 °C and all solvents not miscible in all proportions with water can be evaporated by injecting live, otherwise known as 'open,' steam into the liquid solvent. Thus the great majority of solvents can be steam distilled. Steam distillation has the big advantage, if exotherms may occur in the solvent mixture, that it always operates at a temperature below 100 °C at atmospheric pressure even when the solvent has all been stripped from the feedstock. It is therefore often a solution to the problem posed by a combination of a low-boiling solvent not easily condensible when under vacuum and an exotherm.

If the solvent to be steam distilled were pure and not water miscible, the mixture would distil over when it reached a temperature where the sum of the solvent vapour pressure and steam vapour pressure totalled 760 mmHg (Figure 5.3). At that point, the mole fraction ratio in the distillate would be the same as the ratio of the vapour pressures:

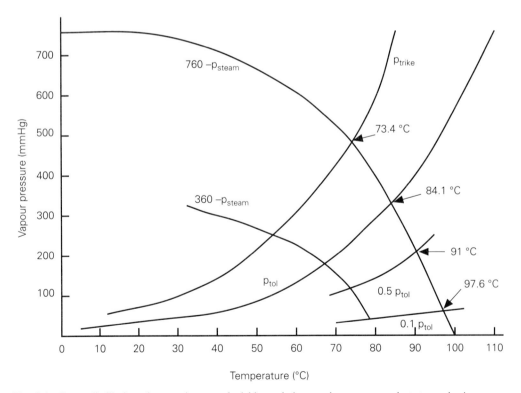

Fig. 5.3 Steam distillation of pure toluene and trichloroethylene under vacuum and at atmospheric pressure

$$\frac{p_{solv.}}{p_{steam}} = \frac{x_{solv.}}{x_{steam}} = 5.1$$

Reading off from Figure 5.3, $p_{solv.} = 335$ mmHg and the toluene–water composition in the distillate would be 19.7% w/w water. Similarly, the trichloroethylene mixture would consist of 7.04% w/w water.

In a more practical system, there would be involatile residue in the solvent and the vapour pressure of the solvent would be $\gamma x p_{solv.}$. Until most of the solvent had been stripped out, γ is likely to be close to unity.

At the point when the toluene mole fraction and the residue mole fraction were equal at 0.5, the steam distillation temperature would be 90 °C and the water content of the distillate is 34.5% w/w.

Eventually when all the solvent in a batch process is stripped out, the vapour would be all steam.

Evaporation in a pot still requires only the sensible heat to raise the solvent to its boiling point plus the latent heat of evaporation. Although in steam distillation the sensible heat is lower because the boiling point is lower (84.1 °C in steam vs. 110.7 °C in the case of toluene), steam is used as a 'carrier' gas and then wasted in the condenser. Further, all the steam used is too contaminated to return to the boiler as hot condensate and its heat is therefore lost. To reduce the amount of steam used, the operation can be run at a reduced pressure (Figure 5.3). Since the lowest temperature at 360 mmHg for the toluene–water system would be 65 °C, there would be no serious problem in condensing at this pressure and the steam saving would be appreciable.

Assuming that steam injected into a batch for steam distillation comes from a boiler system at, say, 10 bar, it will have some available superheat to give up to the charge first to raise it to its boiling point and then to provide the necessary latent heat of evaporation for the solvent. Table 5.1 shows that, even if the heat needed to bring the solvent to its

Table 5.1 Steam consumption for steam distillation of toluene

System pressure (mmHg)	Solvent mole fraction	Boiling point (°C)	Steam in vapour (% w/w)	Live steam per kg toluene (kg)	Condensed steam per kg toluene (kg)
760	1.0	84.1	19.7	0.245	0.17
760	0.5	91.0	34.5	0.543	0.15
760	0.1	97.6	73.3	2.745	—[a]
360	1.0	65.0	17.0	0.205	0.18
360	0.5	62.0	30.5	0.439	0.16
360	0.1	77.5	68.0	2.125	0.04

[a] At this live steam usage there is an excess of superheat and no steam condenses.

boiling point is disregarded, some of the injected steam will be condensed in the batch still. The volume of toluene evaporated is substantially greater than the steam condensed so that only under most exceptional circumstances is there a danger of the volume of liquid in the still increasing and therefore the vessel overfilling.

The comparison of the heat consumption by conventional dry distillation, once again disregarding the heat needed to bring the solvent to its boiling point, shows that dry distillation, using about 0.18 kg steam/kg toluene, is more efficient than steam distillation. The comparison, however, cannot be meaningfully extrapolated to low mole fractions of solvent in the still charge since the temperature of the liquid would need to be raised to 216 °C to make 0.1 mole fraction of toluene boil.

Continuous steam stripping, in which the solvent-rich mixture is fed to the top of the column and steam is injected into the base, has a lower steam requirement than batch steam distillation. To reduce a toluene–involatile mixture from 0.9 to 0.1 mole fraction of solvent will theoretically require 0.315 kg steam/kg toluene, not allowing for the heating up of the mixture, compared with batch steam distillation at about 0.43 kg/kg.

Unfortunately, a continuous operation of this sort is seldom practical. As the solvent is removed from the feed the resin–water mixture becomes difficult to handle and not at all suitable for processing in a packed or a tray column. The only solvent recovery mixture that is likely to lend itself to such continuous processing is one in which the contaminant is a water-soluble salt that can be disposed of after being stripped free of solvent.

In all steam distillation, there is a risk of foam formation and laboratory trials on new mixtures designed to show up a foaming tendency are an essential part of their laboratory screening. It is usually possible to find an antifoam agent which is effective if the carryover of foam spoils the colour of the distillate.

It is normal practice to undertake batch steam distillation at a constant steam input rate since the condenser is likely to be the rate-controlling component of the equipment. Thus the marginal cost per kg of the recovered solvent not only reflects its high steam requirement but also its high plant occupation time.

The residue from the distillation tends to be a wet, lumpy mixture unfit for disposal by landfill and therefore requiring incineration. The operator has to balance the cost of stripping the marginal solvent against the extra costs and hazards of incinerating a highly flammable material. It should be noted that an organic resin, even when thoroughly stripped of solvent and containing considerable occluded water, has a high enough calorific value to be burnt without added fuel.

Up to this point only the steam distillation of sparingly miscible solvents such as hydrocarbons or chlorinated hydrocarbons has been considered. The evaporation, using direct steam injection, of fully water-miscible solvents with atmospheric boiling points below 100 °C is different in principle and is commonly practised. Whereas in the case of immiscible solvents, dry distillation was shown to require less heat and therefore would be more attractive, unless a difficult residue made it difficult to carry out, there is no such advantage in this case. Any used solvent of this sort, whether or not it contains a difficult residue, can be evaporated by injecting steam into it, thus avoiding the need to have a reboiler or evaporator. This can be a useful technique if the solvent contains, for instance, halide salts that would require a heat exchanger made of exotic metals.

The disadvantage of such a course of action is that water builds up in the residue and will be present in the vapour leaving the still. For an immiscible solvent the distillate will separate into two phases after condensing and because of the shape of the vapour–liquid equilibrium (VLE) diagram (Figure 5.4) no fractionating column is needed. However, a water-miscible solvent will have to be freed of water by fractionation or some other means. Further, there are only two solvents in this class that do not form azeotropes with water–methanol and acetone. The latter is difficult to separate from water by fractionation below a level of about 1.5% w/w water so that only methanol can be mixed with water without a considerable penalty. This penalty does not arise if water is already present in the material to be steam distilled.

Fouling of heating surfaces 65

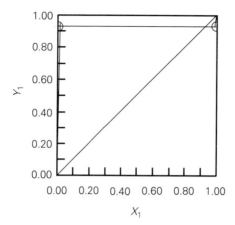

Fig. 5.4 VLE Diagram of methylene dichloride (1)–water (2) at 40 °C. This is typical of the shape of the VLE relationship of all sparingly water miscible solvents (e.g. hydrocarbons, chlorinated hydrocarbons)

A further class of solvents intermediate between the water-miscible low-boiling compounds and the immiscible materials are those which azeotrope with water and form two-phase distillates on condensing. Typical of these are the butyl alcohols, MEK and isopropyl acetate. Each on its own is appreciably soluble in water and the presence of an organic solvent in the water phase makes that phase more attractive to other solvents. In distilling by steam injection a typical mixture of solvents such as are used as thinners and gun cleaners for nitrocellulose lacquers it is not uncommon to lose 6% of the solvent, and about 10% of the active ingredients (alcohols, ketones, esters), into the effluent water.

This presents a difficult disposal problem, in addition to a significant solvent loss, since the waste water has a high BOD and usually a low flash point. One way of eliminating the problem is to recycle the water phase from the decanter to the process. Since it is primarily clean condensed steam, it can be boiled without any fear that it will form scale on the heating surface or throw out resin which will interfere with the heat transfer. The solvents it contains will be evaporated as solvent vapour and returned via the still to the condenser and phase separator. Some water will be lost to the system since the solvent phase from the phase separator leaves the system water saturated. A typical figure for a cellulose thinners distillation would be 5% or less. This water needs to be replaced by make-up (Figure 5.5).

If methanol, and to a less important extent ethanol and isopropanol, are present in the mixture to be steam distilled, they may concentrate preferentially in the water phase. This may result, if their concentration is high enough, in the distillate failing to separate into two layers as the distillation proceeds. A laboratory trial using a Dean and Stark still head will show if, for any given mixture, this is a problem.

Most organic residues release their solvent if steam is sparged into them. There are, however, a few in which the resin cures as soon as its temperature is raised, encapsulating significant amounts of solvent which cannot then be recovered by steaming and are therefore lost. At the same time, the residue forms a mass that is hard to handle and very difficult to remove from the still. To avoid this the solvent-rich residue can be atomized, in a similar way to fuel in a pressure jet, and one or more steam jets are directed into the jet of dirty solvent. The solvent is evaporated and the resin reacts to give small particles which can be easily handled.

66 Separation of solvents from residues

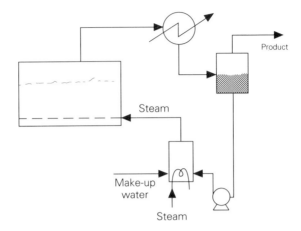

Fig. 5.5 Closed-circuit steam distillation

Vapour distillation

While steam distillation of solvents which are not water miscible produces a recovered solvent that is ready for reuse, or at worst only needs reinhibiting, the wet solvents resulting from steam distillation of alcohols, ketones and esters with boiling points up to about 120 °C are likely to need drying. Pervaporation is a technique particularly well suited to this problem, provided that there are no glycol ethers in the mixture, since it copes with drying from 5% water down to 0.5% water. The equipment is, however, relatively expensive and there is another method for treating solvents contaminated with resins and pigments which avoids the use of steam injection while eliminating a heating surface that is likely to become fouled. This is the Sussmeyer process (Figure 5.6), which

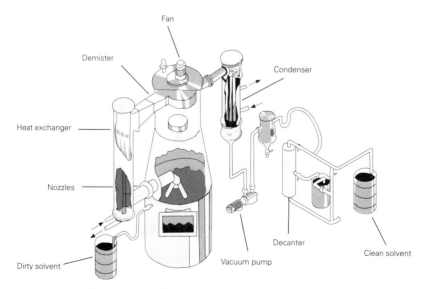

Fig. 5.6 Exploded view of Sussmeyer plant

relies upon superheating the solvent vapour from a batch of contaminated solvent and returning this superheated vapour into the liquid in the still. The heart of the process is a fan set above the still. It draws vapour from the liquid surface in the still up through a demister so that no resin droplets are present to foul heat-exchange surfaces. The vapour is pushed by the fan down through a steam- or hot oil-heated shell and tube exchanger, where the vapour is superheated, and into jets through which the vapour is sparged into the liquid in the still giving up its superheat to vaporize more solvent. As pressure builds up at the still head at the suction side of the fan, surplus vapour not needed for heat transfer duty flows to a standard water-cooled condenser and leaves the plant as recovered product.

While the injection of steam into dirty solvent often gives rise to foaming, which if it cannot be controlled is liable to spoil the product or result in a reduced operating rate, the injection of solvent vapour very seldom gives rise to foam formation. The sludge also is reduced in volume because there is no water present in it and because it is often a hard solid when cold, the possibility exists of having a residue that is acceptable for land fill. Since both steam distillation and 'vapour distillation' are often used for handling solvents that contain paint pigments, there are problems associated with the incineration of residues. Comparatively high concentrations of heavy metals are present in these residues and there is a strong argument in favour of disposing of them in a solid resin to landfill rather than as ash from an incinerator. Incineration of such residues presents a problem in the collection and disposal of toxic dusts.

The whole process is operated under vacuum so that solvents up to 180 °C can be handled without decomposition.

Package plants processing up to 960 l/h of contaminated solvents are available and solvents containing 15–20% of resin are suitable for recovery in such units. Although no water is introduced into the solvent, this does not mean that the system cannot be used to recover water-wet solvents, although their comparatively high ratio of latent heat to sensible heat means that the operating rate is slower than for dry feedstock.

Just as for steam distillation, there is no reason why a fractionating column cannot be inserted between the still and the condenser if mixed solvents are to be separated after removal from an involatile residue, but this also will reduce the operating rate because of the reflux required for the separation in the column.

Hot oil bath

Another method of avoiding a heat-transfer surface that may become fouled is to use a temperature-stable liquid in a 'bath' on to which solvent is fed and from which vapour flashes leaving its residue behind. Such a technique is attractive for unstable solvents which decompose or polymerize if heated to their boiling point over long periods. The liquid being heated in the bath should ideally not dissolve the residue although if the concentration of residue in the feed is small and the liquid is a hydrocarbon fuel, it may be possible to purge contaminated liquid to the fuel system.

Separation of heating and evaporation

Residue tends to come out of solution at the point at which solvent becomes supersaturated. Supersaturation occurs because the solution loses solvent and if this happens

68 *Separation of solvents from residues*

by the formation of vapour bubbles at a heat-transfer surface, this is also where the residue will leave the solution. If boiling does not take place at the heat-transfer surface, it is unlikely that residue will foul the surface unless some other process is also taking place there, such as a further polymerization of the components of the mixture.

In a forced circulation system, the mixture of solvent and residue may be heated under pressure in the heat exchanger but evaporation will not occur until the pressure has been released. This has the disadvantage that the mixture is heated to a temperature significantly higher than if boiling were allowed to take place in the exchanger and undesired chemical changes are more likely to happen. It also means that a considerably greater amount of pumping power is expended in circulating liquid against a head.

If enough vertical room is available, the head may be supplied by the liquid column above the heat exchanger. Provided that the steam pressure or hot oil temperature at the heat exchanger is sufficiently high to provide the heat flux, an evaporation of about 5% of the solvent per pass should be designed for. This requires enough superheat in the liquid at the point where the pressure is released to provide latent heat for one twentieth of the liquid.

At atmospheric pressure the latent heat of toluene is about 87 cal/g and the specific heat is 0.48 cal/g/°C. Hence the required superheat temperature is about 1.8 °C per 1% to be vaporized or 9 °C for 5% evaporation.

The density of toluene near its boiling point is 0.78 so that a liquid head to stop boiling in the heat exchanger relying on static head alone is about 2 m for the above performance. Vaporization will take place in the pipe as the back-pressure diminishes and erosion at the outside of any bend in the pipe, particularly if solid crystals are formed, may be serious. An alternative is to rely not on static head but on the dynamic head generated by a restriction (an orifice plate and/or small-bore pipework) to provide a pressure drop. The vapour can then be allowed to flash off in a large enough chamber, e.g. the bottom of the fractionating column, so that there is no risk of impingement on the vessel wall.

The overall design is similar to that used in salt crystallization and is suitable either for batchwise or continuous operation. In the former case the still contents need to be kept in a form that can easily be discharged. For continuous evaporation the vapour can be fed to the column after passing through a combination of flash vessel and disentrainer. Provided that the latter function is effective, clean side streams can be taken from the column below the feed point.

A forced circulation evaporator depends on the ability of a centrifugal pump to circulate the residue. This sets a limit of about 500 cP at working temperature on the viscosity of the bottoms and often requires that an appreciable amount of solvent must be left unrecovered to keep the residue mobile. In addition, the deliberate superheating of the mixture being circulated is the very opposite of what should be done to avoid exotherms, as discussed earlier in this chapter.

Although if a crystalline salt is present a forced circulation evaporator may be the best choice, it would seldom be chosen for general-purpose solvent recovery operation.

A different approach to avoiding fouling of heat transfer surfaces is available in small (up to 100 l) batch package distillation units (Figure 5.7). These provide the heat energy for boiling the solvent from electrically heated hot oil. The tank holding the boiling solvent has, as an inner liner, a plastic bag capable of withstanding 200 °C. This liner is disposable with its contents of residue leaving the heat-transfer surfaces untouched by

Fouling of heating surfaces 69

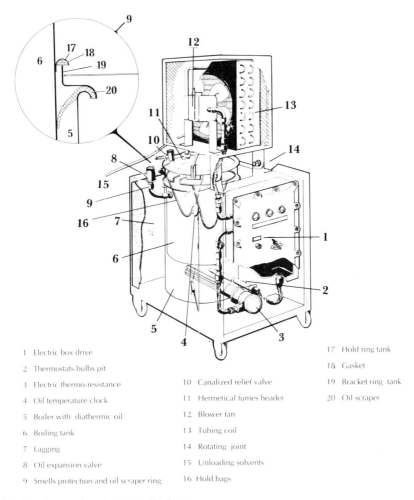

1 Electric box drive	10 Canalized relief valve	17 Hold ring tank
2 Thermostats bulbs pit	11 Hermetical fumes header	18 Gasket
3 Electric thermo-resistance	12 Blower fan	19 Bracket ring tank
4 Oil temperature clock	13 Tubing coil	20 Oil scraper
5 Boiler with diathermic oil	14 Rotating joint	
6 Boiling tank	15 Unloading solvents	
7 Lagging	16 Hold bags	
8 Oil expansion valve		
9 Smells protection and oil scraper ring		

Fig. 5.7 IRAC package unit with removable liner

potentially fouling materials. This equipment is available with vacuum facilities and flame-proof electrics.

Continuously cleaned heating surfaces

To recover the maximum yield of solvent from a mixture, operation at a high viscosity is necessary. This calls for equipment that will attain high heat-transfer coefficients under conditions in which flow would usually become laminar and the discharge of residue well stripped of solvent and too viscous to be handled by a centrifugal pump.

Agitated thin-film evaporators (ATFE) consist of a single cylindrical heating surface jacketed by steam or hot oil. The solvent-rich feed is spread over the heated surface by a rotor turning with a tip speed of up to 10 m/s. There is normally a narrow gap between the rotor tip and the tube wall but if more than 90% of the feed is to be evaporated

70 *Separation of solvents from residues*

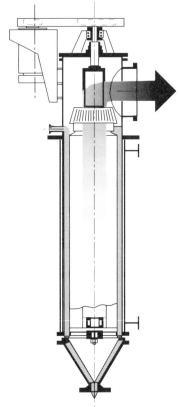

Fig. 5.8 Luwa evaporator, sectional view

movable blades pressed against the wall by centrifugal force can be fitted (Figure 5.8). The spreading process causes constant agitation of the liquid in contact with the heating surface and turbulent flow despite viscosities of up to 30 000 cP toward the bottom of the tube where most of the solvent has been evaporated.

The vapour flows upwards through a separator at the top end of the rotor which knocks out any droplets of liquid carried in the vapour stream. Contact between vapour and liquid provides about 30% more fractionation than the single stage represented by other methods of evaporation.

Usually an ATFE is used in a continuous mode with the solvent being stripped from the residue in a single pass through the evaporator. If it is desired to fractionate the distillate, a column can be fitted between the ATFE and the condenser (Figures 5.9 and 5.10).

It is possible to use an ATFE as an external evaporator on a conventional batch still if more than two solvent distillate fractions are required and if there is a difficult problem of fouling from a resin in solution. Such a problem is conventionally met by using a close clearance impeller within a still, constantly cleaning the jacketed walls of the vessel. However, this solution suffers from the fact that the heat-transfer surface in contact with the batch charge decreases as the volume of the batch is reduced. The combination of a higher boiling point as the volatile solvent is removed and a lower heat-transfer area as the volume in the kettle is reduced results in a high marginal cost of recovery of the last

Fouling of heating surfaces 71

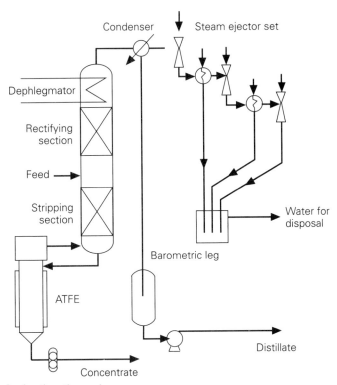

Fig. 5.9 ATFE operation for fractionation under vacuum

Fig. 5.10 Wiped-film evaporator (Raywell)

of the solvent. Provided a pump is available that can feed the batch still contents to the ATFE, the evaporator surface area is maintained throughout the batch and the liquid head over the heat-transfer surface is kept to a minimum.

ATFE have comparatively high overall heat-transfer coefficients due to the agitation of the film in contact with the heating surface. Although, because of their high standard of design and complexity in comparison with other heat exchangers, their cost per unit area is high, ATFE have overall heat-transfer coefficients three or four times those for other evaporators when handling high-viscosity liquids. If exotic, expensive materials of construction have to be used for the heat-transfer surfaces in contact with process liquids, the capital cost of ATFE can be readily justified.

Fluxing residue

Residues from solvent recovery operations are usually materials that have to be disposed of. Although they may, when cold, be solid enough to go to landfill in drums, this method of disposal is likely to become progressively less acceptable, necessitating incineration.

Reference to Chapter 6 will indicate that material for incineration that cannot be handled as a liquid (e.g. nearly solid material in drums) is difficult and expensive to incinerate and it may therefore be necessary to leave sufficient solvent in a residue to make its handling easy. A site where direct transfer may be made from the solvent recovery plant to an incinerator can allow molten solids or liquids pumpable at process temperature to be destroyed without handling problems. However, incinerators tend to need considerable maintenance and to link closely the operation of a solvent recovery plant to an incinerator may not be acceptable.

A solution to this problem is to add to the distillation residue, either at the point of discharge from the solvent recovery unit or by adding to the feed so that it will remain in the residue at the end of recovery, a flux which keeps the residue in a form which allows it to be pumped easily. A commercial solvent recovery firm is likely to have solvents or mixtures of solvents that are of very low value or even unsaleable which can serve this purpose, since they may be used to allow more valuable solvents to be released from a crude mixture.

A firm recovering its own solvents on-site is less likely to have such resources and may need to purchase flux to allow the full recovery of desirable materials. While this flux may be low-quality solvent from a commercial recoverer, it is also possible that it will be liquid hydrocarbon fuel, such as gas oil. The possibility exists that fuel may need to be bought to boost the calorific value of the feed to an incinerator handling high water content waste or the fuel, carrying in solution solvent recovery residues, could be burnt in the steam-raising boilers. In neither case will the fuel be wasted.

Practical experience shows that it is preferable not to allow organic tank residues to come out of solution and then redissolve them in a plant washing step if it is possible to include a solvent in the original charge of feedstock. This is particularly the case if a solvent mixture of, say, acetone, water and a heavy organic is to be treated to recover acetone. As the acetone is removed, the organic tar which is not soluble in water falls out of solution and often adheres to the sides of the vessel and may require large quantities of wash solvent to remove it. If a solvent can be present to hold the tar in solution either in a single phase with the water or as a separate organic phase, this will usually prove more economical. Solvents such as DMF and the glycol ethers will be worth consideration for holding the tar in an aqueous phase while toluene or xylenes may prove effective if the water phase can be separated for, say biotreatment while the organics have to be incinerated.

Adjustment of pH or addition of surface-active agents are other methods worthy of consideration to make residues easier to handle. The springing of amines is often a helpful step. These may have reacted with acids in use and therefore cannot display their solvent properties until regenerated.

Vapour pressure reduction

In addition to the problems of handling residues both during and after solvent recovery, involatile materials cause difficulties by reducing the vapour pressure of solvents, as was shown in the description of steam distillation.

The limitations for a water-free distillation of a solvent from its involatile solute are:

- To condense the solvent with cooling tower water, the solvent vapour should not have a temperature of less than 30–35 °C. This sets, for any particular solvent, a bottom limit for the pressure at which the operation can be run.

- If the solvent is flammable it is usually a requirement that the flash point of the residue be above ambient temperature.

- The temperature of the evaporator will be restricted by the heating medium available, most commonly steam, to about 10 bar at the heating surface, corresponding to 165 °C in the liquid to be processed.

- There will be a pressure drop between the heating surface and the condenser comprising liquid head and pressure loss from the flow of the vapour. Both vary greatly with the equipment available.

The way in which these limitations affect the distillation of solvent–residue mixtures can be seen in Table 5.2.

Table 5.2 Effect of solvent boiling point on recovery from involatile residue

Solvent	Atmospheric pressure b.p. (°C) Col. A	Mole fraction of solvent to give LEL at 23 °C Col. B	Vapour pressure of Col. B mixture at 165 °C (mmHg) Col. C	Pure solvent b.p. at Col. C pressure (°C) Col. D
Xylene	139	1.00	1463	165
Toluene	111	0.44	1272	130
Cyclohexane	81	0.14	819	83
n-Hexane	69	0.075	552	59
n-Pentane	36	0.021	331	12

Column A. This shows that the solvents chosen range in volatility from medium (xylene) to high (*n*-pentane).

Column B. At 21 °C a pure hydrocarbon solvent with a boiling point of about 139 °C will just be within its LEL. As the solvent boiling point is reduced, a lower mole fraction of solvent yields a highly flammable residue.

Column C. At the temperature assumed to be available with a normal industrial steam supply, a high solvent vapour pressure can be generated from a 'safe' toluene–residue mixture. The pressure would be more than ample to allow for pressure drop through fractionating equipment and to condense at atmospheric pressure in this case. Cyclohexane is marginal in this respect but can probably just produce a residue mixture below LEL and be condensed without the use of vacuum in a low-pressure-drop plant.

Column D. n-Hexane will need a reduced pressure operation but even allowing for a pressure drop in the processing equipment, the boiling point of the vapour will not present a condensing problem at 59 °C. n-Pentane, however, cannot be condensed with conventional cooling water at the pressure that must be achieved to make a residue below its LEL.

Although all the solvents listed in Table 5.2 are hydrocarbons, the conclusions are generally applicable to flammable solvents.

While chlorinated solvents do not involve a flash point problem, the extra cost and difficulty in incinerating a solvent–residue mixture containing chlorine are such that low concentrations of solvents in residue are often a requirement for them also. Since the objective will normally be to reduce the chlorine content to a low weight percentage of the residue, the more volatile solvents are the easier ones to strip out to achieve any required specification.

If the priority is not to produce an acceptable residue but rather to achieve the highest recovery of the solvents, it is clearly possible (Table 5.3) to reduce the mole fraction of the less volatile solvents well below the limit set in Column B of Table 5.2.

Table 5.3 Effect of low pressure on stripping toluene from residue

Mole fraction of toluene	Vapour pressure at 165 °C (mmHg)	B.p. at indicated vapour pressure (°C)
0.3	867	116
0.2	578	102
0.1	289	80
0.05	145	62
0.01	29	26

In an appropriately designed plant, there is unlikely to be any insuperable problem to reducing the toluene in the residue below 0.05 mole fraction, but the marginal amount of solvent recovered as the pressure falls is small and the residue is likely to be increasingly difficult to handle.

Another limitation that may have to be considered is the thermal stability of a solvent. Most solvents in widespread industrial use can be expected to be stable at their boiling points, provided their pH is close to neutral but this cannot be assumed when they are undergoing fractionation. This is because a state of equilibrium is continuously disturbed. Two examples will illustrate the problem.

Ethyl acetate forms an equilibrium mixture according to the equation

$$EtOAc + H_2O \rightleftharpoons EtOH + HOAc$$

In the absence of water, no hydrolysis can take place and ethyl acetate is stable. If, however, wet ethyl acetate is fed to a fractionating column, hydrolysis takes place but the equilibrium is not reached because the acetic acid, being much the least volatile component, moves down the column while ethanol, in a low-boiling ternary azeotrope with ethyl acetate and water, moves up the column. The reaction proceeds slowly at low temperature but at the atmospheric boiling point it is fast enough to affect yields seriously and to make an off-specification recovered product. It is desirable to operate at the lowest possible temperature, and therefore pressure, and with a minimum inventory of liquid.

Similarly, dimethylformamide (DMF) decomposes in the presence of water in an exothermic reaction

$$HCONMe_2 + H_2O \rightleftharpoons HCOOH + Me_2NH$$

In this case, the dimethylamine is very much the most volatile component of the system and under fractionation rapidly moves up a column. DMF forms a high-boiling azeotrope with formic acid and this moves to the column base.

In a batch distillation, when the inventory is considerable and the column base is a kettle, a highly acidic condition develops which tends to encourage the reaction in both examples.

A general-purpose recovery unit should therefore have vacuum facilities so that the highest economic yields, the most acceptable residues and the least risk of decomposition can be attained.

The pressure that matters in evaporating solvent from residue is at the heat-transfer surface of the evaporator. This is made up of three components:

1   The absolute pressure at the vent of the condenser.

    This is determined by:

    a   the air-tightness of the plant;
    b   the type of vacuum pump or steam ejector used;
    c   the vapour pressure of the solvent at the temperature of the condenser cooling medium;
    d   the amount of low molecular weight compounds arising from decomposition (cracking) of the feed;
    e   the dissolved air or gas in the feed (in the case of a continuous plant only);
    f   the capacity of the vacuum apparatus to handle incondensables arising because of (a), (c), (d) and (e).

    In general-purpose solvent recovery, a typical pressure at the vent is unlikely to be less than 25 mmHg although lower pressures are achievable with specialist equipment.

2   The pressure drop through the plant.

    Reduced to the minimum of an evaporating surface and a condensing surface placed as close to each other as practicable, a solvent recovery unit can have a very low pressure drop. If any fractionation is needed (Figure 5.9), pressure drop is inevitably introduced in pipework and column packing.

3 The liquid head over the heat transfer surface in the evaporator is, in a batch unit, usually large compared with either (1) or (2) if the plant is designed for a low pressure drop.

A depth of liquid of 1500 mm in the still would be typical of a modest-sized unit and this would exert a liquid head of 100 mmHg at the start of a batch. As the batch progresses, the level will fall and with it the liquid head but this will be offset, in many cases, by the increase in the mole fraction of involatile residue and the decrease in the mole fraction (and therefore partial pressure at a given temperature) of the solvent.

In a unit where the invariant pressures due to (1) and (2) amount to 30 mmHg, the effect of liquid head and solvent mole fraction on the temperature of the solvent mixture are as illustrated in Table 5.4. The initial mixture is 0.85 mole fraction DMF with 0.15 mole fraction of an involatile tar.

Table 5.4 Batch still temperature vs. wiped-film evaporator (WFE) temperature

Head over heating surface (mmHg)	Mole fraction DMF	Vapour pressure (mmHg)	Still temperature (°C)	WFE temperature (°C)
100	0.85	153	102	70
37.5	0.60	112.5	94	77
30	0.50	120	95	82
22.5	0.33	159	103	92
20	0.25	200	109	99
17.6	0.15	317	123	114
16.5	0.10	465	136	127

On a thin-film or wiped-film evaporator (Figure 5.8) (ATFE) where the liquid head over the heating surface is less than 1 mmHg and the operation is continuous, the liquid temperature will rise over the height of the evaporation from top to bottom but the absolute pressure will be constant throughout. Assuming the vacuum system gives a slightly worse performance because of the continuous flow of dissolved air from the feed but the pressure drop over the equipment is the same as for the pot still, Table 5.4 shows a temperature comparison based on a 35 mmHg absolute pressure. It can be seen that not only has an ATFE got a very small inventory and residence time but also it exposes the contaminated solvent mixture to lower temperatures.

Odour

A large proportion of recovered solvents, particularly those recycled through an industrial process, do not have to be judged by the most difficult specification of all—marketability. Solvents, which are incorporated into products being sold for domestic use, must have an odour which is acceptable to all customers. They will be used in the home by people whose noses have not been heavily exposed to 'chemical' odours and who, generally, would rather have no odour at all in their paints, polishes and adhesives, both during and after use.

The custom processor has therefore got to achieve a higher standard than the in-house recoverer as far as a solvent's smell is concerned. If a smell is unavoidable, as it is in most cases, the standard to be reached is that of the virgin unused material, a sample of which is sure to be in the possession of the potential commercial buyer.

Unacceptable smells are often due to decomposition of the solvent itself (e.g. DMF) or of some component of the residue which has cracked to give a low molecular weight product which contaminates the recovered solvent overheads. They usually cannot be masked by reodorants. Indeed, the presence of a reodorant often signals that the smell of the solvent is suspect.

Treatment with activated carbon is sometimes effective if the molecular weight of the contaminant is high. Many unacceptable odours are due to the presence of low concentrations of aldehydes and sodium borohydride can be used to remove them.

It is always better, if possible, to prevent their formation by evaporating at as low a temperature as condensation will allow and to expose the solvent to high temperature for as short a time as possible. If the initial choice of a solvent system is influenced by the smell of a recovered solvent, those with strong odours and good chemical stability (e.g. aromatic hydrocarbons) are less likely than, say, alkanes to become unacceptably contaminated.

One of the most difficult problems associated with smell is its measurement. Not only are individuals very different in their ability to detect odours but they also differ in their preferences. Even in sealed bottles, solvent odours change and usually improve with time. Frequent opening of sample bottles leads to loss of the more volatile components and so alters the overall smell. Exposure in a laboratory to occasional high odour levels can spoil an individual's ability to judge them for the rest of the working day. Frequent comparison of smells involving inhaling significant amounts of solvents is bad for the health. A cold ruins an individual's performance.

For all these reasons, it is useful to develop, when possible, a gas–liquid chromatographic (GLC) headspace analysis for the malodorous compound, although often it is present in very low concentration. Failing a satisfactory laboratory method, an 'odour' panel made up primarily of female non-smoking office workers is likely to be as effective as anything.

6 Separation of solvents

Although all fractional distillation operations rely on exploiting differences in the relative volatility (α) of the components to be separated, this difference can arise in a number of ways. Distillation separation methods are classified in Table 6.1 in the order of occurrence in solvent recovery.

Table 6.1 Fractionation methods

	Continuous	Batch
Atmospheric	*	*
Vacuum	*	*
Steam	*	*
Azeotropic	*	*
Extractive	*	
Pressure	*	*

By careful design a single process unit will be able to carry out every method, although less well than a unit specifically designed for each. However, to be able to design and build a dedicated plant for specific separations is a luxury that a solvent recoverer seldom has. Even less often does he operate a plant solely for the job he believed he had when he designed it. A solvent recoverer should therefore approach a separation problem with a high degree of flexibility in modifying the separation method to the plant and vice versa.

Thus he should ask himself:

1. Is the plant suitable or can it be altered to make it so? The most likely reason that the answer to this query would be negative is that there is a major corrosion problem, but lack of traced lines, the presence of odours and also regulatory barriers are other common problems.

2. Are there any azeotropes in the solvent system that would prevent the required specifications being met by simple means? There are techniques to break azeotropes, so that it is not necessary to abandon hope of using some type of fractionation if there is an azeotrope preventing the achievement of the purity required. Azeotropes are sufficiently frequent among the commonly used solvents to allow the problem they present to be ignored at the early stage of an assessment.

3. Has the column sufficient separating power to achieve the separation required? To answer this it is necessary to know

(a) how many theoretical stages the column contains and
(b) whether this number of stages will be adequate even at total reflux.

Column testing

When a new column is installed it is often covered by a performance guarantee, and even if it is not it is wise for its owner to find at an early stage what separating power it has. This allows a deterioration in performance at some later date to be detected without doubts. A fall-off in performance due to collapsed or displaced trays, blocked distributors or packing that is not wetting properly is difficult to detect without a base from which to make comparisons.

The choice of suitable test mixtures for columns depends on the number of stages that the column may have, since separations that are too 'easy' will call for a very high degree of accuracy in analysing samples.

Test mixtures should be chosen from binaries that have near ideal behaviour ($\gamma^\infty = 1.0$). Also, they must be stable at their boiling point, not exceptionally toxic and inexpensive. They should, of course, not form azeotropes. Suitable binary mixtures are listed in Table 6.2.

Table 6.2 Column test mixtures

Compound	γ^∞	α	Suitable for testing stages
Methanol[a]	0.89	1.7	4–20
Ethanol[a]	1.16		
Monochlorobenzene	1.01	1.13	10–50
Ethylbenzene	0.99		
Toluene	0.96	2.8	2–7
Ethylbenzene	1.08		
n-Butanol	1.00	1.5	5–30
Isobutanol	1.00		
n-Heptane		1.075	20–80
Methylcyclohexane			

[a] A mixture of methanol and ethanol is available as a low excise duty blend as industrial methylated spirits (IMS). It only contains 4% methanol and because almost all of this may fractionate into the column head, more methanol will need to be added to IMS to make a satisfactory test mixture and to avoid excise problems in producing an ethanol that is no longer denatured.

From tests with such mixtures, the number of theoretical stages can be calculated using the Fenske equation:

$$N_{min} \ln \alpha = \ln F \qquad (6.1)$$

where N_{min} is the number of theoretical trays at total reflux and F the separation factor, defined for a binary mixture as

$$F = \left(\frac{x}{1-x}\right)_T \left(\frac{1-x}{x}\right)_B \qquad (6.2)$$

where x is the mole fraction of the more volatile component in a binary mixture and T and B denote the top and bottom of the column, respectively.

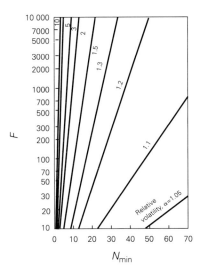

Fig. 6.1 Graphical presentation of Fenske equation

Relative volatility

Figure 6.1 shows this equation in graphical form. It will be seen that for a feed split to make 99% molar tops and bottoms ($F = 9801$), any value of α less than 2.0 will require a very large column. In fact, fractional distillation to produce relatively pure products is seldom the correct choice of technique when the system has a relative volatility of less than 1.5.

For an ideal mixture, such as those listed in Table 6.2, the relative volatility is fairly constant throughout the column and equation 6.1 can be used. It is possible either to predict the number of trays at total reflux needed to achieve a given degree of separation, or the degree of separation that will be achieved by a column whose fractionation power is known. Since the number of theoretical trays at total reflux is the minimum needed for a given separation, it is possible to show, under ideal circumstances, whether a separation is possible or not.

Example 6.1
It is desired to separate a binary mixture of acetone and MEK. These can be considered to be ideal with a relative volatility of 2.0. The acetone at the column top must not contain more than 1 mol% MEK. At total reflux on a column of ten theoretical plates, what will be the composition of the column bottom?

$$10 \ln 2 = \ln\left(\frac{99}{1}\right) - \ln\left(\frac{1-x_m}{x_m}\right)$$

$$x_m = 0.912$$

i.e. the MEK concentration at the column bottom would be 91.2 mol%.

As reference to Appendix 1 will show, there are few binary mixtures of solvents that behave in an ideal fashion, or even close to it, over the whole range of concentrations. In addition, relative volatility changes with temperature and thus will vary over the height of a fractionating column.

However, the large departures from ideality usually occur because of azeotrope formation, which may make a given separation impossible by straightforward fractionation, or at the extreme ends of the composition range. In the latter case, high values of γ^∞ for the more volatile component of a binary mixture make stripping it from the less volatile one easier, whereas to remove the less volatile from the more volatile is more difficult (*see* Chapter 12, equations 12.12 and 12.13). It therefore may make little difference to the overall column requirements of a separation, and can be ignored at the earliest stage of an assessment. This is particularly so since a large safety factor is normal in practice in choosing an appropriate column for a separation.

Tray requirements

At total reflux, a given separation will be achieved with a minimum number of plates (N_{min}). In practice, a recovery operation should not often need to be operated with a reflux ratio of more than about 5:1. Experience shows that about $2.5N_{min}$ theoretical stages are required for the most economical operation. Since an actual tray has an efficiency of 60–70%, this means that $3.8N_{min}$ actual trays would ideally be needed. The reboiler or, in batch distillation, the kettle represents one theoretical tray so, *very roughly*, a column of $2N_{min}$ m between tangents should be sought for a separation based solely on the Fenske equation.

It is also helpful, for the first assessment of a separation project, to estimate a value for α. Figure 6.2 shows a family of vapour–liquid equilibrium curves based on constant values of α for binary mixtures. The compositions are in mol%. Quick comparison, if a curve for the mixture to be separated is available, will provide a value for α.

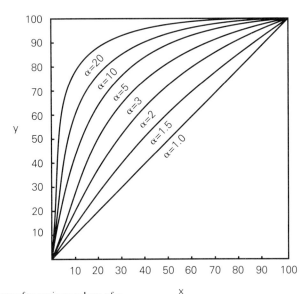

Fig. 6.2 VLE Diagram for various values of α

Alternatively, although it depends on the assumption that γ is equal to 1.0 for both components throughout the composition range, the Antoine or Cox equations can be used to estimate α at any given temperature between the boiling points of the two components.

Example 6.2
Using Cox chart constants, estimate the value of log α for the binary system MEK (1)–MIBK (2).

$$\log p_1 = 7.22242 - \frac{1345.9}{T+230}$$

$$\log p_2 = 7.27155 - \frac{1519.2}{T+230}$$

$$\log \alpha = \log p_1 - \log p_2$$

$$= -0.04913 + \frac{173.3}{T+230}$$

The atmospheric pressure boiling points of MEK and MIBK are 80 and 116 °C, respectively, so at various positions in the column the temperature will lie between the two boiling points. Calculated α values are given in Table 6.3.

Table 6.3 Calculated ideal values of α for MEK–MIBK

T °C	α
80	3.23
90	3.11
100	2.99
110	2.89
116	2.83

As one would expect with a binary mixture of two chemically similar solvents, their behaviour closely follows Raoult's law, with γ_1^∞ being 1.073 and γ_2^∞ 1.0125.

In only a very few cases is the value of α at a lower boiling point.

$$\frac{d \log \alpha}{dT} = \frac{B_1 - B_2}{(T-230)^2} \tag{6.3}$$

Since $B_1 - B_2$ is usually negative, the value of α in a binary system almost always falls as the temperature rises. To aid fractionation it is, of course, desirable to have a high α and operating under vacuum will usually help to provide this.

If the value of α is likely to vary greatly over the height of a column, it is desirable to use the geometric mean of α at the column top and bottom.

Batch vs. continuous distillation

Equation 6.1 can be applied to both continuous and batch distillation. In the latter case, the value of x_B is the mole fraction of the more volatile component in the still kettle, and will vary as the batch proceeds, reducing until the composition of the material in the kettle corresponds to the composition in the reboiler of a continuous column, fractionating the same original feed to the same purity specifications. The number of theoretical trays needed to produce a required distillate from a given feed is thus, at the start of a batch distillation, less than that needed in a continuous one.

Many solvent recovery operations cannot benefit from the long steady-state runs typical of continuous operation, because the necessary quantity of consistent feed is not available. The possibility of achieving a separation with a smaller number of trays makes batch distillation attractive in these circumstances.

Against this advantage there must be set the disadvantages of batch distillation:

- Because operation is not steady state, much more attention must be devoted to running the plant or on-line analysing equipment must be fitted and depended upon.
- Longer residence times for much of the batch at high temperature can lead to decomposition and polymerization of components of the feed. Apart from reducing yield and creating impurities not originally present, this may increase the risk of an exothermic reaction. Because of the larger hold-up in the equipment, the energy and potential damage of such a reaction is liable to be greater in a batch than in a continuous plant.
- Batch distillation tends to produce intermediate fractions because of the unavoidable hold-up in the system. These have to be recycled in subsequent operations reducing the net size of charges.
- The 'housekeeping' involved in batch operation, charging the still and removing quantities of hot residues reduce the available running hours and require operator's time and attention.
- A batch fractionation plant can be considered as a column with a single stripping plate, the evaporation occurring at the surface of the liquid in the kettle. All its actual column consists of enriching plates.

On the other hand, a continuous column provides the flexibility of splitting the available fractionating power into any ratio of stripping to enriching, subject only to the availability of a feed point at the correct position.

Because of the small amount of stripping power available, it will often be difficult to strip the last of a volatile component from the residue in a batch still. In solvent recovery practice, this shortcoming is not as serious a disadvantage as in the production of new solvents. When producing new solvents, one is frequently dealing with a mixture of homologues (e.g. benzene, toluene and xylene) with values of γ^∞ near 1.0. In solvent recovery it is much more common to be dealing with mixtures of chemically dissimilar compounds which have comparatively high values of γ^∞. The vapour pressure of volatile impurities in residues tends to be non-ideal and much higher than Raoult's law would predict. Stripping is therefore much easier and the reliance on a single stripping stage not as restricting as conventional practice derived from solvent production would lead one to expect.

Batch vs. continuous distillation

Once it has been established that a column, whether batch or continuous, is capable of making a separation at total reflux and, therefore, that a simple fractionation is possible, considerations of economics and available capacity need to be made. This involves estimates of the second parameter in making a binary separation, the reflux ratio.

To achieve a separation one needs to use at least a minimum reflux ratio (R_{min}). Using simplifying assumptions of constant volatility and equal molar latent heat for the components,

$$R_{min} = \frac{x_T}{(\alpha-1)x_F} - \left(\frac{1-x_T}{1-x_F}\right)\left(\frac{\alpha}{\alpha-1}\right) \tag{6.4}$$

where x_F is the mole fraction of the lighter component in the feed in the case of continuous operation, or in the still for a batch plant. When a high degree of purity is needed (e.g. $x_T = 0.995$), the equation can be reduced to

$$R_{min} = \frac{1}{(\alpha-1)x_F} \tag{6.5}$$

It has already been shown that, for a high separation factor (F), it is rare to use fractionation with values of α less than 1.5 (Figure 6.1) and this sets a top limit of R_{min} at $2/x_F$.

Thus, whereas separations done continuously need a higher minimum number of plates than do batchwise separations, the latter commence with a value for R_{min} equal to that required for a continuous split, and the required value of R_{min} increases as the composition of the kettle changes and x_F ($= x_B$) becomes lower and lower.

Example 6.3
What are the values of N_{min} and R_{min} required to separate an equimolar mixture of MEK and MIBK into fractions containing 1 mol% of MIBK in MEK and vice versa, assuming $\alpha = 3.0$ throughout the separation?

For continuous operation, $F = 9801$. From equation 6.1,

$N_{min} = 8.36$

$$R_{min} = \frac{1}{(3-1)\times 0.5} = 1:1$$

For batch operation, initially $F = 99$,

$N_{min} = 4.18$

$$R_{min} = \frac{1}{(3-1)\times 0.5} = 1:1$$

Finally, $F = 9801$,

$N_{min} = 8.36$

$$R_{min} = \frac{1}{(3-1)\times 0.01} = 50:1$$

This would indicate that it is impractical to achieve the purity of product at the same yields by batch as it is for continuous distillation when the relative volatility for the system lies in the middle range. Substitution of a higher value for α brings the value of R_{min} at the end of a batch separation down to a more practical level.

Common operating practice usually involves setting the reflux at about $1.25R_{min}$. Once the values of N_{min} and R_{min} are known, the Gilliland correlation between reflux ratio and number of theoretical stages allows the reflux ratio to be worked out for a column with a known number of trays (N):

$$\frac{N - N_{min}}{N + 1} = 0.75 \left[1 - \left(\frac{R - R_{min}}{R + 1} \right)^{0.5668} \right] \tag{6.6}$$

Example 6.4
For a continuous column of 20 theoretical stages, what reflux ratio will be required for the separation in Example 6.3?

$$\frac{20 - 8.36}{20 + 1} = 0.75 \left[1 - \left(\frac{R - 1}{R + 1} \right)^{0.5668} \right]$$

$R = 1.21$

It is very important to note that the above calculation of R_{min} and R is only valid if the column feed in a continuous column is put into the correct position in the column. All trays between the actual feed point and the correct feed point are 'lost'. This is one reason why the nearly complete flexibility that a tray column provides on feed point choice is valuable when the column is to be used for a variety of feedstocks, some of which cannot be specified when the column is designed.

A packed column, whether filled with random or ordered packing, can only have a feed point where there is redistribution and at most every eight or so theoretical stages. The average loss of fractionating power due to malposition of the feed is therefore on average four or more, which may be significant in a short column.

The optimum position of a liquid feed point is where the composition of the feed is the same as that of the liquid leaving the feed tray or, if the feed is a vapour, the vapour leaving the feed tray.

Solvent recovery poses some fractionation problems that are not often encountered when processing unused solvents. It is not uncommon, in both batch and continuous operation (although less in the latter because of the shorter residence time at boiling point), to find that the distillate is contaminated with breakdown products. These may be from hydrolysis or decomposition of the solvents themselves, or from solutes derived from the process in which they have been used. Aldehydes, which impart an unacceptable odour to the recovered solvent, are also sometimes present. Traces of water in hydrocarbons or chlorohydrocarbons may also be found at the column top.

Provided that the column has more than enough separating power for the main fractionation it needs to do, it is often of benefit to take the main product off as a liquid side stream four or five actual trays from the column top. The top trays can be operated at total reflux with occasional purging when the concentration of the light impurities begins to spread down the column or starts to interfere with the effective working of the condenser. In the case of water, a small phase separator will prevent the water returning to the column, although this may need to treat the water as a top or a bottom phase depending on the density of the organic distillate.

The side stream will be in equilibrium with the vapour leaving the same tray. As a result, the product will contain impurities to the extent of the relative volatility and concentration of product and contaminant in the column vapour at the product take-off tray.

The technique may be extended on batch stills to avoid the major disadvantage of their operation. A batch still can, in theory, make a series of pure products whereas a continuous column can only produce at best two pure products, tops and bottoms, although side streams containing concentrates of components can be taken off both as liquids and vapours. However, the column top must have a liquid hold-up in the condenser, reflux drum, phase separator, vent condenser and other vessels, together with their interconnecting pipework. At the point in the batch distillation when one product has almost all been distilled off and the subsequent one is reaching the column top, there is inevitably a mixing of the two leading, if the product specifications require nearly complete separation, to the production of intermediates which have to be recycled to the feed tank. Design of the column top to include a partial condenser (otherwise known as a dephlegmator) to reduce the column top hold-up and eliminate the reflux lines can reduce the volume of the top works at the same time as it adds an additional separation stage to the column. Such a design effectively prevents a phase separator being installed which reduces the column top volume, but also reduces the flexibility of the plant as a whole.

Provided that adequate fractionation power exists for separating a second product from the third, or from the residue if no third product is required, it is attractive to take the second product as a liquid side stream at, say, the column mid-point. The upper half of the column then concentrates any traces of the most volatile product at total reflux. Thus no still time is wasted on taking an intermediate fraction, which usually requires much testing, tank changing and labour-intensive plant operation.

There is no theoretical reason why a third take-off even lower down the column should not be installed for a further distillate fraction, but the increased complication would seldom be justified.

The satisfactory operation of such a system depends on there being no failure of boil-up so that the material held at the column top does not fall down the column and reach the side stream take-off and spoil the product being taken off there. A temperature control linking a point in the column with a stop valve on the side stream is a desirable safety feature (Figure 6.3).

If a series of batches of the same feedstock is planned, the column top will be left at the end of a batch at a suitable composition for turning into a product tank very shortly after the commencement of the next batch, since no 'heavy' material ever reaches the column top.

The other major operational disadvantage of a batch still is its lack of stripping plates, although this, too, can be partially overcome with the use of a connection at or near the column mid-point. The conventional way of starting a batch charge is to fill the kettle with feedstock and to commence boiling. If the feed is pumped not to the kettle but to, say, the column mid-point and boiling is commenced in the kettle as soon as the coils are covered or the circulating pump can be primed, the vapours meeting the feed descending the column will strip out the most volatile components of the feed. Not only will this

88 Separation of solvents

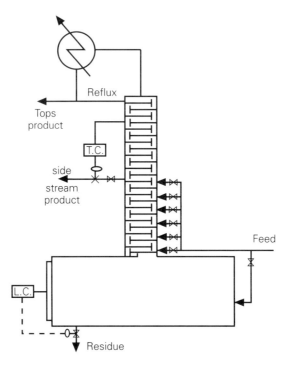

Fig. 6.3 Hybrid batch/continuous still

provide a stripping action for producing the first fraction, but it will increase the size of the batch, since accommodation for most of the first fraction need not be found in the kettle. Thus, if the first fraction is a large proportion of the feed, a batch may be doubled or enlarged even more in size. Here again, safe operation requires a level control on the kettle to cut off the feed when the kettle is full.

The combination of batch and continuous operation that these two techniques provide is particularly applicable to a solvent recovery plant where the separation requirements may vary widely and be difficult to predict.

Vacuum distillation

The choice of fractionation under vacuum as the means of carrying out a separation, whether continuous or batchwise, can be made for three reasons:

- to achieve improved α in comparison with α at atmospheric pressure;
- to keep the highest temperatures needed in the solvent recovery operation as low as possible for economic reasons;
- to avoid damaging materials being processed which may have a tendency to decompose, polymerize or exotherm at higher temperatures.

Improved relative volatility (α)

As equation 6.3 shows, the temperature dependence of $\log \alpha$ is proportional to the difference between the values of Cox chart B for the two components in a binary mixture.

Using the Cox equation,

$$\log \alpha^* = (A_1 - A_2) - \frac{B_1 - B_2}{T + 230}$$

where α^* is Raoult's law perfect relative volatility, T (°C) is the system temperature and the subscripts 1 and 2 refer to the more and less volatile components of the mixture, respectively.

Both A and B increase with increasing boiling point of the solvent so that $A_1 - A_2$ and $B_1 - B_2$ for two solvents in the same class will always be negative. By differentiating with respect to temperature it is clear that the value of α^* increases as the temperature is reduced:

$$\frac{d(\log \alpha^*)}{dT} = -(B_1 - B_2)(T + 230)^{-2}$$

However, the effect of reducing temperature is small if the value of $B_1 - B_2$ is small, as will be the case for two solvents of the same chemical class if their atmospheric pressure boiling points are close together.

Whatever the value of $B_1 - B_2$, the advantage of working at the lowest practical temperature is clear (Table 6.4). A reduction in T from 200 to 100 °C will increase the value of $\log \alpha^*$ by 0.70×10^{-3} $(B_1 - B_2)$ whereas reducing T from 200 to 50 °C will increase $\log \alpha^*$ by 1.24×10^{-3} $(B_1 - B_2)$. There is a practical limit below which the temperature in vacuum distillation cannot be reduced owing to the difficulty of condensation and this usually is in the range 40–50 °C.

Larger values of $B_1 - B_2$ are obtainable when the components of a mixture are from different chemical classes (Table 6.5 and Figure 6.4.) This is particularly true if one, but not both, is an alcohol. The value of $B_1 - B_2$ for solvents boiling at 150 °C can then be up to 400, and the advantage of operating at 50 °C means that $(B_1 - B_2)/(T + 230) = 0.376$. Therefore, α^* is about 2.4 times greater at 50 °C than at 150 °C.

Table 6.4 Effect of absolute pressure on ratio of vapour pressures of esters

Atmospheric pressure b.p. of ester (°C)	Vapour pressure at 100 °C (mmHg)	Relative volatility at 100 °C	Vapour pressure at 50 °C (mmHg)	Relative volatility at 50 °C
100	760	1.00	120	1.00
110	554	1.37	81	1.48
120	403	1.89	55	2.18
130	291	2.61	37	3.24
140	210	3.62	25	4.80
150	150	5.07	17	7.06
160	107	7.10	11	10.91

90 *Separation of solvents*

Table 6.5 Values of A and B in the Cox equation for compounds with an atmospheric pressure boiling point of 150 °C

Chemical class	A	B
Cyclopentanes	7.15762	1625.2
Aromatic hydrocarbons	7.23646	1655.2
Aliphatic hydrocarbons	7.24133	1657.0
Haloaliphatics	7.26002	1664.1
Aliphatic nitriles	7.28081	1672.0
Aliphatic ketones	7.31923	1686.6
Aliphatic ethers	7.33844	1693.9
Nitroalkanes	7.38423	1711.3
Aliphatic esters	7.52344	1764.2
Aliphatic alcohols	8.27396	2049.4

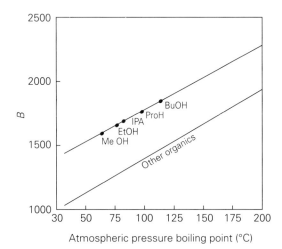

Fig. 6.4 Value of Cox Chart B for alcohols and other organics as boiling points.

The case of a practical separation which relies on low pressure for a fractionation illustrates the effects of temperature and chemical class:

	B.p. (°C)	Chemical class	B
Cyclohexanone (1)	156	Ketone	1716.5
Cyclohexanol (2)	161	Alcohol	2110.6

Hence $B_1 - B_2 = 394.1$. Of this difference in B, chemical class contributes about 360 and difference in boiling point only about 30. The relative volatility of the mixture at its atmospheric pressure boiling point is about 1.17 and, since high purity is needed for both components in the mixture, fractionation at 760 mmHg is not practicable. At 100 mmHg α is about 1.7 and the throughput of a 30 theoretical stage column where reflux ratio and the vapour handling capacity of the column must both be taken into account is optimum at this pressure.

It should be noted that the cyclohexanone–cyclohexanol system obeys Raoult's law almost exactly so that the effect of lowering the pressure and temperature does not depend on any variation of activity coefficient. It is not safe to rely on ideality except when components belong to the same class, but the effect of low temperatures on the values for activity coefficients is not very large in the solvent recovery range and atmospheric pressure boiling point values can be used with caution.

Improved ΔT in reboiler

Distillation tends to be a very large consumer of heat in any chemical factory and may be the deciding factor in determining at what pressure steam must be generated or to what temperature hot oil or other heating medium must be raised and distributed. This can have a fundamental effect on the thermal efficiency of the factory calling for thicker insulation, return mains, increased costs of leaks, problems with flash steam, etc.

The fundamental equation that governs the reboiler or evaporator in a distillation operation is

$$q = hA(T_H - T_P) \tag{6.7}$$

where
- q is the amount of heat transferred;
- h is the overall heat-transfer coefficient;
- A is the heat-transfer area;
- T_H is the temperature of the heating medium;
- T_P is the temperature in the process.

If T_H is to be kept low for economic or operational reasons, the steps that can be taken are as follows:

a Minimize q at points of heat use where T_P is high. If heat can be put into a distillation column by heating the feed rather than the residue, the presence of the volatile components of the feed will keep the boiling temperature low, and since the feed is injected some way up the column the pressure at which the feed boils will be less than the pressure at which the reboiler operates.

b Operate at as low a pressure as possible. This will be determined by the ability to condense the column top vapour. Ample cooling tower capacity and heat transfer area on the condenser are needed. A vent condenser fed with chilled brine or glycol–water should be considered. There is a considerable difference in the pressure drop per theoretical stage between various tower internals, and if those with low pressure drop can be used for other operational reasons, their extra cost may be justified. If vacuum is provided with a liquid ring pump, a low vapour pressure circulating medium (e.g. glycol, gas oil) should be used rather than water.

c Have ample heating surface in the reboiler and choose a system that does not foul and has an intrinsically high overall heat transfer coefficient.

Avoiding chemical damage

This topic is covered in Chapter 5.

Against the advantage of fractionating under reduced pressure must be set some important negative aspects.

First, the diameter of a fractionating column is determined by the vapour load it can carry and flooding will take place if, for any particular design, a certain value of G^2/ρ_G is exceeded, where G is vapour velocity in weight/s/unit of cross-sectional area and ρ_G is vapour density. Since the value of G sets the rate at which a column can produce distillate, the higher the value of ρ_G the greater is the productive capacity of the column and therefore the lower the column diameter needed. For a given solvent, ρ_G is higher when the absolute pressure of the system is higher. This effect is partially offset by the increase in boiling temperature at a higher system pressure.

For the separation of an alcohol from a ketone, an example suitable for vacuum operation, the factors listed in Table 6.6 determine the effect of fractionating at low pressure.

Table 6.6 Product rate comparison of cyclohexanol–cyclohexanone separation under vacuum and at atmospheric pressure[a]

	p (mmHg)	
	760	100
T (K)	432	378
$p/T\ (\propto G^2)$	1.76	0.26
$(p/T)^{1/2}$	1.33	0.51
α	1.17	1.75
R_{min}[b]	11.76	2.67
$R\ (= 1.25 R_{min})$	14.7	3.34
$(p/T)^{1/2}/1+R$[c]	0.085	0.118

[a]This simplified comparison assumes sufficient fractionating stages to achieve the chosen product purity. With a low value of α for the atmospheric pressure case, a large number of stages would be needed.
[b]Assuming a high-purity distillate and using equation 6.5 for $x_F = 0.50$.
[c]$(p/T)^{1/2}/1+R$ is proportional to the rate of product.

As has been shown (Table 6.5), the separation involving an alcohol is particularly suitable for low-pressure operation and, even so, the product rate on a given column is not remarkably different from that at atmospheric pressure.

Second, a distillation plant operating at atmospheric pressure can often discharge product to storage without a pump, since the height of the column provides sufficient head. If the condenser on a high-vacuum column is not high enough above the ground to provide a barometric leg, a pump is needed to transfer product, and it will not have a positive pressure over its suction. It will therefore be liable to leak air into its suction in the event of a seal failure. Similarly, any pump handling residue or column bottoms will always be under vacuum and here an expensive double mechanical seal or a glandless pump will be needed for reliable operation.

Steam distillation

The disadvantages of steam distillation are that many of the lower boiling solvents form water azeotropes which are difficult to dry, and an appreciable amount of contaminated water can arise from it. These matters are covered in Chapters 5 and 7.

For solvents that are not appreciably water miscible, steam distillation can achieve the same sort of advantages in improving relative volatility as vacuum distillation. The steam can be viewed as an inert carrier gas that allows the solvent mixture to boil at a temperature below its atmospheric boiling point. Thus at 84.5 °C toluene has a vapour pressure of 333 mmHg, and would boil under a vacuum of that level. Water at 84.5 °C would contribute a vapour pressure of 427 mmHg, so that a water–toluene mixture would boil at atmospheric pressure.

While toluene would have a relative volatility with respect to ethylbenzene at atmospheric pressure of 2.1, it would have a relative volatility of 2.23 at 84.5 °C as expected from the boiling point effect of the Cox chart B values of toluene and ethylbenzene.

However, this gives the same effect as a very modest reduction in pressure. For a substantial improvement in α, mixtures should be sought in which the steam contributes the major part of the combined vapour pressure. This will correspond to a temperature of, say, 99 °C, at which the vapour pressure of the solvents will be about 30 mmHg. This corresponds to a solvent mixture with an atmospheric pressure boiling point of about 160 °C, at which (as Table 6.4 showed) an appreciable beneficial effect on the relative volatility of the components of the mixture might be expected.

The drawback of obtaining an improved relative volatility in such a way is that very large amounts of steam may have to be used. The moles of steam used per mole of solvent can be calculated.

$$\frac{n_w}{n_s} = \frac{p_w}{p_s} = \frac{P - p_s}{p_s} \tag{6.8}$$

where

n_w = moles of steam;
n_s = moles of solvent;
p_w = partial pressure of steam;
p_s = partial pressure of solvent;
P = total system pressure.

At 99 °C and atmospheric pressure:

$$\frac{n_w}{n_s} = \frac{730}{30} = 24.3$$

Typically a solvent with a boiling point of 200 °C will have a molecular weight of about

94 Separation of solvents

120, although this can vary widely, so that about 3.5 kg of steam is needed for each kilogram of solvent.

If the relative volatility required for the separation is provided at a temperature of 99 °C and a system pressure of 100 mmHg can be achieved, then only a 70 mmHg contribution is required from steam and the steam injected to achieve boiling is only

$$\frac{n_w}{n_s} = \frac{100 - 30}{30}$$

i.e. 0.35 kg/kg solvent. On the other hand, if the highest value of α is required, the system temperature should be reduced to the lowest achievable figure. This will correspond to the limit of the combined condenser and vacuum-inducing system and may typically be about 40 mmHg in a general-purpose plant fitted with a liquid ring vacuum pump.

For the separation of two aliphatic esters boiling at 180 and 200 °C and of the same chemical class (in Table 6.5), the figures listed in Table 6.7 would be typical.

Table 6.7 Interaction of steam and vacuum on α of aliphatic esters boiling at 180 and 200 °C

System temperature (°C)	System pressure (mmHg)	α	Steam injected (kg/kg solvent)
190	760	1.30	0
100	38	2.03	0
99	760	2.03	3.5
99	100	2.03	0.35
50	38	2.50	1.75

Steam distillation can be carried out in both batchwise and continuous plant and, since the equipment is very simple and can be used on a routine basis for preparing the plant for maintenance or for cleaning between campaigns, steam injection facilities should be fitted even when process application is not immediately foreseen.

If the amount of steam used per unit of overheads is large and the steam pressure at the injection point is high, the superheat available may be enough to provide the latent heat needed for the distillation. However, heating steam should normally be controlled and measured separately from 'live' steam to give full control and flexibility to the system.

Azeotropic distillation

Azeotropic distillation is a commonly used solution to a fractionation problem in which, at whatever pressure, the value of α is too low for the techniques described so far. It is particularly valuable for breaking apart the components of existing azeotropes in a system, but it can also be used when the required separation is a very difficult one (i.e. $\alpha < 1.5$).

It involves adding a further component, an entrainer, which forms an azeotrope with one of the members of an existing mixture and not the others (or the other, in the case of a binary mixture). Usually it is desirable for the entrainer to form an azeotrope with the small component rather than with the majority of the mixture since this, in most cases, reduces the amount of entrainer to be recycled. However, the entrainer's most important property is ease of separation from the solvent it is removing.

In a few cases, this does not present a problem. Dimethylformamide (DMF) forms low-boiling azeotropes with heptane and xylenes. The addition of water in the fractionation system as an entrainer allows the formation of azeotropes with these hydrocarbons, but not with DMF. The hydrocarbon–water azeotrope at the column top splits into two liquid phases, so that the hydrocarbon can be removed and the water recycled. Similarly, in drying ethanol and isopropanol the unwanted water forms a separate phase which can be removed while the entrainer is recycled to pick up more water.

Figure 6.5 shows the likely range in which immiscibility may be found and it will be seen that it is comparatively rare in the absence of water among alcohols, esters, ethers and glycol ethers.

Since water tends to be the key to phase separation, it is often necessary to add it to the overheads outside the column. Thus, to separate methanol and acetone, which form an

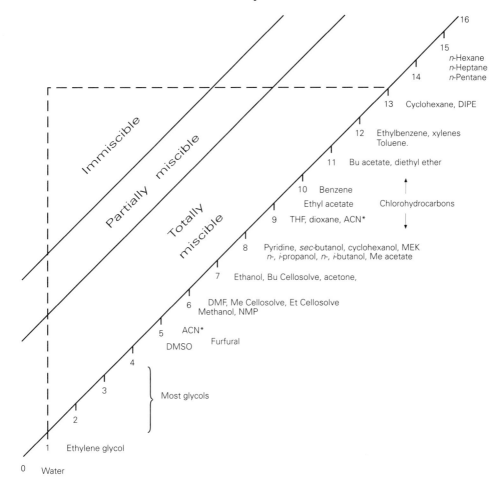

Fig. 6.5 Mutual miscibility of common solvents. The example shows that ethylene glycol is immiscible with diisopropyl ether
* ACN shows unusual properties with miscibility at both the top and bottom of the 0–16 scale

azeotrope at 55 °C with 12% methanol, it is possible to add methylene dichloride (MDC) to the system. MDC forms an azeotrope with methanol (boilling point 38 °C, 7% w/w methanol) but not with acetone, so the MDC–methanol may be taken as an overhead. Addition of water to this azeotrope results in a two-phase system with the methanol partitioning strongly in favour of the water phase. The MDC phase is recycled to the column while the methanol can easily be separated from water by fractionation or, because it is a very cheap material and only 12% will have been present in the original acetone azeotrope, it may be more economic to dispose of it.

A similar application of the addition of water to recover the entrainer is in the separation of methanol from THF. Even if THF and methanol did not form an azeotrope, they have boiling points so close that there would be little expectation of separating them by ordinary fractionation. Using n-pentane as an entrainer, methanol can be removed from the methanol–THF azeotrope with comparatively few fractionation stages and, as can be seen in Figure 6.5, methanol is very close to being immiscible with pentane. Hence the addition of a small amount of water will break the mixture into two phases, allowing an almost pure pentane to be returned to the column.

The drawback to this as a recovery method is that pentane carries very little methanol out of the system (pentane–methanol azeotrope, 8% w/w methanol) so that if the original mixture of methanol and THF is rich in methanol, a preliminary concentration is desirable to reach the THF–methanol azeotrope (31% methanol). Since the relative volatility between methanol and the azeotrope is about 2.0, this should not be a difficult separation.

Because it is both cheap and easy to water-wash from hydrocarbons and chlorohydrocarbons, methanol will often prove to be a useful entrainer. It has the additional advantage that it displays azeotropism with a large number of solvents and that, while it has a low boiling point, it is not so low as to cause problems in condensation.

Azeotropic distillation can be done continuously or batchwise. In the latter case, enough entrainer should be used so that while it is removing the solvent with which it azeotropes, there should be a low but positive concentration of it in the still kettle. Azeotropic distillation done batchwise is particularly well suited to a hybrid unit (Figure 6.3), since small amounts of entrainer can be held in the column top while the second product is being removed at the column mid-point.

For continuous azeotropic distillation, most of the entrainer will usually need to be returned to the column as reflux and the remainder may be mixed with the column feed, possibly throwing water out of solution from the feed. If there is no separation of this sort, the 'spare' entrainer should be fed continuously above or with the feed depending on whether the entrainer is less or more volatile than the feed.

Phase separations, whether involving water or not, can be considerably affected by the temperature at which they take place. It may be economic to carry them out at low temperature with heat exchange between the hot overheads leaving the condenser and the reflux which is cold as it leaves the phase separator. The reflux should then be reheated before being returned to the column.

Column behaviour will be non-ideal for both batch and continuous azeotropic fractionating. Although calculation of trays and reflux required for a separation is possible when the necessary data are available, it is usually advisable, once a likely entrainer has been identified, to carry out laboratory trial fractionation of the system. Using a Dean and Stark column head, this is easy for both modes of operation.

It is possible to identify azeotropic entrainers on a theoretical basis, but the very comprehensive tabulation of azeotropes, their compositions and azeotropic temperatures available makes it quicker and easier to search for them (*see* Bibliography).

Extractive distillation (ED)

While the application of azeotropic distillation is restricted by the small choice of effective entrainers, it is a technique that is comparatively easy to apply since the equipment required is very similar to that for standard atmospheric pressure fractionation, both in a batch and a continuous mode.

There are plenty of entrainers for ED but the equipment needed is specialized. It is, however, a way of separating by distillation that has great potential in solvent recovery, where mixtures of solvents with similar boiling points, but differing chemical characteristics, are common.

Most of the published theoretical and practical applications of ED have been in the separation of hydrocarbons (alkenes from alkanes, toluene from naphthenes, benzene from paraffins, etc.) where, even in the absence of azeotropes, the relative volatilities are very low.

The number of solvent recovery applications is small primarily because ED cannot, except in a few specialized cases, be carried out batchwise. This tends to limit ED to comparatively large streams. Such streams do not arise in solvent recovery until a process approaches maturity, at which point there is likely to be resistance to changing solvent recovery methods.

To understand how ED works, it is necessary to consider how low values of relative volatility arise.

$$\alpha = \frac{\gamma_1}{\gamma_2} \times \frac{P_1}{P_2}$$

where P is the vapour pressure of a pure substance, γ is the activity coefficient and subscripts 1 and 2 refer to the component with the higher and lower boiling point at atmospheric pressure, respectively.

If α is low (i.e. <1.5), this can be for three reasons:

Case 1. P_1 and P_2 may be substantially different (e.g. ethanol and water, $P_1/P_2 = 2.29$), but the ratio γ_1/γ_2 may be sufficiently lower than 1.0 that an azeotrope ($\alpha = 1.0$) can form.

Case 2. P_1 and P_2 can be very close (e.g. ethanol and isopropanol) but, because of their chemical similarity, the values of both γ_1 and γ_2 are very close to 1.0.

Case 3. P_1 and P_2 can be very close but, despite their chemical dissimilarity, γ_1/γ_2 is close to 1.0 (e.g. benzene and carbon tetrachloride, $P_1/P_2 = 1.13$).

It has been shown (equation 6.3) that, unless one of the components is an alcohol, solvents that have similar boiling points have a substantially constant P_1/P_2 value throughout the range of solvent recovery pressures and temperatures. ED is based on introducing an entrainer that modifies γ_1/γ_2 to increase the value of α.

In considering the appropriateness of ED for a separation, it is important that it is known why α is low. The chances of finding an entrainer that will raise α to a point at which fractionation will be easy is good for Case 1, provided that γ_{1E} (the activity coefficient of component 1 in the extractive distillation solvent) is greater than γ_{2E} since here the activity coefficient ratio is reinforced by P_1/P_2.

When P_1 and P_2 are very close as in Case 3, it is not important to reinforce the P_1/P_2 effect if an entrainer can be found that gives $\gamma_{2E} \gg \gamma_{1E}$. Both products from ED are distillates, so that even if one of the products is due to be discarded or burnt, there is no advantage to be gained by making it the column bottoms of the extraction column.

A special case may exist if ED is applied to a binary mixture, only one component of which is to be recovered, and where water can be used as an entrainer. For such a separation, the entrainer recovery column can be dispensed with and a very dilute solution of bottoms product in water can be sent to effluent disposal. Such possible special cases are listed in Table 6.8. For all these cases, a mole ratio of water to feed of

Table 6.8 Effect of water on relative volatility

Product component	Disposal component	Normal relative volatility	ED relative volatility
Isopropanol	Ethanol	1.16	1.5
sec-Butanol	n-Propanol	1.10	1.3–1.6
tert-Butanol	Isopropanol	1.03	1.4
MEK	Ethanol	1.05	3.0
tert-Butanol	Ethanol	1.2	2.0
MEK	Isopropanol	1.1[a]	2.1
MEK	tert-Butanol	1.1[a]	1.7
DIPE	Ethanol	1.04	5.7

[a] Systems in which the relative volatility is not reversed by the presence of water.

9:1 has been assumed and, in all but two systems, the lower boiling component has been transformed into the more volatile one by the presence of water.

The product component when water is used as an entrainer is, of course, a water azeotrope in many cases and complete recovery will involve an extra refining process. However, ED does provide a possible solution to very difficult separation problems. All the separations listed in Table 6.8 could also be done using other very polar entrainers apart from water but, because of their much higher molar volumes (9:1 entrainer to feed molar ratios, Table 6.9), they are unlikely to be practicable in general-purpose plant.

Table 6.9 Molar volumes of some potential ED entrainers

Entrainer	MW	Liquid density (g/cm^3)	Volume per mole (cm^3)
Water	18	1.00	18
MEG	62	1.11	56
Phenol	94	1.06	89
NMP	99	1.03	96
Furfural	96	1.16	83
Xylene	106	0.87	122
Cresol	108	1.03	105
MCB	112	1.11	101
DMF	73	0.95	77

- Theory indicates that ED will be more efficient the lower is the operating temperature but the higher the entrainer concentration. The higher the entrainer concentration, the higher is the operating temperature in Sections B and C, so these factors work against each other but, other things being equal, the lower the boiling point of the entrainer the better.

On ground of cost, toxicity, molar liquid volume and potentially high values of γ, water can be a very attractive entrainer for low-boiling solvents, but there are few such solvents with which it does not azeotrope requiring a further processing step after ED.

Figure 6.6 shows a typical relationship of γ versus concentration for an entrainer that will have useful potential for ED. It is clear that the increase in γ is not very considerable

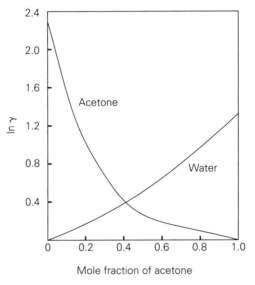

Fig. 6.6 Activity coefficients for the binary system acetone–water showing the effectiveness of a high concentration of water in increasing the relative volatility of acetone

at low entrainer concentrations and that, to be effective, a minimum mole fraction of about 0.80 will be needed. At this level and higher, the inter-reactions of a ternary system can be ignored, and it is adequately accurate at early stages in the design to consider two binaries, component 1 and the entrainer and component 2 and the entrainer, in arriving at values of activity coefficients.

Inspection of Figure 6.7 will show that the ED columns can be divided into five sections. Of these, two (B and C) can be considered as 'active' in that the entrainer is present in these along with components 1 and 2. The remaining three sections, A, D and E, are performing 'cleaning-up' operations. Consideration of the functions of all these sections will reveal the criteria for choosing a suitable entrainer.

Section A
The entrainer must be easily separated from the tops product, both so that the tops product reaches its purity specification and so that the entrainer, which is often an expensive chemical, is not lost from the system.

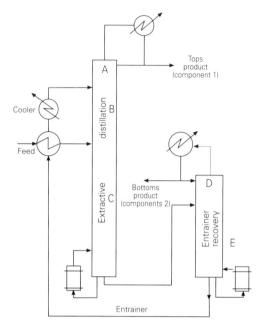

Fig. 6.7 Extractive distillation plant

The reflux returned to this section should be minimized since when it reaches Section B, it will need to be mixed with at least 4 mol of entrainer.

A high value of P_1/P_E coupled with a low value of γ_{E1}^∞ (the activity coefficient of the extractive distillation entrainer in low concentration in component 1) is ideal for these purposes.

Section B
The entrainer being fed into the column at the top of this section should be at a temperature close to the boiling plant of component 1. This is likely to require the entrainer, which leaves the bottom of Section E at its boiling point, to be cooled further even after it has been through the heat exchanger by which the cold feed enters.

The action taking place in this section requires that γ_{1E}/γ_{2E} should be high. Since there will be effectively no component 2 at the top of Section B, this is the place where the entrainer may not be fully miscible with the solvent system to be separated.

To be effective, the entrainer must be present in high concentration in the liquid containing both components of the mixture to be separated. If the liquid forms two phases, the concentration of the entrainer in the solvent will be diminished. High values of γ_{1E} are associated with immiscibility and $\ln \gamma_{1E}^\infty$ in the range 7.0–7.5 or above is likely to fail in this respect. This is seldom met if water is not part of the system in combination with hydrocarbons or chlorohydrocarbons.

Section C
At the same time as a high value of γ_{1E} is wanted, a low value of γ_{2E} is required. If the feed does not contain equal molar concentrations of components 1 and 2, it will follow that at the top of section B or the bottom of Section C, the mole fraction of the entrainer will be appreciably higher than the average.

It is usually true that the addition of an entrainer does more to improve separation by increasing γ_1 to γ_{1E} than by reducing γ_2 to γ_{2E}. If, therefore, there is a large difference between the concentrations in the feed and the mixture is a Case 2 or Case 3, it should favour taking the smaller fraction as overheads in the extractive distillation column, whatever slight advantage may be derived from P_1/P_2.

Section D
It is usual to find that in ED a column with up to 40 theoretical stages is needed to accommodate Sections A, B and C, whereas the column comprised of Sections D and E is short.

Section D, like Section A, is primarily involved in making a component free from extrainer for cost and product quality considerations. There is no reason for a particularly low reflux ratio on Section D, unlike Section A.

Section E
It is most important to strip all component 2 from the entrainer since little fractionation will take place in Section A, and the work done in Sections B and C can be thrown away if the returned entrainer is not pure, so γ_{2E}^∞ should be high.

This requirement clashes with the need for optimum operation of Section C, where γ_{2E}^∞ was preferentially low. Since there is no great disadvantage in having extra trays and/or extra reflux in the entrainer recovery column, it will usually be better to have, as an objective,

γ_{1E} high for sections B and C
γ_{E1} low for Section A
γ_{2E} low for Sections B and C (but not E)
γ_{E2} low for Section D

Any residue that is brought in as part of the feed is not normally eliminated in the entrainer recovery column and will build up in the entrainer. Vacuum facilities should therefore be considered for this column so that, between campaigns, it can distil over the entrainer leaving behind any residue. The residue is likely also to include any inhibitor that may be present in the feed.

For the merchant solvent recoverer whose operating pattern cannot justify a plant dedicated to ED, the possibility may exist to use a pair of batch stills in an ED mode. The capacity of the batch kettles makes the operation easy to control, at the cost of an investment in entrainer. However, the ED column will normally require at least 30 theoretical stages with an entrainer feed six stages from the column top. The remaining stages are then split equally above and below the feed for typical ED operation. If a tray column is used, it is very important to consider the liquid handling capacity of the downcomers, as this is likely to control the plant throughput. The liquid rate is, of course, very much greater than for ordinary batch distillation where it cannot exceed the boil-up.

ED entrainer selection

Some of the properties required of an entrainer have nothing to do with their effectiveness in the system but are, nonetheless, essential:

- Stable at all temperatures in the system at which the highest is found in the lower part of the entrainer recovery column.
- Compatible with all the materials of construction of the equipment.
- Acceptable as far as toxicity and handling (e.g. low freezing point) are concerned.
- Economic. This is of importance to merchant recoverers who may not be able to offset entrainer costs over very long ED runs.

If the above four criteria are met, the operational characteristics of the potential entrainers can be compared.

- Easy separation of entrainer from both products. This will, in general, demand fairly high values for P_1/P_E and P_2/P_E, particularly if γ_{2E} is low.
- A high value of γ_{1E} subject to its miscibility with component 1.
- A small liquid volume per mole of entrainer so that the capacity limitation of liquid downcomers is minimized.

It should be noted that, apart from the power involved in pumping entrainer from the base of the recovery column to the top of Section B and the amount of sensible heat removed to waste from the entrainer, there is no cost penalty for increasing the entrainer/feed ratio until the columns can no longer handle the liquid load. In a plant specifically designed for ED, additional heat exchange between Section E bottoms and Section C bottoms can recover much of the heat otherwise wasted, so that entrainer/feed ratios of up to 19:2 could be used.

The values of γ^∞ show the absolute maximum effect that might be obtained when the concentration of solute approaches zero, and are therefore of only qualitative use in screening possible entrainers. They demonstrate that, if the presence of water in the methyl acetate product at the azeotropic concentration of about 3% w/w is not acceptable, other possible classes of entrainers exist (Table 6.10).

Those systems marked a in Table 6.10 have a Raoult's law relative volatility reversed by the presence of the entrainer. Since for methyl acetate–methanol P_1/P_2 is 1.43, it is

Table 6.10 Possible entrainers for methyl acetate (1)–methanol (2)–entrainer (3) system.

Entrainer	γ_{E1}^∞	γ_{1E}^∞	γ_{2E}^∞	γ_{E2}^∞	$\gamma_{1E}^\infty/\gamma_{2E}^\infty$	Azeotropes with entrainer	
						1	2
Water	8.52	23.6	2.2	1.93	10.6	Yes	No
Ethanol	1.86	1.52	0.89	1.13	1.7	No	No
n-Propanol	2.79	3.59	1.06	1.27	3.4	No	No
Benzene	1.42	1.25	7.42	9.73	0.17[a]	No	Yes
Cyclohexane	3.59	3.39	18.34	20.87	0.15[a]	Yes	Yes
MEK	2.07	2.02	1.02	1.02	2.0	No	Yes
MCB	1.32	1.17	7.76	4.94	0.15[a]	No	No
Butyl acetate	1.21	1.24	3.38	2.89	0.4[a]	No	No

[a] See text.

worth considering as a suitable entrainer one which yields a value of $\gamma_{1E}^\infty/\gamma_{2E}^\infty$ of 0.15.

Using the information in Table 6.10, the following conclusions may be reached for short-listing entrainers for further evaluation:

- *Alcohols.* The values of $\gamma_{1E}^\infty/\gamma_{2E}^\infty$ appear to be rising as the polarity of the alcohol decreases, and it would be worth screening n-butanol and MEG. The latter has a very low molecular weight for its volatility. It also is easy to separate from both products.

- *Aromatic hydrocarbons.* These will all reverse the natural volatility of the products and therefore need to have better values of $\gamma_{1E}^\infty/\gamma_{2E}^\infty$ than entrainers that do not reverse the volatility. Ethylbenzene or xylenes would be the lowest boiling hydrocarbons not to form azeotropes with methanol, but monochlorobenzene looks to be worthy of further testing.

- *Paraffins (alkanes) and naphthenes.* These both form azeotropes with methanol up to a boiling range at which pure hydrocarbons are not readily available.

- *Ketones and esters.* These do not look likely to give large or small enough values of $\gamma_{1E}^\infty/\gamma_{2E}^\infty$ to be of interest.

Table 6.11 shows a similar exercise for the system acetonitrile (ACN)–toluene, which indicates that a C_9 aromatic hydrocarbon would be a suitable entrainer.

Table 6.11 Possible entrainers for ACN (1)–toluene (2)–entrainer (E) system

Entrainer	γ_{E1}^∞	γ_{1E}^∞	γ_{2E}^∞	$\gamma_{1E}^\infty/\gamma_{2E}^\infty$	γ_{E2}^∞	Azeotrope of component with entrainer
Butyl acetate	1.84	2.26	0.91	2.5	0.93	No
n-Butanol	4.76	4.08	2.31	1.8	3.51	Yes (2)
n-Heptane	9.88	20.31	1.20	16.9	1.35	Yes (1)
MCB	2.82	3.27	0.97	3.3	1.00	No
Ethylbenzene	2.46	4.95	0.96	5.1	1.08	No
Xylenes	2.16	5.69	0.83	6.8	1.05	No

Gas–liquid chromatography (GLC) works in a similar way to ED and provides a suitable method of testing a short list of entrainers for a separation. A GLC column packed with an inert solid works in a similar way to a batch distillation column, so that the injected sample eventually leaves the column as a series of peaks in boiling point order. Azeotropes occupy their boiling point position.

Normal GLC operation involves treating the column with a stationary phase which plays the same role as the entrainer in an ED operation, so that azeotropes are split because there is a very high concentration of 'entrainer' to 'solvent'.

Potential entrainers can be tested by coating a GLC column with them and observing the effective separation on a given solvent mixture. A more rough and ready method for screening is to compare the composition of the first 8% of an Engler distillate of the solvents to be separated, with the first 2% distillate when the distillate is from a mixture of 20% solvents and 80% potential entrainer.

Table 6.12 shows the results of this test on a variety of systems that are typical of

Table 6.12 Results of screening test for various entrainers

Mixture (% w/w)	Entrainer	First distillate[a]
THF–n-hexane (20:80)	None	22/78/–
	NMP	7/92/1
	2-Nitropropane	6/85/9
	EC 180	20/68/12
MDC–Freon 113 (50:50)	None	50/50/–
	NMP	33/59/8
	2-Nitropropane	26/48/26
	EC 180	55/31/14
Diisopropyl ether–acetone (50:50)	None	39/61/–
	EC 180	48/42/10

[a] The numbers in the third column are the percentages by weight of the first 8% of an Engler distillation of the mixtures in the first column with or without the entrainer in the second column.

solvent recovery operation. EC 180 is a mixture of C_{12} isoalkanes and Freon 113 is a trade-name for trichlorotrifluoroethane. All the results conform to the expectation that the more polar entrainer reduced the concentration of the more polar solvent in the distillate and vice versa. The results also give an indication of the difficulty likely to be experienced in Section A of the extractive distillation column in keeping entrainer out of the tops product.

Once suitable entrainers have been identified, it is very desirable to be able to carry out a laboratory trial of the plant-scale extractive distillation.

The three functions which together form an ED operation are best carried out separately batchwise to prove the practicability of the operation. Without sophisticated controls, not often available in the laboratory, the complication of performing these simultaneously is too great and yields no information that cannot be obtained by doing them separately.

Function 1
Carrying out the operation done in Sections B and C provides material for doing the other two functions, so this should be done first.

Because the liquid capacity of an Oldershaw and other laboratory plate columns is small, it is best to carry out the trial in a packed column.

Hot entrainer at the temperature of the column top should be fed continuously at the column head, while a conventional batch distillation takes place. Product, inevitably containing some entrainer, should be withdrawn at the column top and the batch should be continued until there is no further component 1 in the system. The still should be very large in comparison with normal laboratory batch distillation practice, since it will have to hold almost all the entrainer fed into the column head during the course of the batch.

Function 2
The crude component 1 can be batch distilled off the entrainer in the usual way to obtain information on the difficulty of operating Section A.

Function 3
Similarly, the residue from the ED run can be stripped free from component 2 to check what problems may arise in Sections D and E.

7 Drying solvents

In organic solvent recovery the commonest separation is the removal of water. Water has many harmful effects upon a solvent. It can spoil its solvent power and this may call for a reduction of water content to, say, 1.0%. It can slow down a reaction as one finds, for instance, in esterifications. Here the water content of the solvent should be as low as possible but often 0.1% will be an economically acceptable limit. It can also destroy a reagent (e.g. a Grignard reagent or a urethane) and a water content as low as 100 ppm or even less may be demanded by process economics.

At the same time, solvents range from those miscible with water in all proportions to those in which water is very sparingly soluble, although always detectable (Table 7.1).

Table 7.1 Solubilities

Solvent	Solubility of water in (% w/w)	Solubility in water (% w/w)
Hydrocarbons		
Pentane	0.012	0.036
Hexane	0.011	0.0138
Heptane	0.005	0.0052
Octane	0.005	0.0015
Nonane	0.004	0.0006
Benzene	0.063	0.18
Toluene	0.033	0.052
Xylene	0.05	0.0076
Ethylbenzene	0.035	0.02
Cyclohexane	0.01	0.0055
Chlorinated hydrocarbons		
Methylene dichloride	0.15	1.85
Chloroform	0.072	0.82
Carbon tetrachloride	0.01	0.077
1,2-Dichloroethane	0.15	0.81
Trichloroethylene	0.033	0.11
Perchloroethylene	0.008	0.015
1,1,1-Trichloroethane	0.05	0.02
Monochlorobenzene	0.033	0.049
Ketones		
Acetone	Completely	Miscible
Methyl ethyl ketone	12.0^a	27.5
Methyl isobutyl ketone	1.9	1.7
Cyclohexanone	8.0	2.3

Table 7.1 Continued

Solvent	Solubility of water in (% w/w)	Solubility in water (% w/w)
Ethers		
Diethyl ether	1.3[a]	6.9
Diisopropyl ether	0.62	1.2
Dioxane	Completely	Miscible
Tetrahydrofuran	Completely	Miscible
Alcohols		
Methanol	Completely	Miscible
Ethanol	Completely	Miscible
n-Propanol	Completely	Miscible
Isopropanol	Completely	Miscible
n-Butanol	20.0	7.7
Isobutanol	15.0	8.7
sec-Butanol	36.3[a]	15.4
Cyclohexanol	11.8	4.3
Esters		
Methyl acetate	8.2[a]	24.5
Ethyl acetate	3.3	7.7
n-Propyl acetate	2.9	2.3
Isopropyl acetate	1.8	2.9
n-Butyl acetate	1.64	0.67
Amyl acetate	1.15	0.17
Miscellaneous		
Pyridine	Completely	Miscible
Acetonitrile	Completely	Miscible
Furfural	6	8.3
Nitrobenzene	0.24	0.19

[a] Solvent whose azeotrope with water is not two phase despite the fact that they are not completely water miscible [Class B (i); see below].

Further, there are a number of solvents less miscible with water as their temperatures rise, e.g. n-butanol and butyl Cellosolve, although the majority are more miscible at higher temperatures.

When faced with the range of chemical types of solvents and the requirements for their water content, it is not surprising that there are many drying methods to choose from, and it is the purpose here to try to indicate which may be applicable in various situations.

The methods of water removal covered are:

- Fractionation
- Azeotropic distillation
- Extractive distillation
- Pressure distillation
- Adsorption
- Membrane separation (pervaporation)
- Liquid–liquid extraction
- Hydration, reaction and chemisorption

- Salting-out
- Coalescing
- Fractional freezing

Fractionation

For solvents that do not form an azeotrope with water, fractionation should always be considered as a possibility for drying provided that very low water contents are not required. Even then the use of fractionation followed by, say, molecular sieve or desiccant contacting may be the best method.

For solvents more volatile than water (e.g. methanol, acetone), fractionation may serve the double purpose of drying and removing non-volatile impurities or colour bodies. It also removes the diacetone alcohol formed during boiling acetone. However, taking all the solvent as an overhead by distillation, often at a significant reflux ratio because of the minimum reflux necessary, can be very expensive in the heat required if the solvent–water mixture being dried is already a distillate and does not need 'cleaning up'. Thus one would consider evaporating 98 kg of acetone at a 1:1 reflux ratio to remove it from 2 kg of water as a bad operation in terms of the heat required.

Most medium volatility solvents, boiling between 70 and 140 °C, form azeotropes with water and cannot be dried satisfactorily by straightforward fractionation.

Water has a relatively high volatility with respect to high-boiling solvents such as DMF, DMSO, DMAC, NMP and monoethylene glycol have high relative volatilities with respect to water and they can be dried easily by fractionation, although it can be difficult to remove traces of solvent from the distillate water and the implications that this may have on the plant effluent should be borne in mind.

Although theoretically batch fractional distillation may seem an adequate means of drying a solvent to a low water level, it is often the case that the hold-up of material in industrial-sized equipment makes the task more difficult than would be expected. The addition of a desiccant to the batch kettle when it is nearly water-free makes the drying quicker and more economical. Desiccants for this purpose are listed in Table 7.13.

Azeotropic distillation

For considering the drying of solvents that form azeotropes with water, it is convenient to divide the solvents into four classes.

Class A
Solvents that have water azeotropes that form two liquid phases on condensing.

Class A (i). Solvents very sparingly miscible with water so that the distillate, in either a batchwise or continuous distillation, splits into two phases, of which the water phase can be rejected to waste. All hydrocarbons and chlorinated hydrocarbons belong to this group and, depending on the value of the solvent or the cost of disposal of the contaminated water, many high-boiling oxygenated solvents can also be treated in this way.

108 *Drying solvents*

Class A (ii). Solvents partly water miscible so that the water phase is too costly to dispose of and must be processed to recover its solvent content. *n*-Butanol is typical of this group and the method of recovering the solvent from the water is illustrated in Figure 7.1.

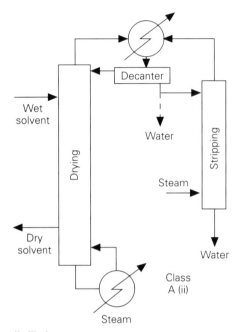

Fig. 7.1 Drying *n*-butanol by distillation

Class B
Solvents that have water azeotropes that are not two phase.

Class B (i). These solvents are not fully water miscible and form two phases, often close to the azeotropic composition. Three are marked in Table 7.1. They can be dried azeotropically by using an entrainer as if they fell into Class B (ii). Alternatively, it is possible to place, between the condenser and decanter, a contacting column to saturate the azeotrope with an inorganic salt. The presence of salt causes a phase split and the water phase (saturated with salt) can then be discarded. MEK can be dried in this way (*see* Salting-out section).

Class B (ii). Solvents miscible with water in all proportions. It is necessary to add an entrainer when distilling these wet mixtures to produce at the column top a two-phase mixture of which one phase is low in water and the other high. All such entrainers fall in Class A (i) and preferably they should form a binary azeotrope with water and no ternary azeotrope involving the solvent to be dried.

Such a system is the use of pentane to dry THF–water (Figure 7.2). Even the low carrying power of pentane for water (1.4% w/w) can be turned to good effect since not all the pentane in the overheads is needed to be returned as reflux to the column top. The

Azeotropic distillation 109

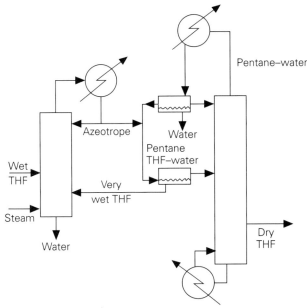

Fig. 7.2 Drying THF by distillation

very low solubility of water in pentane can be used to partition out the water present in the THF–water azeotrope being fed to the drying column.

Other solvents in Class B (ii) which can be treated using a similar flow sheet include pyridine being dried using benzene or cyclohexane and *n*-propanol using cyclohexane.

However, in many cases there is no suitable entrainer that does not form a ternary azeotrope. Tables 7.3 and 7.4 show the entrainers possible for use in drying ethanol and IPA using a flow sheet as in Figure 7.3.

The properties that must be considered in selecting the optimum entrainer are listed in

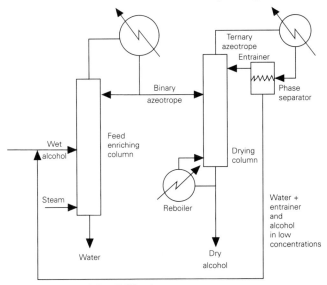

Fig. 7.3 Flow diagram for drying of class B(ii) solvents

110 *Drying solvents*

Table 7.2. If a potential entrainer does not give a satisfactory performance in the first three criteria it should usually be disqualified from further consideration.

Table 7.2 Criteria for selecting an entrainer

1 Toxicity
2 Corrosion
3 Stability
4 Ability to form an effective azeotrope
5 Phase separation
6 Fractionation
7 Boiling point
8 Latent heat
9 Ease of handling
10 Availability and price

Toxicity

Much work on the composition of azeotropes in the reference literature was done in the 1930s. Entrainers such as benzene, chloroform and carbon tetrachloride had a lot of research done on them. Today such materials are considered very undesirable in industrial situations and will present major handling problems as well as being unacceptable, even in the lowest concentrations, in the finally recovered product. Full consideration should therefore be given to the difficulties of solvents with TLVs of 10 ppm or less before including them on a list of 'possibles'.

Consideration of the toxicity of entrainers should be made in a comparison with the toxicity of the solvent to be dried.

Ethanol has a low toxicity (TLV 1000 ppm), so that all the entrainers worth considering present a greater hazard, as Table 7.3 demonstrates, but in the case of a more toxic solvent such as pyridine (TLV 15 ppm) there is no advantage to be gained in rejecting any entrainer with a greater TLV than this.

Table 7.3 Ethanol entrainer TLVs (ppm in air)

Entrainer	TLV	Entrainer	TLV
Benzene	1	Toluene	100
Carbon tetrachloride	10	Diisopropyl ether	250
Chloroform	10	Cyclohexane	300
Ethylene dichloride	10	Trichloroethylene	350
n-Hexane	50	n-Heptane	400

Corrosion

Just as the fractionation column may not have the power to make a separation, it may also not be able to handle certain corrosive materials. An entrainer is treated very severely, possibly being held at its boiling point for many days, and materials often thought of as stable can deteriorate and possibly yield corrosive agents such as hydrochloric acid or acetic acid under such long-term stress.

Stability

An entrainer must not react with the materials to be separated, nor must it polymerize under the column conditions. Particularly when an aqueous phase is being discharged from the system there is a possibility that any inhibitor supplied in the commercial grade of an entrainer (but not necessarily present in the reagent grade material) will be lost from the system, leaving it unprotected.

Ability to form an effective azeotrope

In general, it is preferable to operate a Class A (ii) rather than a Class A (i) drying process, so the search for an entrainer that forms only a binary azeotrope with water in a given system should not be abandoned lightly. If no such entrainer exists, as is the case for ethanol–water, it is very attractive to use an entrainer which minimizes the amount of solvent and entrainer partitioning into the water phase (see diisopropyl ether in Table 7.4). It may be economic to dispose of this water phase and avoid the complexity of recycling it to recover its solvent content (Figure 7.3).

The very large difference in performance of various possible entrainers is illustrated by the 'loss' of ethanol and IPA in Tables 7.4 and 7.5.

Phase separation

If the azeotrope does form two phases, the separation is usually done in a gravity decanter. It is therefore important that the phases should have a substantial difference in density. Other matters to be borne in mind are that in a general-purpose plant, the water phase to be rejected may be either the upper or lower one, and that the volumes of the phases are likely to be very different, which may cause problems with residence times and settling.

For all these reasons, the separation stage is important when selecting an entrainer. The various settling characteristics of possible ternary azeotropes formed in the ethanol–water–entrainer combination illustrate the problems (Table 7.6).

While the well designed decanter may minimize the difficulty of low density difference (e.g. toluene) and varying residence times (e.g. DIPE vs. chloroform), large liquid hold-up at the top of a distillation column is not helpful to good fractionation and high density differences, leading to small decanter volumes, are therefore desirable.

Sometimes an existing plant may be unable to make a clear phase separation and the choice has to be made of reducing the residence time of one phase to improve the quality of the other.

Most of the solubility data quoted in the literature are for conditions at 20 or 25 °C. In a minority of cases mutual solubility increases with reducing temperature (e.g. MEK–water and diethyl ether–water) and it is best to make the phase separation near the boiling point. Usually, however, it is better to consider cooling the condensate before it reaches the decanter and reheating the entrainer phase before returning it to the column. Since the entrainer phase is almost always very much larger than the rejected aqueous phase, this interchange can often be done between the two streams without external sources of cooling or heating.

Table 7.4 Ethanol drying entrainers

	Entrainer									
	CCl$_4$	Benzene	Chloroform	EDC	Heptane	Cyclohexane	DIPE	Trichloroethylene	Hexane	Toluene
	Ternary azeotrope (% w/w)									
EtOH	10.3	18.5	4.0	15.7	33.0	17	6.5	16.1	12.0	37
Water	3.4	7.4	3.5	7.2	6.1	7	4.0	5.5	3.0	12
Entrainer	86.3	74.1	92.5	77.1	60.9	76	89.5	78.4	85.0	51
B.p. (°C)	61.8	64.6	55.5	67.8	68.8	62.1	61.0	67.0	56.0	74.4
	Water-rich phases (% w/w)									
EtOH	48.5	52.1	18.2	41.8	75.9	64	20.2	48	75.0	54.8
Water	44.5	43.1	80.8	46.6	15.0	31	78.0	38	19.0	20.7
Entrainer	7.0	4.8	1.0	11.6	9.1	5	1.8	14	6.0	24.5
Density	0.935	0.892	0.976	0.941	0.801	0.95	0.967	0.95	0.672	0.86
	Entrainer-rich phase (% w/w)									
EtOH	5.2	12.7	3.7	12.5	5.0	2.5	5.9	14	3.0	15.6
Water	0.1	1.3	0.5	2.3	0.2	0.5	1.2	2	0.5	3.1
Entrainer	94.8	86.0	95.8	85.2	94.8	97.0	92.9	84	96.5	81.3
Density	1.52	0.866	1.44	1.17	0.686	0.78	0.737	1.35	0.833	0.849
Closest boiling binary	EtOH–CCl$_4$ 65.0°C	EtOH–benzene 67.8°C	CHCl$_3$–water 56.1°C	EtOH–EDC 71°C	EtOH–heptane 72°C	EtOH–cHex 64.9°C	DIPE–water 62.2°C	EtOH–TCE 70.9°C	EtOH–hexane 58.7°C	EtOH–Tol 76.7°C
EtOH loss[a] to water phase (% w/w)	5.7	6.4	1.2	4.7	26.6	10.8	1.4	6.6	20.8	13.9

[a] Assuming feed to be 95% ethanol–5% water and aqueous phase not reprocessed.

Table 7.5 IPA drying entrainers

	Entrainer						
	DIPE	Toluene	Benzene	DIB[a]	Isopropyl acetate	Cyclohexane	EDC
	Ternary azeotrope (% w/w)						
IPA	6.6	38.2	19.8	31.6	13	18.5	19.0
Water	3.1	13.1	8.2	9.3	11	7.5	7.7
Entrainer	90.1	48.7	72.0	59.1	71	74.0	75.3
B.p. (°C)	61.8	76.3	65.7	72.3	75.5	64.3	69.7
	Water-rich phase (% w/w)						
IPA	15.0	38	14.4	55.2	11.5	41.6	20.7
Water	84.6	61	85.1	39.4	85.6	55.7	78.4
Entrainer	0.4	1	0.5	5.4	2.9	2.7	0.9
Density	0.976	0.930	0.966	0.88	0.981	0.92	1.117
	Entrainer-rich phase (% w/w)						
IPA	6.3	38.2	20.2	26.9	13	17.2	18.8
Water	0.3	8.5	2.3	3.1	5.6	1.0	3.1
Entrainer	93.4	53.5	77.5	70.0	81.4	81.8	78.1
Density	0.732	0.845	0.895	0.74	0.870	0.78	0.968
Closest boiling binary	DIPE–water	IPA–water	Benzene–water	DIB–IPA	IPAc–water	cHex–IPA	EDC–water
	62.2 °C	80.6 °C	69.3 °C	77.8 °C	76.6 °C	69.4 °C	72.3 °C
IPA loss[b] to water phase (% w/w)	2.9	10.1	2.8	22.8	2.2	12.2	4.3

[a] Diisobutylene
[b] Assuming feed to be 86% IPA–14% water and aqueous phase not reprocessed.

114 *Drying solvents*

Table 7.6 Settling characteristics of water entrainers in ternary azeotropes with ethanol

Entrainer	Density of		Density difference	Relative volume	
	Entrainer phase	Water phase		Top	Bottom
Chloroform	1.44	0.98	−0.46	6	94
Benzene	0.87	0.89	+0.02	86	14
Hexane	0.67	0.83	+0.16	90	10
Heptane	0.69	0.80	+0.11	65	35
EDC	1.17	0.94	−0.23	13	87
DIPE	0.74	0.97	+0.23	97	3
Toluene	0.85	0.86	+0.01	47	53

Fractionation

Whether an azeotrope is a binary or a ternary it is desirable that it should be fractionated easily from the other component(s) of the system. In the absence of vapour–liquid data the boiling point gap is the best indication of how easy the split is. The comparative complexity of the column contents can be illustrated by the ethanol–water system with cyclohexane added as a dewatering entrainer (Table 7.7).

Table 7.7 Boiling points of components in the water–cyclohexane–ethanol system

	B.p. (°C)	Composition (% w/w)
Ethanol–water–cyclohexane	62.1	17:7:76
Ethanol–cyclohexane	64.8	30:70
Cyclohexane–water	69.5	91:9
Ethanol–water	78.2	96:4
Ethanol	78.4	100
Cyclohexane	80.7	100
Water	100	100

The effects of poor fractionation will be to allow some of the binary, ethanol–cyclohexane, to reach the column top. It has a higher ethanol to cyclohexane ratio than the ternary (18:82 vs. 30:70) so it will increase the ethanol concentration in the tops and therefore the solubility of water in the entrainer phase. This in turn will result in more of the water which had reached the column top being returned to the system and less being rejected. From this point of view it is instructive to examine the other entrainers for ethanol–water dehydration (Table 7.8).

In a situation in which fractionating power is known to be barely adequate, the two solvents (DIPE and chloroform) with low-boiling binary azeotropes including water rather than ethanol have the advantage that it is positively helpful to have their water binaries admixed with the ternary in the decanter (Table 7.8).

Boiling point

The water-containing binary or ternary azeotropes will always have lower boiling points

Table 7.8 Potential effect of insufficient fractionation on column top composition in drying ethanol

Entrainer	Binary b.p. (°C)	Second component	Ternary b.p. (°C)	Difference (°C)
Cyclohexane	64.8	Ethanol	62.1	2.7
Benzene	67.8	Ethanol	64.6	3.2
Chloroform	56.3	Water	55.5	0.8
Hexane	58.7	Ethanol	56.0	2.7
EDC	71.0	Ethanol	67.8	3.2
DIPE	62.2	Water	61.0	1.2
Trichloroethylene	70.9	Ethanol	67.0	3.9
Heptane	71.0	Ethanol	68.8	2.2
Toluene	76.7	Ethanol	74.4	2.3

than the solvent they are being used to dry. The chance of a ternary azeotrope being formed decreases as the boiling point difference between solvent and entrainer increases. Low-boiling entrainers such as pentane or methylene chloride can thus be used in a Class A (ii) mode whereas hexane or trichloroethylene can only be employed in a Class A (i) way. However, the difficulty of condensing, at atmospheric pressure and in a general-purpose plant, below 40 °C may negate the attraction of Class A (ii) operation.

Latent heat

Since the entrainer in an azeotropic distillation is continually being evaporated and condensed with its latent heat being wasted, it is important that the quantity of heat involved should be considered. For the removal of water, as Table 7.9 shows, the amount of heat needed can be modest compared with the reflux ratios involved in straightforward fractionation.

This may be offset in economic terms by the fact that almost all useful azeotropes in

Table 7.9 Comparison of heat requirements for azeotropic drying and heat needed for fractionation under reflux

Water entrainer	B.p. (°C)	Azeotropic b.p. (°C)	Water (% w/w)	Equivalent reflux ratio
n-Pentane	36	34.6	1.4	11.0
Chloroform	61	56.1	2.8	3.8
DIPE	69	62.2	4.5	2.9
n-Hexane	69	61.6	5.6	2.5
Trichloroethylene	87	73.4	7.0	1.4
Benzene	80	69.4	9.0	1.8
Cyclohexane	80	70.0	9.0	1.6
n-Heptane	98	79.2	12.9	1.0
DIB	101	81.0	13.0	0.7
Perchloroethylene	121	83.5	17.2	0.4
Toluene	111	85.0	20.0	0.6
n-Octane	126	89.6	25.5	0.4
MCB	132	90.2	28.4	0.4
Ethylbenzene	136	92.0	33.0	0.3
n-Nonane	151	95.0	39.8	0.2

solvent recovery are low boiling. As a result, they reduce the temperature difference over the condenser and hence reduce its capacity. Thus, to use perchloroethylene to dehydrate DMF at atmospheric pressure requires 781 cal/g of water removed at a column top temperature of 83.5 °C whereas ordinary fractionation might need 950 cal/g at 100 °C. With cooling water at 20 °C the load on the condenser would be harder to handle for the lower heat input.

It is frequently true in solvent recovery that time used on the plant is much more expensive than the energy costs per unit of output. This may mean in cases like that above that the lower energy route is not the most cost effective.

No account has been taken in Table 7.9 of the specific heat of the entrainer in the reflux ratio calculation. Although the solubility of water in the entrainers increases with rising temperature, the effect is often so small that it is not worth cooling the condensate to improve the separation. If cooling is necessary, or cannot be avoided, the sensible heat that must be put in to bring the refluxing entrainer back to its boiling point can be appreciable. For instance, the sensible heat to raise perchloroethylene from 20 to 83.5 °C is over 25% of its latent heat. In such cases heat exchange between condensate and refluxed entrainer may be justified.

The general approach should be to use the highest boiling entrainer that is suitable from other operational considerations and to use the one with the lowest latent heat among those with equal performance.

Ease of handling

In addition to toxic hazards, it should be noted that many entrainers are highly flammable and may call for precautions not involved with the handling of the original components of the mixture (e.g. toluene to remove water from DMF). The inventory of entrainer used on a 1 Te/h plant will amount to perhaps 1000 l. This therefore requires drum handling and storage. The densities of trichloroethylene and perchloroethylene make drums of them difficult to handle without mechanical equipment. Both benzene and cyclohexane have freezing points above zero and heating may be required for winter operations for both drums and pipelines.

Availability and price

There are very many azeotropes reported in the literature which involve entrainers that are not easily obtainable on an industrial scale.

Extractive distillation

Extractive distillation is a very effective technique for performing certain difficult separations, but it has the following drawbacks:

- It needs special equipment unlikely to be available in a general-purpose plant.
- Since the extraction solvent is never evaporated, it can only be used to dry wet solvents that are clean distillates (Figure 7.4).

Extractive distillation 117

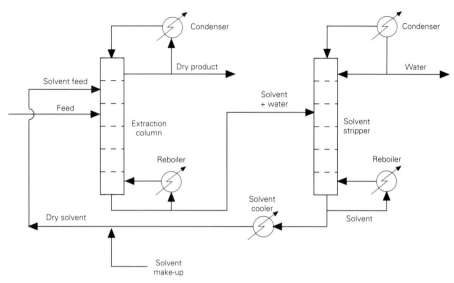

Fig. 7.4 Drying by extractive distillation

- The solvents useful for water removal such as glycerol and monoethylene glycol are not outstandingly stable and any low-boiling decomposition product ends up in the dried solvent product.
- Because the mole fraction of the extracted water needs to be low in the water–extraction solvent mixture, extractive distillation tends to be limited to wet solvents with a low initial water concentration.

Pressure distillation

The composition of an azeotrope varies with absolute pressure. In water–solvent mixtures, such as MEK and THF where this effect is industrially important, the water content of the azeotrope increases with increasing pressure. Thus, if two columns at different pressures are run in series (Figure 7.5), a dry solvent can be made without using an entrainer to break the azeotrope. This can also be done in a batch still, but for both continuous and batch operations the equipment is specialized and the hazard of handling flammable solvents at high pressures must be borne in mind.

Adsorption

A number of highly porous solids adsorb water preferentially when contacted by wet solvent mixtures and can remove water to very low concentrations. While they can be used on a once-through basis they are capable of being regenerated for many cycles of reuse by heating and such regeneration is economical for long-term operations.

Molecular sieves are available in a range of pore sizes and this allows solvents with larger molecular sizes to be excluded from the pores (Table 7.10). The larger the pore size, the greater is the water capacity of the molecular sieve, so it is desirable to use the largest pores that will not be taken over by solvent.

118 Drying solvents

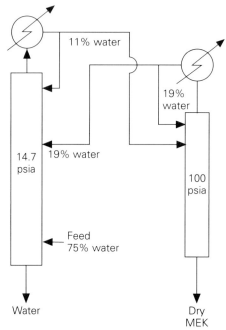

Fig. 7.5 Using pressure distillation to dry MEK

Table 7.10 Properties of molecular sieves of different pore sizes

Pore size (Å)	Adsorbs	Excludes
3	Water	All other solvents
	Methanol	
4	Ethanol	Butanol
5	n-Alkanes	All iso compounds
		Benzene and all aromatics

Thus all solvents except methanol can be dried, although the level of water content that can be achieved may vary. Subject to a laboratory trial, 50 ppm of water is a reasonable target.

Silica gel and alumina have similar properties to molecular sieves but with larger pore sizes and therefore a higher loss of solvent, although this can be recovered during regeneration (Table 7.11). They also have less favourable characteristic curves.

Table 7.11 Comparison of properties of molecular sieve, silica gel and activated alumina

	Molecular sieve	Silica gel	Activated alumina
Pore size[a] (nm)	0.4	20–140	1–7.5
Regeneration temperature (°C)	300	150	180–250
Water content adsorbed[b] (% w/w)	22	30	15
Heat of adsorption (cal/g)	1000	222	333

[a] The diameter of a water molecule is 0.265 nm.
[b] This is the water content when there is no competing solvent present. Although silica gel adsorbs water preferentially it may well pick up less water in a given application than the more precisely 'tailored' molecular sieve.

The capacity of the molecular sieve is also fairly constant whatever the water content of the solvent whereas the capacity of silica gel is proportional to the water content of the solvent over the range 1–30%.

Regeneration in each case needs a hot, dry gas, preferably nitrogen. In most industrial applications molecular sieve regeneration needs electric or flue gas heating since no normal heating medium (steam, hot oil) will attain the regeneration temperature. Nitrogen or some other inert gas must be used because of the necessary consideration of the solvent autoignition temperature. There is some evidence to suggest that autoignition temperatures are lowered when solvents are adsorbed on active surfaces so the risk of an explosion if oxygen is present may be more than would be estimated.

The lower regeneration temperatures for silica gel and, less so, for alumina help to make up for their poorer other properties.

In all cases some solvent will be present in the regeneration gases in the early stage of the heating process and it would be desirable in most cases to pass this gas through a carbon bed adsorber to reduce solvent losses and environmental pollution.

The inorganic adsorbents are resistant to almost all solvents although heating for regeneration may cause reactions leading to blockage of the pores.

Organic adsorbents of the ion-exchange resin type are less inert and may be attacked by some solvents. They are, however, attractive for dehydrating:

- Ion-exchange resins can be regenerated by heating to 120 °C and may be damaged if this temperature, easily achieved from industrial steam sources, is exceeded. Lower temperatures can be accepted if the regeneration takes place under vacuum. Air is an acceptable gas for drying in most cases.

- Non-polar solvents can be dried to less than 50 ppm. This can be particularly useful for drying chlorinated solvents.

- Capital cost of adsorbent per unit weight of water adsorbed is about half that of molecular sieves. The type of resin suitable for this application is Rohm and Haas Amberlite IR-120 and Dowex 50W-X8. Both are sulphonic-type exchange resins in their sodium and potassium form, respectively.

Solvent drying by adsorption cannot easily be made into a continuous process. It is usually either a single-bed batch process or a twin-bed process with one bed on adsorption while the other is being purged, heated and cooled. It does not lend itself to adsorbing large quantities of water. It does, however, have advantages over distillation when very low concentrations of water in product need to be achieved. Probably its best area of application economically and operationally is from 1% water to the 100 ppm level with beds changing from adsorption to regeneration mode once every 24 h.

If a one-off drying operation has to be carried out, possibly owing to an accidental contamination, the reagent cost of removing water will be about £5000–10 000 per tonne of water removed. Since an adsorption bed capable of removing 200 l of water can easily be moved by crane and/or fork-lift truck, it is possible to dry a tank's contents by recirculating *in situ* without the need for a 'wet' and a 'dry' tank.

120 *Drying solvents*

Membrane separation (pervaporation)

Just as solids can be designed and made to adsorb water and reject solvents, so membranes can be designed to pass water and retain solvents. This is the basis for a relatively recently commercialized solvent drying process that has the advantages of:

- being a continuous process (unlike adsorption);
- not being affected by azeotropes (unlike distillation);
- not being affected by the solvent boiling at a lower temperature than water (unlike distillation);
- only needing an electricity supply to make it operate.

Called pervaporation, its name implies the combination of permeation through a membrane followed by evaporation from the downstream membrane surface (Figure 7.6). For solvent drying the membrane is chosen so that water is very soluble in it whereas the solvent to be dried is sparingly soluble.

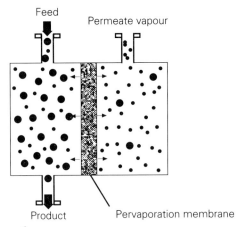

Fig. 7.6 Principles of pervaporation

The driving force to effect the transport of water through the membrane is proportional to the difference in partial pressure of the water on the upstream and downstream sides of the membrane and to achieve a satisfactory operation the upstream should be at least five times the downstream partial pressure. Since the liquid on the downstream side is effectively pure water, it must be evaporated at a very low pressure. On the upstream side, the highest temperature that the membrane will withstand is one factor in limiting the partial pressure of the water.

The other controllable factor is the water concentration, which falls as the feed passes through the system and which, if the product must be reduced to a very low water level, can result in a low driving force in the latter plate modules. This may be offset by the high activity coefficients that water has at low concentrations in many solvents (Appendix 1).

The evaporation of the permeate needs latent heat and this is provided by sensible heat from the feedstock conducting through the membrane. The membrane currently available will only stand about 100 °C and the feed, heated to this temperature, cools as it

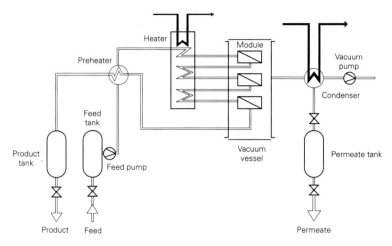

Fig. 7.7 Flow diagram of pervaporation plant

gives up its heat and needs to be reheated in a series of heat exchanges of up to eight stages (Figure 7.7).

The prime restrictions on this drying process are related to the materials of the membrane which are soluble in atropic solvents such as DMF, glycol ethers and DMSO. Since the membrane and its installaton are the most costly part of the equipment, it is vital to test unknown mixtures before drying them by pervaporation on an industrial scale, but the most attractive applications tend to be the drying of mixtures, particularly those containing an impurity that would stop phase separation (e.g. acetone–isopropanol–water) or where water can be removed from, say, acetone without the need to evaporate the acetone.

The membrane is selective for water with only the exception that some methanol will be passed through it if a methanol–water mixture is pervaporated. Water contents of 1–15% in the feed and 0.1–1% in the product define the optimum range for pervaporation.

Liquid–liquid extraction

As is clear from Table 7.1, there are large differences in water miscibility between various classes of solvents. Some solvents, such as hydrocarbons and chlorinated hydrocarbons, are so hydrophobic that they can be used in liquid–liquid extraction processes to drive the water out of a more hydrophilic solvent.

Thus it is possible to separate an ethyl acetate–water mixture using nonane or a similar highly paraffinic hydrocarbon. The ethyl acetate shows a partition coefficient strongly in favour of the hydrocarbon phase. Since the other impurities present in ethyl acetate recovered from a carbon bed absorber (ethanol and acetic acid) are strongly hydrophilic, the quality of the ethyl acetate distilled off the nonane is good (Figure 7.8).

Similarly, DMF in dilute aqueous solutions, which would be difficult to dehydrate economically by fractionation, can be extracted with methylene chloride. The low reflux ratio required for removing the methylene chloride plus its low latent heat makes the

122 *Drying solvents*

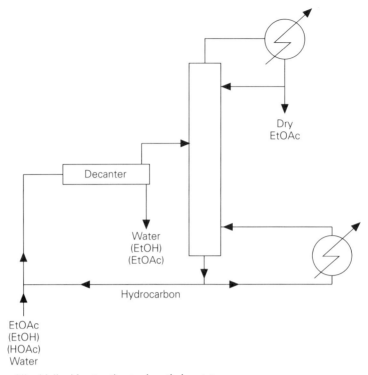

Fig. 7.8 Use of liquid–liquid extraction to dry ethyl acetate

subsequent distillation economic and the small amount of water dissolved in the organic phase distils off as the methylene chloride–water azeotrope.

In considering the possibilities of removing water from a solvent which is completely water miscible, it is useful to know the relative attraction of the water for the solvent. This can be done by considering the activity coefficient of the solvent at infinite dilution in water. Solvents partially miscible with water tend to have relatively high γ^∞ values (e.g. MEK 27.2, methyl acetate 23.6), whereas, as Table 7.12 demonstrates, solvents completely miscible with water usually have lower values of γ^∞. It would therefore be easier, using a solvent with low water miscibility combined with an affinity for the dissolved solvent, to extract the dissolved solvent from water if its γ^∞ is high. By the same reasoning, it is unlikely that there is an extraction solvent to remove methanol from water economically.

A major disadvantage of liquid–liquid extraction is that the aqueous phase will be

Table 7.12 γ^∞ in water values

	γ^∞ in water		γ^∞ in water
Methanol	2.15	Pyridine	11.2
Ethanol	5.37	Isopropanol	11.5
Ethyl Cellosolve	6.9	Butyl Cellosolve	14.8
Acetone	8.86	Dioxane	15.8
Acetonitrile	9.48	Methyl Cellosolve	19.4

saturated with the organic solvent introduced into the system and may be unfit to discharge as effluent, thus requiring incineration or further treatment.

A method for overcoming the relatively high attraction of a solvent to water in liquid–liquid extraction is to employ a pair of extraction solvents, one with a very strong affinity to water and the other with a great affinity to the solvent being separated from water, a technique known as fractional liquid extraction (FLE).

The choice of FLE solvents should be guided by the activity coefficients of water and the solvent to be removed from water in them at low concentrations. Thus, to separate water and ethanol one seeks solvents in which the values of their γ^∞ are low in the phase which they should partition into and high in the phase from which they should be absent. Thus a possible pair of solvents to separate water from ethanol could be MDC and monoethylene glycol:

	γ^∞
ethanol in MDC	1.25
ethanol in MEG	2.05
water in MDC	311
water in MEG	1.04

The FLE solvents must also be very sparingly miscible in each other for satisfactory performance and normally several extraction stages will be required.

A further 'exotic' method of extraction for drying solvents is the use of supercritical fluids such as carbon monoxide, propane and butanes. This approach has been demonstrated in the laboratory for alcohols except methanol, and would seem also to be effective for other oxygenated solvents although no industrial plants have been announced.

Hydration, reaction and chemisorption

In general, the use of chemicals to dry solvents is most common for small-scale operations or for a final stage of dehydration once the major part of the water has been removed by some other means. Because of solution effects or reactions, there is no chemical that is suitable for drying all organic solvents and, particularly for solvent mixtures, laboratory trials are always needed. The desiccants listed in Table 7.13 are far from being a comprehensive list of those which can be used industrially for dehydration.

The capacity to remove water using some of the desiccants varies widely, as Table 7.14 shows, and obtaining their full effectiveness often poses difficult problems of chemical engineering design. Of those listed in Table 7.14, only potassium carbonate is commonly regenerated, requiring temperatures of about 200 °C. The others are relatively cheap chemicals and, if they are used to remove only low levels of water often on a small batch basis, are uneconomic to process.

Caustic soda is sometimes used both as a desiccant and to remove peroxides from solvents, particularly ethers, where their presence in a still is dangerous, but because pellets of NaOH tend to fuse together it is especially difficult to get good solid–liquid contact with them.

The combination of distilling solvent from a still kettle holding desiccant is often practised when small quantities of very dry solvent are required and the products must

124 Drying solvents

Table 7.13 Potentially useful desiccants

Solvent	Desiccant												
	$CaCl_2$	CaO	CaH_2	$CaSO_4$	Na	NaOH	Na_2SO_4	KOH	K_2CO_3	P_2O_5	B_2O_3	$LiAlH_4$	$MgSO_4$
Hydrocarbons	A	N	A	C	B	N	C	N	N	B	N	A	N
Alcohols	X	C	C	C	C	X	X	N	C	X	N	X	X
Glycol ethers	B	X	B	N	B	X	N	X	N	X	N	B	N
Chlorinated hydrocarbons	B	X	B	B	X	X	B	X	B	B	N	N	B
Ketones	X	X	X	B	X	X	B	X	B	X	B	X	B
Ethers	C	B	N	C	B	B	N	B	C	X	N	B	N
Esters	X	X	B	B	X	X	B	X	B	B	N	X	B
Miscellaneous:													
DMF/DMAc	N	X	B	N	X	X	N	X	N	B	N	X	N
DMSO	X	X	B	X	X	X	N	N	N	N	N	X	N
Pyridine	B	D	B	N	D	B	D	B	N	N	N	B	D
Acetonitrile	C	X	B	C	X	X	C	X	C	B	B	X	B
Aniline	X	C	B	N	B	C	N	C	N	X	N	X	N

A Dryness to 1 ppm.
B Dryness to good industrial standard of about 50 ppm.
C Can achieve industrial standard but not very efficient.
D Can remove some water.
X Ineffective and can react with solvent, possibly dangerously.
N No information.

Table 7.14 Capacities and relative cost of desiccants

Desiccant	Capacity (%)	Cost[a]
$CaCl_2$	20	Moderate
$MgSO_4$[b]	20–80	High
CaO	30	Low
Na_2SO_4[b]	120	Low
K_2CO_3	20	Moderate/high
$CaSO_4$	20	Moderate
NaOH	c	Low

[a] Low cost £500/Te of water, high cost £10000/Te of water.
[b] Anhydrous salts.
[c] Very dependent on application.

not contain any inorganic salts in solution. For such an operation Table 7.15 sets out desiccants that may be used provided appropriate safety precautions are taken.

Table 7.15 Desiccants suitable for producing very dry and pure solvents under batch distillation conditions

Compounds	Desiccant
Hydrocarbons	Na or $LiAlH_4$
Alcohols	MgI_2
Chlorinated hydrocarbons	P_2O_5
Ethers	Na or $LiAlH_4$
Esters	P_2O_5
Nitriles	K_2CO_3

Salting-out

This involves bringing the wet solvent into contact with a solid, usually an electrolyte, which has the power to withdraw some of the water present to form a second phase that can be removed by decantation. The dehydrating substance may be either a solid or a saturated aqueous solution. The latter is more easily adapted to counter-current operations.

The solid chosen, as in the case of drying by hydration, must not react with the solvent and, since this method is almost always followed by a distillation step, the problems of corrosion, e.g. from chlorides, must be borne in mind. The solid is also not normally recoverable so its cost is an important factor.

Hydration of the salt may also take place if, say, calcium chloride is used.

The dehydrating power of salts in any salting out operation in which there is a solid salt phase present is, at any given temperature, in inverse relation to the vapour pressure of water over the salt's solution in pure water (Table 7.16).

Thus lithium chloride is the most effective of those listed in producing a dry solvent, but it is very water soluble and therefore large quantities are needed to produce a saturated solution. It is also one of the more expensive of the solids listed and, in any particular combination of solvent purchase cost, water present and other drying means available, NaCl or Na_2SO_4 is likely to be the most economic for an industrial process.

The drying of MEK and pyridine is among commonly used applications of salting-out for binary mixtures of solvent and water. The MEK–water azeotrope is just single phase

126 *Drying solvents*

Table 7.16 Relationship between salt solubility and water vapour pressure at different temperatures

Salt	Solubility (g/l at 20 °C)	Water vapour pressure of saturated solution (mmHg)			
		15 °C	20 °C	25 °C	30 °C
NaCl	36.0	9.0	13.0	18.0	24.0
MgCl$_2$	54.5	4.5	6.0	8.0	10.0
NH$_4$Cl	37.2		13.8	18.6	24.4
LiCl	67[a]	1.8	2.1	2.7	3.6
CaCl$_2$	74.5	5.0	6.1	7.08	7.1
Na$_2$SO$_4$	19.4		16.1		
NH$_4$SO$_4$	75.4		14.1	19.1	25.6
Na$_2$CO$_3$	21.5		14.6	20.9	

[a] At 0 °C.

at ambient temperature and the addition of a salt produces two liquid phases and a solid–salt phase. The aqueous phase contains 4% MEK and is seldom worth recovering. The MEK-rich phase is easily split into the azeotrope and a dry MEK fraction.

The pyridine–water azeotrope, containing 43% of water, is also single phase but can be split into two phases using sodium hydroxide or sodium sulphate, again leaving so little pyridine in the aqueous phase that it is not economically worth recovering, subject of course to the cost of disposal of the aqueous effluent.

Coalescing

The majority of processes defined above involve phase separation, often of two phases with modest density difference. Since most solvents dissolve less water at low than at high temperatures, it is worth operating at as low a temperature as is practicable without running a risk of freezing either solvent or aqueous phase (Figure 7.9). Under cooling, the water leaving the solvent phase forms a fog of droplets too small to precipitate quickly and a coalescing pad or an electrostatic field is needed to remove these droplets. Such an addition reduces the volume required in the decanter, which is particularly desirable in batch distillation operation.

Fractional freezing

A small number of solvents with freezing points above 0 °C can be dried by batchwise fractional freezing, but this is a technique more useful in the laboratory than in plant-scale operations where it needs unusual special-purpose equipment.

Conclusion

This review has been directed at the removal of water from pure single organic solvents. In industrial systems, even when theoretically this is the position, there can be traces of impurities which can arise from inhibitors (e.g. ethanol in chloroform), denaturants (e.g. methanol or diethyl ether in ethanol) or plant rinsing (e.g. acetone) and either in batch

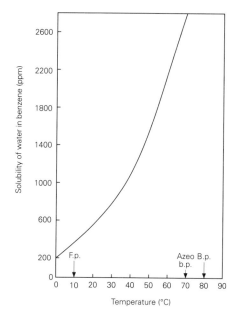

Fig. 7.9 Relationship between temperature and solubility of water in benzene

or continuous operations they may build up in concentration at the column top. Clearly, if such small concentrations of impurities in the feed can cause problems, ternary mixtures including water can be even more difficult to dehydrate, particularly when using azeotropic distillation techniques.

Examination of Figure 7.1 shows a typical problem. If a feed of n-butanol and water contains a very small concentration of methanol or acetone, these volatile components will accumulate in the column tops. Both water and butanol products, leaving the plant as column bottoms, will not carry away such light materials. Modest concentrations will change the mutual solubility of n-butanol and water so that the azeotrope does not form two phases and the decanter will cease to operate.

With the large variety of dehydration methods available (Table 7.17) there is usually more than one which will be effective for any mixture, but it is important to be aware of the effects that apparently trivial concentrations of impurities may have. It is also important in evaluating methods of water removal to consider the cost of drying per ton of water removed. This can vary between less than £100 per ton for an easy fractionation to over £10000 per ton for an adsorption or chemical method where there is no recovery of the reagent. Very often it is economic to use two methods, the first to get rid of large quantities of water followed by a second method to reach water contents in the 1000 ppm or lower range.

Since many solvents are hygroscopic, it is often best to separate these stages, holding the partly dried solvent in storage until it is about to be used and carrying out a final 'polishing' stage as it is transferred into process for reuse.

128 *Drying solvents*

Table 7.17 Useful dehydration methods for various common solvents

Solvent	Fractionation	Azeotropic distillation	Extractive distillation	Pressure distillation	Adsorption	Pervaporation	Liquid–liquid extraction	Hydration	Salting-out	Coalescing	Fractional freezing
n-Pentane					×	×		×	×	×	
n-Hexane		×			×	×		×	×	×	
n-Heptane		×			×	×		×	×	×	
Benzene		×			×	×		×	×	×	×
Toluene		×			×	×		×	×	×	
Xylenes		×			×	×		×	×	×	
Cyclohexane		×			×	×		×	×	×	
Methanol			×		×	×		×			
n-Propanol	×	×			×	×		×			
Isopropanol	×	×			×	×		×			
n-Butanol	×	×			×	×		×			
Isobutanol	×	×			×	×		×			
sec-Butanol	×	×			×	×		×			
MEG					×	×		×			
MDC					×		×	×		×	
Chloroform		×			×	×	×	×		×	
1,2-EDC		×			×	×	×	×		×	
Trichloroethylene		×			×	×		×		×	
Perchloroethylene		×			×	×		×		×	
Acetone	×				×	×		×			
MEK	×			×	×	×	×	×	×		
MIBK					×	×	×	×			
Diethyl ether				×	×	×	×	×	×	×	
Dioxane	×	×			×	×	×	×			
THF	×	×		×	×	×	×	×			
Ethyl acetate		×			×	×	×	×		×	
Butyl acetate		×			×	×		×		×	
DMF	×	×			×	×	×	×			
Pyridine	×	×			×	×		×			
Acetonitrile	×				×	×		×	×		
Furfural	×				×	×		×		×	
Aniline					×	×		×			

8 Options for disposal

In any consideration of the relative economics of solvent recapture and processing for reuse against the purchase of new solvent, the cost of environmentally acceptable disposal must be included.

Liquid solvent thermal incinerators

For many years, dumping or land filling was considered to be the standard method of solvent disposal by all solvent users, from the smallest organizations to large international firms. Ideally almost any hole in the ground which could be used for getting rid of dry domestic rubbish could also receive drums and cans of solvents and chemicals. Even tanker loads of used solvent which could be absorbed by refuse might be accepted.

When damage to the local groundwater by such actions was observed and the amount of waste solvents increased dramatically, selected tips such as rock quarries, pits lined with clay, salt mines and underground coal mines that did not leak were still accepted for the disposal of toxic and flammable wastes.

The practice became a matter of public interest when dumping sites which had been filled with toxic waste and capped with a supposedly impervious layer were used for housing and other purposes involving unrestricted access. By this time, in the developed countries, many of the geographically convenient holes with suitable geological characteristics had been filled or had relatively short futures. It was then clear that a combination of neighbourhood concern and lack of usable sites demanded a great change in the disposal of used solvents, and that the new methods would cost much more than land filling, however sophisticated the best operated of the disposal sites might be.

The initial change was to move towards using the sea instead of the land for disposal. Relatively lightly contaminated water that was unacceptable in rivers and tidal estuaries could be discharged in the sea from sea-going barges and small tankers. These discharges were usually in designated areas not used for fishing, and a high proportion of the volatile solvent in contaminated wastes evaporated from the sea in a very short time to add to VOCs. Concern about the condition of the North Sea and other shallow seas with restricted contact with oceans is now such that the future of sea dumping is likely to be very restricted, and any new solvent-using operation being considered in the 1990s should not be planned on the assumption that this form of sea disposal will continue to be available.

Incineration at sea also became available in the 1970s. The scale of incineration is such that, while landfill disposal was available, there were few places where enough waste could be accumulated to support an economic incineration operation. As a result, a series of ships were fitted with on-board incinerators and plied the world collecting accumulations of liquid waste at sea-served tank storage installations. The waste was then burnt in areas of the sea where fishing did not take place. Incineration at sea was particularly attractive for handling chlorinated hydrocarbons, since the sea is alkaline and hydrochloric acid produced in burning was neutralized as the exhaust gases made contact with the sea surface. The extra cost of scrubbing the exhaust gases, generally necessary if chlorinated materials were burnt in land-based incinerators, could thus be avoided. Problems of supervision, with ships operating from countries where they were not registered, finally led to restrictions on sea-going incinerator ships while the quantity of material requiring incineration, displaced from landfill disposal by legislation, helped to improve the economics of fixed land-based plants.

The situation in the 1990s is that sites with large quantities of internally generated solvent-based wastes are likely to be able to justify the installation of in-house incinerators, often with biodegradation plants coping with solvent-contaminated water. Since chemical incinerators require considerable maintenance, even such sites may need to use commercial incineration as a fall back if their own unit's capacity is overloaded by arisings or is down for maintenance.

Arising from smaller operations will be sent in drums or bulk to commercial plants with capacities in the range of 20000–50000 Te/yr. While the operators of in-house incinerators have control over the material they have to burn and can specify the form in which it is delivered to them, merchant incinerators are often obliged to take 200 l drums containing material which has set solid in the drum. Since, under these circumstances, the drum and its contents must be charged directly to the incinerator as a single parcel, the size of the unit must be sufficient physically to accommodate the drum and to cope with the heat load imposed by a cold drum suddenly fed into it. This sets a minimum size of about 15000 Te/yr.

Many of the dedicated incinerators have capacities of less than 5000 Te/yr and can be much less complex than commercial ones. If they do not have to handle halogen-containing solvents or chemicals, they do not need to scrub their waste gases with alkali. Similarly, if there are no inorganics in their feed, they do not need to scrub out dust. Since, in addition, there is no need to transport waste off site, it is possible that in-house incineration can be very much less costly than commercial incineration, even when the scale of operation is a great deal smaller.

Thermal incineration relies on a high combustion temperature and an adequate residence time to achieve its effect. Since complete combustion is necessary, excess air of 30–50% is normally used and if the chlorine content of the incinerator feed is high, methane may need to be added to ensure that all the chlorine has reacted to hydrochloric acid.

A minimum temperature of 1100 °C, combined with a residence time of 4 s, is needed in the high-temperature zone to ensure satisfactory destruction of organics. To achieve the minimum temperature a lower calorific value of about 5000 kcal/kg (9000 BTU/lb) is needed for the feed. This is easily reached for hydrocarbons and other solvents containing little water. However, if much water is present, additional support fuel may be necessary.

Lower calorific value of 100% acetone 6808 kcal/kg

Lower calorific value of 75% acetone	5106
Lower calorific value of 25% water	−136
Lower calorific value of 75:25 mixture	4970 kcal/kg

It will be seen that absence of solvent has a far larger effect than the negative effect of the water.

There are restrictions on the effluent gas from incinerators which vary from country to country. A typical requirement would be:

VOC	5 mg/m^3
Carbon monoxide	1 mg/m^3
Hydrochloric acid	4 mg/m^3
Sulphur oxides	30 mg/m^3
Nitrogen oxides	180 mg/m^3

In addition, regulatory authorities will require that the key hazardous constituents, usually chosen because they are hard to burn, will be destroyed or removed to 99.99% (referred to as 'four nines') of the amount in the feed.

Normally the best practice calls for the plume of waste vapour, which would otherwise be apparent after scrubbing the effluent, to be eliminated by reheating. This can usually be done by heat exchange between unscrubbed and scrubbed gas.

It is important that gases leaving the very high temperature section of the incinerator should not spend any appreciable time at temperatures between 250 and 400 °C. This is because dioxins can be formed between these temperatures, and it is customary to use water quenching after heat transfer with the scrubbed effluent to prevent this.

Clearly, the costs of incineration can vary depending on the calorific value of the liquid waste, its chlorine content, the haulage involved and whether it is in bulk or drums. Prices in the range £150–500 ($250–850)/Te for bulk waste are likely.

Liquid solvent to cement kilns

Cement kilns have most of the requirements for satisfactory destruction of waste solvents. In particular, they have very high operating temperatures of about 1500 °C. Unless this temperature is reached, the cement clinker is not formed, which effectively guarantees temperatures well above those necessary for effective incinerator operation. Further, the gas residence time at high temperature is of the order of 30 s. Dust is removed from the gases being discharged from the stack by electrostatic precipitators, which are very effective compared with scrubbing, and the normal cement kiln has a stack several hundred feet taller than is normally fitted to an incinerator.

Finally, the conditions in the kiln itself are highly alkaline and turbulent so that, if halogens are included in the waste solvent fuel, they are reacted very quickly and form part of the cement clinker.

There are two major manufacturing processes for making cement: the wet and the dry process. The heat requirement for the former is about 6×10^6 BTU/Te of cement while the dry process needs about half this amount of heat. To remain competitive, the wet process has to use low-cost fuel wherever possible. The benchmark price is that for coal at about \$40/Te with a calorific value of 20×10^6 BTU/Te.

The capital cost of equipping a cement kiln to burn waste solvent covers

- tank storage and blending facilities;
- kiln firing equipment;
- solvent testing laboratory.

Since only about one-third of the fuel used on a kiln can be waste solvent, the possible cost saving can at best be 10% and a typical cost reduction is 6%. To achieve this, the charge for disposal of waste solvent of 5500 kcal/kg (10000 BTU/lb) with 3% maximum chlorine content is about \$35/Te, which shows a great saving over the cost of merchant incineration. In the USA, about 10^6 Te/yr of waste solvent is disposed of via cement kilns, and about 30 kilns have the necessary facilities. This is more than the quantity disposed of by incineration. In Europe a much smaller number of kilns are able to handle used solvents.

It is, however, not a route for disposal that will take any solvent regardless of composition. It is vital that the quality of the cement produced is not adversely affected.

The following specification limits are typical:

Solid particles	3 mm diameter maximum
Heat content	5000 kcal/kg minimum
Ash	10% maximum
Sulphur	3% maximum
Fluorine	1% maximum (fluxes kiln lining)
Chlorine* Bromine	4% maximum
pH	5–10
Viscosity	100 cP maximum

Metals:
  Lead
  Zinc
	1000 ppm (stop cement setting)

Chromium
  Cadmium
  Arsenic
  Mercury
	200 ppm (toxic)

Solvents:
  Carbon tetrachloride
  Benzene
	Non detectable (toxic)

Other highly toxic PCBs	50 ppm maximum

Since the quality of individual loads of waste solvent will vary very widely, it is important to screen incoming materials on arrival and to have ample storage and blending capacity to maintain a consistent quality of fuel to the kiln. Stirred storage tanks

---

*In order to avoid the excessive formation of calcium chloride, sodium and/or potassium must be present in the system.

that do not allow pigments to settle to the tank bottom or water-immiscible chlorinated solvents to form a separate bottom phase are desirable. One or more 'quarantine' tanks to hold loads that can only be bled slowly into blends are also necessary.

Some of the most attractive solvents to dispose of in cement kilns are washings from paint mills. These contain substantial quantities of resins and pigments but, in general, have a high calorific value (8500 kcal/kg) and are primarily composed of low-cost solvents, which makes them unattractive to recover. They tend to contain relatively high concentrations of iron and titanium, neither of which is harmful to the properties of cement. Because paint must not contain highly toxic metal compounds or solvents, it would be unlikely that paint mill washings would present toxicity problems.

Steam raising with waste solvents

Other processes for making use of the heat derived from used solvent are the generation of steam in specially equipped boilers and the drying of road stone in coating plants. Since the majority of small boilers are not designed to handle fuels with a high ash content, the selection of used solvents for the scale of operation typically covered by package boilers must be done with care. If the quantities of solvent arising on a site are substantial, it may be economic to process them to produce three disposal streams:

- high ash residue for destruction off-site in a merchant incinerator;
- distillate high calorific value fuels to burn usefully to raise steam on-site;
- water for biological treatment on-site.

If the biological treatment plant is already in existence to handle dilute aqueous wastes, the cost of such a strategy is well worth considering in comparison with sending the whole waste stream to a merchant incinerator who will penalize the negative calorific value of the water in the waste.

The presence of chlorine in solvents for incineration makes the problem of recovering the heat generated a difficult one. To avoid the production of dioxins or furans, temperatures greater than 1100 °C must be attained in the combustion. Additionally, the rapid quench down to 280 °C does not allow an easy opportunity for heat transfer. The more complex an incinerator is made, the greater are the possibilities of problems and breakdowns. Thus the cost of chlorinated solvents, if they cannot be recovered, can be very substantially increased to the user by the cost of disposal of the chlorinated solvent itself, and by the effect it will have on the cost of disposal of any other solvents mixed with it. If the presence of an organochlorine compound is unavoidable, it may be possible by, for instance, treatment with finely divided sodium, to remove it chemically and thereby make substantial cost savings in the overall effluent disposal bill.

Thermal and catalytic vapour incineration

Chlorine is also harmful in most cases in which catalytic incineration rather than thermal incineration is used. It clearly makes little sense in the removal of VOCs and the process odours sometimes associated with them if the organic molecules are recaptured from the air by carbon adsorption, only for the material that is removed from the bed during regeneration to be incinerated to waste. Thermal or catalytic incineration of the contaminated air is an effective way of cleaning it, and the solvent vapour present in the air makes an appreciable contribution to achieving the temperatures required.

Catalytic incineration usually runs at about 500 °C, depending on the solvent to be destroyed and the concentration of the solvent in air. This is unlikely, for safety reasons, to exceed 30% of LEL and may be much less. The percentage destruction will depend on the allowable limits of discharge which normally take account of the odours involved. In most cases, the catalyst is platinum based and will be specified for a given solvent mixture. The presence both of high dust levels and of halogens would influence the choice strongly against catalytic incineration. The higher capital cost and the lower fuel requirement of catalytic incineration against thermal incineration can only be compared for a specific duty.

Adsorption vs. incineration

MEK, a solvent which causes problems when adsorbed on activated carbon and which is difficult to recover from water, will increase the air temperature by about 550 °C if burnt at a concentration of 30% of LEL. With heat recovery, this will be sufficient to make both a catalytic and a thermal incinerator self-supporting in fuel.

As Table 8.1 shows, there is little pattern discernible in the heat rise that will be obtained from particular solvents. The high heats of combustion and molecular weights of hydrocarbons are offset to some extent by their low LEL. All systems with good heat recovery between incoming polluted air and outgoing 'incinerated' air should be self-sustaining at a 30% LEL concentration.

Table 8.1 Heat content of contaminated air at 30% solvent LEL

Solvent	LEL (ppm)	Heat of combustion (cal/g)	Temperature rise (°C)
Hexane	12000	10692	424
Toluene	12000	9686	411
Acetone	26000	6808	394
Methanol	60000	4677	345
n-Butanol	14000	7906	315

It would be under very unusual circumstances that expensive solvents (e.g. THF) should not be recovered even when the recovery costs are high. Cheap solvents, such as the hydrocarbons, are also cheap to recover after activated carbon adsorption, unless the odour of the recovered solvent presents a problem. Methanol, usually the cheapest of all the solvents, needs straightforward fractionation after desorption from activated carbon, but will often prove a marginal case for recovery.

It should not be forgotten that when very high percentage recovery is practised, as is now required in advanced industrial countries, there is a tendency for impurities to build up in systems where, in times past, they were purged away in vapour losses. The complexity and cost of re-refining solvents for reuse therefore tends to become greater, and disposal in an environmentally acceptable way is thus made more attractive.

Biological disposal

Solvent recovery processes, both for recapture from air and water and for working up to a reusable condition, give rise to large amounts of solvent-contaminated water. Part of this will arise from the processes themselves but, particularly in a general-purpose plant where tanks and process equipment have to be decontaminated frequently, much will arise from housekeeping activities.

If the solvents handled are sparingly water miscible, much of the contaminated effluent will need phase separation, which will remove a large part of the solvent present. If the separation can take place at an early stage of the flow through the site, such separation will prevent the whole aqueous effluent from being saturated with all the solvents being handled (*see* Chapter 4).

Ignoring water used for cooling, which should only be contaminated if condensers or coolers leak, a typical water discharge for a general-purpose solvent recovery plant may lie in the range of 1–5 Te per tonne of solvent processed. Whether this water is processed for return to the environment at a municipal treatment facility, where great dilution with domestic and other trade effluents may be expected, or on site, where bacteria particularly effective in dealing with specific contaminants may be used, a disposal of solvents of 2% of the organic part of the incoming solvent may well be achieved in this way. Provided the treatment is truly biological and not a thinly disguised air stripping process, such disposal is environmentally acceptable and is likely to be a great deal cheaper than 'incinerating' water.

9 Good operating procedure

There is no reason why a solvent recovery operation, whether independent or in-house, should not achieve the same high standard of safety and good housekeeping as any other chemical plant. There are, however, problems and potential hazards that are more commonly found in reprocessing solvents than in the generality of chemical manufacture and they warrant special consideration in the design and management of such facilities.

While the solvents used in industry are documented and information is available on the dangers involved in their use, this cannot necessarily be assumed to be true of solvent mixtures and is certainly not true of the mother liquors which are frequently worked on to recoup their solvent content.

Unusual dangers also stem from the attitudes of generators of used solvent who, too frequently, regard them as 'waste' rather than as raw material for the recovery process. Such attitudes affect, among other things, the labelling of drums, the quality of drums used for storage and the care devoted to avoiding cross-contamination. Indeed, far too often, used gloves and unwanted sandwiches are found in consignments of solvent for recovery.

While in-house recovery usually requires expertise in the hazards of a limited range of solvents, a commercial recovery firm is likely to need to handle safely a greater number and diversity of solvents than any user or producer. To do this calls for a very high standard of management and a well trained labour force. Indeed, a case may be made, on the grounds of safety, for restricting the number of different solvents handled on a single site or in a single self-contained unit.

In making recommendations for good operating procedures there is little advantage in trying to isolate the hazards arising specifically because used solvents are the feedstock of a recovery operation from those relevant to the everyday safe processing and handling of toxic and flammable solvents in general. It is useful, however, to consider the problems in a methodical way particularly if one has the luxury of building an operation from scratch. Such a consideration will form the basis for standard operating procedures.

A general code of practice for solvent recovery, except in the case of a very large commercial solvent recovery organization, will contain much that is not applicable to individual sites or firms.

The headings for consideration are not necessarily listed here in order of importance, although the requirement for an able and well trained staff supported by adequate laboratory facilities cannot be stressed too highly.

- Staff
- Laboratory

- Installation design and layout
- Principal hazards
- Storage and handling of solvents
- Feedstock screening and acceptance
- Process operations
- Maintenance
- Personal protection
- First aid
- Fire emergency procedure

Staff

Matters to be considered include the following.

1   Educational standard

A solvent recovery plant is potentially a hazardous environment and for his own sake as well as for that of his fellow workers, an adequate standard of literacy and numeracy is vital for every employee no matter how humble his or her role may be.

2   Colour blindness

Colour coding of drums, pipelines, etc., is a common and useful aid to operation. Many people, especially males, are colour blind and are therefore at increased risk.

3   Physique

Although the handling of full drums by fork lift truck is the normal practice, there will always be occasions when items of this size have to be moved by hand and above-average height and weight make such an operation safer and easier.

4   Sex

Particularly when processing solvent residues from the pharmaceutical industry there exists a danger of contact with materials having teratogenic properties. The wisdom of employing females, either in the laboratory or the plant, who may become pregnant needs careful consideration.

5   Skin complaints

Some people's skins are particularly susceptible to dermatitis and other complaints, however careful they are with protective clothing, barrier cream and personal hygiene. They should not be employed in solvent handling.

6 Health inspection

Before engagement, all employees should have blood and urine tests to check for abnormalities and to provide a datum for subsequent tests.

7 Liver function

The combined load on the liver of heavy alcohol intake with exposure to solvents can be harmful.

Laboratory

Access to an adequately equipped and staffed laboratory is essential for the safe operation of a solvent recovery unit. Its function can be divided into various areas.

1 Development of process

This requires laboratory-scale equipment that will allow a simulation of the conditions attainable on the plant.

2 Monitoring of goods in and out

It is highly desirable for both feedstocks and recovered solvents that methods jointly agreed with supplier or customer are used and that similar equipment is employed. It should be appreciated by both parties that specifications appropriate to virgin solvents are not necessarily sufficient for recovered solvents, since impurities never present in the manufacture of virgin solvents may be found in a recoverer's feedstock and product.

3 Storage of samples

The laboratory will normally be responsible for the taking and keeping of samples of goods in and out samples. Retention for a year or more may be required and safe housing in a ventilated and fireproof building, ideally separate from the laboratory itself, must be provided.

4 Monitoring of process

Samples, possibly taken hourly, will need to be checked to follow the progress of plant batches.

5 Quality control of products

After processing and blending have been completed, the product must be passed fit for sale. It is very important that this test is done on a sample that truly represents the tank contents.

6 Certification of equipment

Meters to test explosive and toxic atmospheres must be totally reliable in use since a person's life may depend upon the test results. Before use in the field, instruments should be checked against standard vapour mixtures in the laboratory.

7 Ventilation of laboratory

In a laboratory handling solvents, good ventilation is essential and much of the work done can be carried out with advantage in fume cupboards. Ventilation at a low level in the laboratory removes heavy vapour most effectively and avoids drawing vapour upwards where it is more likely to be inhaled.

8 Minimum solvent inventory

The amount of flammable solvent in the laboratory at any time should be kept to a minimum since this is probably the highest fire risk area within the whole solvent recovery operation.

Installation design and layout

Hazards, particularly from fire and explosion, can be reduced by careful layout of a site. Future expansion should not be forgotten in settling the initial layout. The various points to be considered are the following.

1 Segregation

Storage should be segregated from process plant and dangerous processes from less dangerous ones. Thought should be given to the way a fire may spread to involve other areas, the slope of the ground and the natural drainage system.

2 Routine site access

Road transport accidents are a common cause of death or injury at work and internal site roads should be designed to avoid blind corners and junctions. Consideration should also be given to access for lifting equipment used in maintenance and construction.

3 Emergency site access

Fire appliances should have easy access to hazardous areas with allowance being made for variable wind direction. The siting of fire hydrants should reflect this. The layout of road tanker loading/discharge bays should allow for a tanker to be driven out forwards and without difficult manoeuvring in the event of fire.

4 Tanker parking

It will be necessary to have an area, preferably close to the laboratory, where samples can be safely taken from a tanker and where the tanker can then stand while the sample is being tested.

5 Incident control rooms

Particularly in the circumstances when there are few people on a site (night shifts, weekends) consideration should be given to them giving each other the maximum mutual support in an emergency.

Assembly points for staff should be close to control points but not on vehicle access routes.

Secure communication for obtaining outside help is vitally important.

6 Flammable inventory

Operators of solvent recovery plants never admit to having enough tank storage both as regards tank size or numbers. However, both the cost of the storage and the usually greater cost of the tank contents must be set against the easier operation which a multiplicity of tanks provides.

Commercial recoverers who must be able to cover the requirements of their customers should have a minimum storage capacity of a delivery (say 20 Te) plus a week's production plus ullage (5%) for each of their products sold in bulk. If their plant is a multi-purpose one the frequency of production of batches or campaigns may increase the production period they must accommodate. It is more difficult to set a target for the storage of used solvent since this may not be set by factors under the processor's control. A reasonable first estimate, in the absence of more precise information, would be for crude storage of a campaign or batch plus two deliveries plus 5% ullage.

Thus, for the refining of a solvent stream arising at 10 Te/week and being processed at 5 Te per batch on a dedicated plant:

- Crude storage   47.25 Te
- Product storage 23.625 Te

A modest-sized solvent recovery operation processing five different solvents may thus need 350–500 Te of solvent storage in ten different tanks as a minimum.

Principal hazards

The majority of hazards on a solvent recovery site and accidents arising therefrom are those involving handling, climbing, lifting, vehicles, etc., which are common to most heavy industry operations.

The hazards which require special attention in solvent recovery are explosion, fire and toxic risks.

142 *Good operating procedure*

1 Explosion

Many solvents and their solutes can decompose, polymerize or react very rapidly with oxygen or water, thereby creating a cloud of gas or vapour. If confined, this vapour will cause a high pressure, which may lead to the confining vessel bursting.

1.1 Laboratory investigation

Such reactions, as far as solvent recovery is concerned, tend to take place at elevated temperatures and are seldom, if ever, triggered by the material (steel; copper) of which the plant is constructed.

It should therefore be an unbroken rule, when handling used solvents, to carry out in every detail on a laboratory scale any operation that is planned to be done subsequently on a larger plant scale. Protective screens should be used in the laboratory for such experiments and liquid, not vapour, temperatures should be observed. Some decomposition reactions have an induction period so the laboratory experiment should cover a time at least as long as the plant work is expected to last.

1.2 Preventing exotherms

It is difficult to design a venting system for a plant to cope with the energy released by a decomposition of this sort and every effort should be made to prevent an exotherm occurring. Some methods for coping with the problem are considered below.

1.3 Inventory reduction

If an unstable material is being evaporated, the plant inventory and material residence time should be kept to a minimum, e.g. use a thin-film evaporator in a continuous process.

1.4 Heat removal

Monitor the plant for a temperature rise or, in a temperature controlled process, a fall-off of heat (e.g. steam) input with maintenance of temperature. If an exotherm is detected, automatically cut off heat and remove heat from system (e.g. water douse).

1.5 Low-temperature heating medium

Avoid direct heating on which skin temperature control is difficult.

Use the lowest temperature heat source practicable.

Lower the boiler operating pressure to ensure that a set input temperature cannot be exceeded.

1.6 Low operating temperature

If an exotherm has been found either by replicating a process in the laboratory or by differential thermal analysis, the temperature at which the exotherm has been detected should be at least 20 °C higher than highest temperature authorized for plant operation.

1.7 Restabilizing

Because inhibitors and stabilizers can be fractionated out of distillates from, and the hold-up of, a fractionating column it may be necessary to add such materials continuously to the system to ensure that all its contents are protected.

1.8 Inert atmosphere

Use low-oxygen inert gas for breaking vacuum. Although nitrogen with an oxygen content of up to 8% is sufficient to inhibit fires, it may well not be satisfactory for preventing a fast reaction and 99%+ nitrogen should be used.

2 Fire

Fires can only take place if three components are available:

- oxygen;
- a source of ignition;
- a combustible material.

2.1 Oxygen

No solvent or solvent mixture can burn with less than 8% oxygen so that by reducing the oxygen content of air from 21% it is possible to create a safe gas for blanketing tanks and venting vessels. Normally, to give an adequate margin of safety, inert gas generators make a 3% oxygen product. If such a gas is being used, care must be taken before entering tanks and vessels that the atmosphere in them is fit to breathe.

2.2(a) Ignition source—autoignition

Solvent vapours can be ignited by contact with a sufficiently hot surface without any flame or spark. Materials especially dangerous in this respect are listed in Table 9.1.

Table 9.1 Autoignition temperatures of dangerous solvents

Solvent	Autoignition temperature (°C)	Equivalent steam pressure (psig)
Carbon disulphide	100	0
Diethyl ether	160	75
Dioxane	180	131
Ethylene glycol dimethyl ether	201	216
Ethylene glycol diethyl ether	208	250
Dimethyl sulphoxide	215	290

The dangers of handling these materials when hot oil, high-pressure steam or electric heating are used are obvious and spillage on the outside of imperfectly lagged steam lines can cause a fire. When materials have to be handled above their autoignition temperatures, so that by definition both an ignition source and a combustible material are present, there is no safe alternative to the use of an inert gas to prevent the presence of oxygen and hence a fire.

2.2(b) Ignition source—electrical apparatus

Electrical equipment can provide a source of ignition either by producing a spark or by having a surface hot enough to cause autoignition.

The solvent recoverer may not be able to predict the solvents handled on his plant in the future, but must specify a temperature classification for the electrical equipment based on an appropriate material code and be sure that this danger is understood by his staff. Table 9.1 shows that 200 °C is an adequate limit for covering the majority of solvents. Protection against the sparks produced by electrical equipment is also covered by appropriate national codes.

2.2(c) Ignition source—flames

Hot work on the plant should not take place without a permit to work signed by a properly qualified person after a thorough survey (*see* Maintenance Clause 1 in Maintenance section later). This should also apply to hot work done on equipment (e.g. a defective heat exchanger) removed from the plant to a safe area (e.g. a maintenance workshop) but possibly still containing flammable liquid. No equipment should be sent to an outside contractor without certification as gas free.

2.2(d) Ignition source—smoking

The boundaries of the hazardous areas of a plant should be marked and it should be clear to strangers (e.g. contractors) where smoking is and is not permissible. Matches and lighters, portable radios, camera flash equipment and other portable sources of ignition should not be brought into the danger area without a written permit.

2.2(e) Ignition source—petrol-driven engines

These should not be allowed in the hazardous area and in particular petrol-driven sump pumps and other contractors' tools, liable to be left running without the constant attendance of an operator, should be banned.

2.2(f) Ignition source—static electricity

Static electricity is generated in solvents when new surfaces are formed under such circumstances as pumping a two-phase mixture (air–solvent or water–solvent) down a pipeline or into a tank. The faster the rate of pumping, the greater is the charge generated. The higher the electrical conductivity of a liquid, the more quickly the charge will dissipate. The least conductive flammable solvents are the hydrocarbons and it is good practice not to pump these at more than 1 m/s even if pipework and receiving tankage are fully earthed.

2.2(g) Ignition source—hot oil leaks

Spillage of flammable solvent on lagging or, in the case of hot oil heating systems, leakage of oil at flanges and valve stems onto lagging produces a high fire risk. Where spillage is likely, good sheet metal cover of the lagging is desirable and solvent or oil-soaked lagging should be stripped from heated equipment. In oil-heated systems there should be a minimum of flanges, and valves should be installed with spindles horizontal or vertically downwards where possible.

2.2(h) Non-conductive containers

Care must be taken to earth drums of flammable liquid when filling them and to use electrically conductive hoses. Filling plastic jugs and other non-conductive vessels such as glass bottles should, if it has to be done, be performed very slowly to allow the static charge to leak to atmosphere.

2.2(i) Lightning

In a well earthed plant this should not prove a problem, but if the tallest building on a site houses a plant which might have a flammable atmosphere it should be fitted with a lightning conductor.

2.3 Combustible material

The third component of a fire is the vapour, which can mix with air, over the surface of a flammable liquid. Solvent vapours will only burn in air over a restricted concentration range bounded by the UEL (upper explosive limit) and LEL (lower explosive limit). Table 9.2 sets out for a typical range of flammable solvents their UEL and LEL values

Table 9.2 Flammability properties of various solvents

Solvent	LEL (%)	Lower flash point (°C)	UEL (%)	Upper flash point (°C)	B.p. (°C)
Toluene	1.2	4	7.0	37	110
Heptane	1.0	−4	7.0	29.5	98
Octane	1.0	13	6.5	50.5	126
Benzene	1.3	−11	7.9	14	80
n-Hexane	1.2	−22	7.7	8	69
m-Xylene	1.1	21	6.4	59	132
n-Nonane	0.9	31	2.9	55	151

and their flash points, which are effectively the temperatures at which the solvent-saturated air attains the LEL.

If the solvent vapour is mixed with a gas other than air (e.g. oxygen), different limits, and therefore flash points, would apply.

The chance of ignition from static electricity is especially high when handling liquids that have vapours above their LEL and below their UEL, since the sparks tend to take place near the liquid surface where the vapour will neither be too rich nor too lean to catch fire. At ambient temperatures toluene, heptane and octane are particularly liable to electrostatic ignition, whereas benzene, *n*-hexane, *m*-xylene and *n*-nonane at normal ambient temperatures are outside their explosive range.

Initial boiling point is a good guide to the most dangerous hydrocarbons and 95–130 °C is the most dangerous range.

3 Toxic hazards

3.1 Vapour inhalation—acute

Death, or long-term damage to health, can occur in a relatively short time with some solvent vapours when they are well below the solvent's LEL.

Table 9.3 sets out IDLH (immediately dangerous to life and health) values alongside the maximum concentrations that may be attained due to a spillage in an unventilated room or solvent evaporating from the surface of a pool in a tank. The only solvents in the list in which the saturated vapour is below the IDLH are those which are relatively involatile and relatively non-toxic (e.g. *n*-butanol, white spirit).

3.2 IDLH definition

This is defined as the maximum vapour level from which one could escape within 30 min without symptoms that would impair one's ability to escape and without irreversible health effects. This would be relevant to lifesaving emergencies.

3.3 Toxicity of chlorinated solvents

Table 9.3 underlines the hazardous properties of the chlorinated solvents. Being relatively low boiling they have high vapour pressures at 21 °C. Since they produce a very heavy vapour, ventilation needs to be unusually powerful to displace their vapours and they have relatively low IDLH values.

3.4 Vapour inhalation—chronic

Using the faculty of smell is a very crude method of detecting solvent vapour, but it is valuable to know for which solvents it is useless as a protection against harmful long-term exposure. These are the solvents which have an odour threshold higher than their TLV (e.g. chloroform). The odour threshold varies between individuals and tends to increase with length of exposure (i.e. one becomes used to a smell). It also can be affected by the presence of other solvents which can mask a smell.

3.5 Solvent mixtures

For much more accurate determination of the level of vapour in air, proprietary equipment exists but since mixtures of solvents may contain components which reinforce each other's harmful effects, care is needed in using even the most accurate results.

3.6 Neighbourhood nuisance

While odour thresholds much lower than the TLV may assist in reassuring nearby

Table 9.3 Toxic hazard properties of various solvents. All figures in ppm

Solvent	Odour threshold	TLV	IDLH	Solvent saturated vapour at 21 °C
Acetic acid	1	10	1000	16 000
Acetone	100	750	20 000	250 000
Acetonitrile	40	40	4000	94 000
Aniline	0.5	2	100	340
Benzene	5	10	2000	105 000
n-Butyl acetate	10	150	10 000	14 000
n-Butanol	2.5	50	8000	6300
Carbon tetrachloride	10	5	300	127 000
Monochlorobenzene	0.2	75	2400	13 200
Chloroform	250	10	1000	220 000
Dichloromethane	250	100	5000	500 000
Diethyl ether	1	400	10 000	100 000
Dimethylformamide	100	10	3500	3700
Ethyl acetate	1	400	10 000	100 000
Ethanol	10	1000	20 000	60 000
Heptane	220	400	19 000	610 000
Isopropanol	90	400	20 000	46 000
Methanol	100	200	25 000	130 000
Nitrobenzene	6	1	200	270
Nitroethane	160	100	1000	210 000
Nitropropane	300	25	2300	22 000
n-Octane	4	300	3750	16 000
n-Pentane	10	600	5000	580 000
n-Propanol	30	200	4000	18 000
Pyridine	0.02	5	3600	22 000
Perchloroethylene	5	50	500	22 000
Toluene	0.2	100	2000	31 000
Trichloroethylene	50	50	1000	80 000
White spirit	1	200	10 000	3400
Xylenes	0.05	100	10 000	9200

communities that a smell may not be harmful, they also indicate to a solvent recoverer what concentrations must be achieved to avoid causing a nuisance. The ratio between odour threshold and solvent saturated vapour gives a measure of the dilution problem posed by each solvent, though a judgement as to which smells are acceptable and which are not is very subjective.

3.7 Adsorption through skin

Some solvents, such as dimethyl sulphoxide and dimethylformamide, are very readily adsorbed through the skin and have the ability to carry solutes through the skin with them. Such solvents when present in feedstocks need to be treated with great care, particularly when they have a pharmaceutical origin. Quoted figures for TLV are irrelevant when considering the handling of such solvents in an unrefined state.

Storage and handling of solvents

1 Regulations

In most industrialized countries regulations exist governing the storage, in bulk or in drums, of highly flammable liquids. The storage of solvents which are not flammable (e.g. most chlorinated hydrocarbons) or which have a flash point above normal ambient temperature (e.g. DMF) are unlikely to be regulated unless they pose a very serious environmental hazard.

2 Drum storage—principles

The principles to be adopted are as follows:

- external storage wherever possible;
- storage area to have an impervious surface and means for retaining spillage and leaks, e.g. by a retaining sill;
- area to be separated from buildings, boundaries, fixed sources of ignition or tank bunds by at least 4 m;
- drums to be stacked for easy access and inspection;
- if weather protection is needed, this should consist of a lightweight roof and open sides. Such protection minimizes the risk of contaminating rain water with leakage from drums.
- if internal storage cannot be avoided, then the building should be constructed of half-hour fire-resistant materials, unless separated from other buildings, boundaries or tank bunds by at least 4 m;
- storerooms should incorporate permanent natural ventilation by a substantial number of low- and high-level air bricks, means for retaining spillage within the room, electrical equipment (where necessary) to explosion-proof standard and a self-closing door.

2.1 Drums in workrooms

The number of drums in a workroom should be as small as possible. Closed drums can be stored temporarily outside process buildings provided that:

- the building wall has at least half-hour fire resistance;
- they are not within 2 m of a door, plain glazed window, ventilation opening or other building opening, or directly below any means of escape from an upper floor regardless of distance;
- their siting and quantity do not prejudice the safety of any means of escape from the building.

2.2 Drum storage—layout

While the purchaser and user of drummed solvents will normally minimize his inventory, a solvent recoverer may need to hold a large stock of raw material in drums. Further, these drums are very seldom new.

Drums will normally be received in 80 drum loads with occasional loads of up to 100 drums and a storage layout which allows for access around such a load for inspection for leakage, stocktaking, etc., is desirable.

2.3 Drum storage—details

Full drums should be stored in a vertical position since, in the event of a fire, drums normally fail at their ends and a vertical drum will retain much of its contents if its head fails. In addition, since the majority of drums are head fillers, their bung and titscrew washers will not be liable to leak in storage. Leaking drums can be located without difficulty in a stack two pallets high and two pallets wide, so a block five pallets long by two wide by two high will accommodate a standard 80 drum load. A 0.5 m access passage around such a block is adequate for inspection purposes.

Despite the risk of reignition of a solvent fire from smouldering wood, it is on balance safer to store drums on pallets. A stack of palletized drums is more stable, particularly if the ground is at all uneven. Palletized drums are less prone to damage in handling by fork-lift truck than when handled loose, even if specialized drum handling attachments are used and in an emergency pallets can be handled more quickly. Finally, in a bunded area where rainwater may accumulate the base of a drum may be corroded whereas it is rare for a pallet to rot under the same conditions.

2.4 Drum storage—compatibility

Drums of incompatible chemicals should not be stored together. Incompatibility can be due to the potential for a dangerous reaction if two chemicals come into contact or if a leakage of a corrosive chemical affects the integrity of a flammable liquid receptacle.

2.5 Drum labelling

Used solvents for recovery are rarely stored in new drums. It is most important not only that used solvent drums are clearly labelled on their sides with their contents, but also that old markings referring to the drum's previous use are totally obliterated.

For solvents that are not being recovered in-house, internal code names or numbers (e.g. X12 Mother Liquor) are not a sufficient marking and internal abbreviations (e.g. IPA for isopropyl alcohol or isopropyl acetate) must be avoided.

3 Handling and emptying drums of feedstock

3.1 Opening drums

The use of a standard drum key for opening drum bungs should present no safety problems, but if a drum is 'bulged' with the contents possibly under pressure, care is needed. Bulging may be due to warming after liquid overfilling but may also be due to a chemical reaction or corrosion taking place inside the drum after filling. In the latter case a quantity of gas under pressure may violently blow out the bung and some of the drum contents when the bung is unscrewed. If a drum is suspected of being under pressure the bung should be loosened and the drum vented. Only when there is no flow of vapour should the bung be fully unscrewed.

3.2 Eye protection

This should always be worn when opening drums in case the contents are under pressure.

3.3 Opening difficult drums

A drum suspected of containing flammable solvent should never be chiselled open if the bung cannot be unscrewed. If penetrating oil does not free it and the titscrew also is seized, the destruction of the drum top using acid is a possible method of getting at the drum contents safely.

3.4 Sucking out drums

To protect operators from solvent fumes while drum emptying, the use of a vacuum receiver is effective. If a centrifugal pump is used, spillage from its suction hose and pipe is hard to avoid, whereas a vacuum will suck the suction pipe clear. A liquid ring vacuum pump has the additional advantage of scrubbing the exhaust air from the system to minimize the flammable or toxic vapours generated.

3.5 Safe handling of empty drums

An emptied drum is potentially full of flammable vapour which, if ignited, can present an immediate explosion hazard greater than that from a full drum. This vapour will be heavier than air and a drum that is stored bung hole uppermost will continue to be an explosion hazard for a considerable time. It can be steamed out, but this can expose the operator to high vapour concentrations.

Storage of empty drums on their side with bungs removed and bung holes at the lowest point is a crude but effective way of removing heavy vapour. This must be done in an area where no-one is exposed to high vapour concentrations and where all equipment is flameproof. Before disposal each drum should be checked by explosimeter.

3.6 Storage of clean empty drums

Whereas full drums should be stored standing on their ends, clean empty drums should be stored on the roll. This aids fire-fighting teams who need to know the hazards they are coping with. It also avoids the possibility of water standing on the heads of empty drums and infiltrating into the drums before they are filled. This could spoil the drum contents and may cause a foam-over hazard if material over 100 °C is filled into the drum.

4 Drum filling

4.1 Ullage

It is important that drums are not overfilled since a moderate temperature rise can cause an overfilled drum to leak or burst owing to the pressure caused by liquid expansion. For use in the UK a 5% ullage should be allowed, and this means that a standard drum will hold 205 litres or 45 Imperial gallons. Drums for use in hotter climates may need extra ullage.

4.2 'Remade' drums

Some drum reconditioners 'remake' drums and do not differentiate between remakes and standard drums, although remakes are smaller. Particular care should be taken when filling to a standard weight or volume that such drums have enough ullage.

4.3 Drum strength

There is a wide range of density (from pentane 0.63 to perchloroethylene 1.63) in the solvents handled by recoverers and heavier gauge drums are required for those with densities over 1.0.

4.4 Earthing

Drums being filled with flammable liquids should be earthed at all times with a flexible electrical lead to a 'proved' earth.

4.5 Drumming hot materials

Drums filled with hot products or residues should be allowed to cool before being closed, since otherwise they may distort or even collapse.

4.6 Plastic drums

The suitability of plastic containers should be considered in the light of possible degradation if exposed to UV radiation. Special care is necessary when filling plastic or plastic-lined drums to conduct away static electricity.

4.7 Personnel safety measures—eyewash bottles

Eyewash bottles should be available when drums are being filled with solvents. Eye protection should be worn at all times.

4.8 Ventilation

An operator filling drums is potentially exposed for long periods to the vapour of the solvent being handled. A ventilated hood over the drum or a drum-filling lance with built-in ventilation are appropriate methods of protection.

5 Forklift trucks

5.1 Training

The use of lift trucks requires particular care. Truck drivers should attend appropriate training courses and be formally licensed. Many accidents with trucks are caused by misuse, e.g. allowing persons to ride on the truck, using the forks as a means of access to heights without adequate protection and driving carelessly. Adequate supervision is necessary to detect these practices and take appropriate action.

5.2 Drum lifting attachments

These should be firmly clamped to the forklift trucks' tines when in use and should be regularly inspected for wear.

5.3 Emergency stopping

If a forklift truck is not fully flameproofed, the driver should stop the engine at once if he drops a drum of flammable solvent or penetrates a drum with the tines.

5.4 Pallets

When handling and stacking drums on pallets, only sound pallets should be used. Protruding nails on pallets can puncture drums.

5.5 Loads on pallets

The specification of forklift trucks for use in drum handling should allow for the possibility of a four drum pallet weighing a maximum of 1200 kg.

6 Bulk storage

6.1 Tanks above or below ground

Tanks should be above ground and in the open air. This facilitates cleaning, repairs, examination, painting, leak detection and the dispersal of vapour from vents and leaks. If there is no alternative to underground tankage, it is important that leaks from both tanks and their associated underground pipelines are detected and that leakage into the surroundings is contained and does not contaminate the water table. Burying tanks in concrete cells and using washed sand as back fill is one method of reducing the risk of external corrosion, but internal corrosion of mild steel tanks is a greater risk in solvent recovery than it is in the storage of unused solvents.

6.2 Separation distances

Tanks for storing flammable liquids should be separated from buildings, site boundaries, process units and fixed sources of ignition by the distances laid down in national codes.

6.3 Bunding

Tanks for storing flammable liquids should be surrounded by bund walls high enough to contain 110% of the largest tank in the bund. Bund walls over 1.5 m can make fire fighting difficult and may interfere with ventilation of the bottom of the bund.

6.4 High-density products

If high-density products (e.g. chlorinated hydrocarbons) may be stored in the bund, it must be designed to withstand the hydraulic pressure that may be exerted by them.

6.5 Impermeable floors

The floor of the bunded area, including the area beneath the tank bases, should be impermeable. The surface between the tanks should be laid to fall so that no spillage can form a puddle in the bund which can be a health as well as a fire hazard.

6.6 Bund drainage

From the low point in the bund a suitable means of draining rainwater, spillage, overflows, etc., must be provided. If this takes the form of a valve or penstock, steps must be taken to prevent it being left open.

6.7 Calibration

Tanks must be calibrated in litres so that an operator can tell accurately their available volume. Sight glasses are not satisfactory for this if a two-phase mixture may be stored in the tank.

6.8 Plastic tanks

Tanks should be designed and constructed to a recognized national standard. Even if their contents are not flammable, plastic tanks should not be located in the bunds containing tanks of flammable liquids. Repairs to metal tanks using fibreglass or other non-metallic materials are not satisfactory for tanks in 'flammable' bunds.

6.9 Identification

All tanks should be prominently numbered and these numbers should be visible to fire fighters. A schedule should be kept so that fire fighters can find out the contents of each tank. To avoid confusion, the tank number should also be visible at the tank's dip hatch.

6.10 Painting

In difficult climatic conditions (seaside, chemical works), mild steel tanks can be severely pitted by corrosion even when most of their paintwork is satisfactory. A high standard of repainting is desirable under such conditions.

6.11 Splash filling

Splash filling tends to generate static electricity and filling tanks from the top should be avoided if possible.

6.12 Earthing

All metal parts of the tank installation should be continuously earthed to eliminate electrostatic sources of ignition. The earthing efficiency should be proved and recorded annually.

6.13 Ullage

In operating tanks, consideration should always be given to allowing sufficient ullage. Factors to be considered include:

- changes in ambient temperature;
- mixing by air or inert gas;
- filling from tankers using air or inert gas;
- heating of tank contents with coils including the possibility that a thermostat will fail or a valve not close tightly;
- change of volume during blending.

6.14 Drain valves and sampling

It is undesirable to have single valves opening to the atmosphere at the bottom of tanks unless such valves are normally blanked off. Therefore, sampling the bottom of a tank should be via a dip hatch in the tank top with a bottom sampler. A self-closing dip hatch is recommended.

6.15 Pipeline labelling

Pipelines at loading/unloading points, whether they are points for hose connections or are solid pipelines with valves, should be clearly marked with the number of the tank to which they give access.

6.16 Valve closing

Whenever an operation is stopped for an appreciable time (e.g. 1 h), the valve on the tank should be closed. Reliance should not be placed on a valve distant from the tank, particularly if a hose forms part of the unprotected system.

6.17 Valve types

Bottom phases of water are frequent in solvent recovery tanks and valves, particularly drain valves, at the bottom of tanks are liable to freeze up. It is very important that under these conditions the valve body does not crack, leading to a serious leak when the ice thaws. Cast iron valves are therefore not suitable for such service. Cast steel valves are to be preferred.

Dirty solvents containing solids can make it difficult to keep the seats of gate valves clean and plug or ball valves where the seats are wiped in operation are preferable.

Diaphragm valves are difficult to specify because of the variety of solvents to which the diaphragm itself must be resistant. The only multi-purpose diaphragm material, PTFE, is liable to be damaged by the solids that dirty solvents may contain.

6.18 Pump selection

Pumps should be installed on plinths and, particularly in vehicle discharge areas, be protected from vehicles by curbs or safety railings. When handling flammable solvents, mechanical seals should be standard fittings and their specifications should reflect both the solvents being handled and the suspended matter in them that can jam the seal spring of an unsuitable seal. Glandless pumps are also suitable.

Because of the undesirability of using gland packing, reciprocating pumps are not suitable for pumping solvents and, if a positive pump is needed, double-diaphragm air-operated pumps are worth considering.

Rotary pumps, because of their close clearances (unsuitable for suspended solids) and because of the poor lubricating properties of most solvents, are seldom a viable choice.

The high vapour pressure of many solvents means that care in designing suction hoses and pipework is needed and low NPSH (net positive suction head) may be a necessary specification for a centrifugal pump particularly if run at 2900 rpm.

6.19 Pipeline blockage

When handling spent solvents there is a greater than normal risk that a pipeline may become blocked by tarry substances. A centrifugal pump running against a blocked delivery may become extremely hot and thermal decomposition of the liquid in it may occur.

6.20 Expansion of pipeline contents

Solvents have high coefficients of thermal expansion and pipelines heated by the sun can develop very high internal pressures.

In designing pipework systems, long lengths of line with tight shut-off by valves at both ends should be avoided if they incorporate:

- pumps with cast iron bodies that can fail under high pressure;
- hoses;
- ball valves with plastic seats that can be forced out of position.

If there is no alternative, relief valves must be fitted on such pipelines with discharge to a storage tank.

7 Tanker loading and unloading

7.1 Standing instructions

These should be clearly written instructions governing the loading/unloading operation. These should include:

- precautions against the tanker moving during the operation;
- checks on the hose connections and valve settings before pumping;
- exclusion of sources of ignition during pumping;
- earthing using a system well maintained and dirt and grease free at both the tank and installation ends;
- precautions against overfilling; adequate ullage is essential to cater for any expansion of tank contents due to temperature increase—standards on appropriate ullage space are contained in ADR and IMDG codes;
- vehicle inspection to ensure that dipsticks correspond to vehicle compartments, the vehicle is clean enough for the job and manhole gaskets appear in good condition.

7.2 Vent emissions during loading and unloading

The operation of filling or emptying a road tanker will inevitably lead to flow at tank vents. This flow may be flammable, toxic and/or environmentally unacceptable and consideration should be given to its discharge to a safe place. Linking vents of the filling and emptying vessels is the ideal solution. If the vapours vented from a tank wagon being filled are toxic or narcotic, consideration must also be given to the safety of the operator dipping the contents of the tanker.

7.3 Tanker unloading

Tankers can be unloaded by:

a   a static or mobile pump based at the installation;

b   a pump on the vehicle driven by the vehicle's diesel engine either directly or via a hydraulic system;

c   an air compressor on the vehicle driven by the vehicle's engine;

d   compressed air or compressed inert gas produced on the installation.

The tanker driver should be present throughout the operation if methods (b) and (c) are used. When handling highly flammable materials, method (a) is much to be preferred over method (b) and methods (c) and (d) should only be used in exceptional circumstances.

If method (c) or (d) is used there is a need for considerable ullage in the receiving tank at the end of unloading when a slug of gas will enter the base of the tank and carry some of the tank's contents out of the vent or overflow. The possibility of generating static electricity when air is blown through low electrical conductivity solvent mixtures is considerable.

7.4 Avoiding runback during loading

The risk, particularly in the event of a centrifugal pump stopping, of material from a tank running back and overfilling the tanker being unloaded should be guarded against with a non-return valve or syphon-breaker in the storage tank fill pipe.

7.5 Weight of tanker load

While it is the tanker driver's responsibility to avoid having an overweight vehicle, the staff at any loading installation without a weighbridge must be able to provide information on the density of the materials being loaded. The possibility of a storage tank, and hence a tanker loading from it, having a dense lower phase should be borne in mind.

7.6 Loading hot materials

Before loading a clean product it is standard practice to check the internal dryness and cleanliness of a tanker, but it is also important if loading a hot water-immiscible material (e.g. a distillation residue) to ensure that a tanker is dry, since a foam-over can occur if water beneath such a material boils.

7.7 Containment

A roll-over bund is desirable at a tanker loading bay particularly if a spillage at this point could spread over a large area.

7.8 Adsorbents

To deal with small spillages, particularly where the material spilt may make the surface dangerously slippery, a small ready-for-use stock of adsorbent should be available at tanker loading/unloading bays where the contents of a hose may be spilt.

7.9 Detection of water

Tanker loads of used solvents may contain an aqueous phase either above or below the solvent. To find the interface when the aqueous phase is the lower phase, water-finding paste that changes colour in the presence of water should be applied to the dipstick or dip tape. When the aqueous phase is on top, grease will usually be washed as off the dipstick by the solvent phase but not by the aqueous phase.

7.10 Tank vents

If pressure discharge of tank wagons or the clearance of pipelines of contaminated solvent with inert gas is routinely practised, there is a risk that droplets of solvent will be caught by flame traps or gauzes on the tank vent. As the solvent evaporates any residue will be left behind, restricting or blocking the vent.

Feedstock screening and acceptance

1   Information

A prerequisite for the safe handling of a chemical is a detailed knowledge of its properties. The transfer of adequate information between suppliers and users of a chemical is essential, and in this transfer solvent recoverers will be involved as providers and recipients.

2   Information from producers

Producers of raw materials for recovery, termed 'feedstock', cannot always identify its exact chemical composition and the composition may vary from batch to batch. Feedstock therefore needs to be described in terms of its general nature and properties.

It is essential, when a new feedstock is being considered, to obtain from its producer:

- the process from which the feedstock is generated;
- the feedstock's important components;
- the Health and Safety data for these important components;
- information on any known hazards associated with handling the feedstock;
- a definitive sample of the feedstock.

3   Producer's duty on changes

Once a recovery process has been fixed, based on a definitive feedstock sample, it is important to make clear to the feedstock producer that he is responsible for informing the recoverer of any significant changes in the feedstock's composition, including accidental contamination while under the producer's control.

4   Pre-acceptance tests

Before accepting and discharging a bulk consignment of feedstock, a sample should be checked to ensure that it broadly corresponds with the definitive sample. Further more detailed checking may be required before processing.

5   Testing of incoming drums

In the case of feedstock in drums the drums should be held in quarantine until a sample taken at random from the square root of the number in the consignment has been tested.

6 Producer's standards

Knowledge of a feedstock producer's standards is important when handling his materials. It is unrealistic to assume that all producers have the highest standards of technical competence.

Process operations

1 Minimum manning

When handling toxic and flammable materials, it is not good practice for an operator to be on his own for long periods and hourly contact with another person on-site should be a minimum standard. Fire alarms to give warning of an emergency should be available around the plant whether or not an operator is on his own.

2 Special orders

A process operator should be informed in his written instructions of the hazards associated with the materials he is due to handle, along with the appropriate precautions to take and protective clothing to wear.

3 Standing orders

For routine operations (e.g. still charging), a set of standing orders should be readily available for the operator. These should be supplemented by the special orders for the particular operation to be carried out. Between them, standing and special orders should contain a procedure for the safe shut-down of a plant. It should be possible to implement these if the plant operator himself is absent or incapacitated. All orders should be signed by the person taking responsibility for their accuracy and correctness. It is the management's task to ensure that the plant operator is sufficiently trained to understand and carry out any orders issued to him and that he can obtain advice and assistance whenever he needs it.

4 Charging stills

A still can be charged with feedstock from a tanker, a feedstock storage tank or drums. Assuming that the tanker or storage tank is suitably calibrated, only as far as the drums are concerned is the quantity charged to the still not accurately ascertainable.

4.1 Vacuum charging

Since the solvent recovery unit should be designed to withstand vacuum, one method of charging is to suck feedstock into the still. This removes air from the unit during charging and reduces the amount of flammable vapour that may be discharged when air is displaced through the vent early in a batch. It is the most effective way of emptying drums without spillage, since the suction hose can be sucked dry after each drum.

4.2 Vent scrubbing

If a liquid ring pump is used to make the vacuum, its circulating liquid can be chosen to absorb or react with vapours that might be environmentally objectionable or toxic.

5 Charging on top of residues

Charging on top of the residue of a previous batch is not good practice and should only be done if there is no doubt of the residue's stability. Air introduced into a still between batches can cause peroxide formation, leading to an unstable residue in a subsequent batch.

6 Residues

The handling properties of a residue are some of the most important properties revealed in a laboratory trial distillation. This may show, for instance, that in order to reduce viscosity so that residue may be pumped, it has to be handled above its flash point or in metallic hoses.

7 Laboratory checks on residues

Among other properties that should be checked are:

- acidity;
- peroxide presence and concentration;
- flash point;
- pour point if intended for landfill disposal in drums;
- odour;
- water miscibility.

Residue can be discharged into drums, a tanker or a receiving tank. Its ultimate disposal may determine which should be used.

7.1 Transfer of residues

Transfer can be by sucking into a tanker or tank, by blowing out of the still with air or, much to be preferred, inert gas, by pumping from the base of the still or by gravity. If residue is put directly into drums, gravity filling is usually the safest method since no high pressures are involved. Since hot residue will contract on cooling, drums should not be sealed until they have cooled, to avoid sucking in.

8 Still washing

If water-soluble organic or inorganic residues have to be dealt with, water boil outs may

Process operations 161

be necessary to maintain the heat transfer surfaces and this is especially so with external forced-circulation heat exchangers, since if a tube becomes blocked it will not wash clean and will behave as a stay tube, under stress when the reboiler is heated.

9 Venting at end of batch

Column packing entails the creation of a very large area of metal. During fractionation a thin liquid film is spread on this. At shut-down this film is hot and is particularly susceptible to reaction with oxygen. If such a reaction occurs with the accumulation of heat in the lagged column, fire may break out.

In a hot state, packed columns should never be flooded with air but only with inert gas.

Maintenance

1 Permits to work

Even in a very small organization there can be misunderstanding between individuals and when handling toxic and flammable materials the handover of plant for maintenance is a point of particular risk.

1.1 Handover

It is important that the plant operator knows what is planned to be done and prepares his plant accordingly and that the craftsman knows the limits of the preparation in both extent and degree (e.g. isolated or drained or steamed out).

1.2 Handback

It is similarly important when engineering work is completed that the plant operator is fully informed by the craftsman of anything relevant to the operability of the plant that may have been changed.

The exchange of information before and after the work should be on a formal written basis embodied in a Permit to Work procedure with signatories, when equipment is handed over and returned, by both parties.

2 Cleaning

2.1 Cleaning procedures

Good initial design of pipework and vessels facilitates preparation of equipment for maintenance and repair. The positioning of flanges so that blanks can be inserted in pipelines and the provision of drain cocks or plugs at low points are typical of items to be considered at the design stage.

2.2 Cleaning standards

It should be the intention that a craftsman working on a plant need not wear chemical

162 *Good operating procedure*

protective clothing apart from eye protection because the plant operator can wash or blow through piplines and drain off vessel or pipe contents as part of the plant preparation. The likely exception to this would be the clearance of blockages.

3 Gas-freezing plant

3.1 Principles

There is a wide difference between solvent concentrations in air that are flammable and those that are toxic, e.g. for toluene:

	ppm
UEL (sat. vap. 37 °C)	70 000
Sat. vap. at 21 °C (70 °F)	31 000
LEL (sat. vap. at 15 °C)	12 700
IDLH	2000
TLV	100
Odour threshold	0.2

It is therefore most important to know for what purpose a vessel is being gas freed. For a tank which has to be entered for desludging there is a negligible chance of attaining an atmosphere which will not call for a breathing mask. An explosion set off by an accidental spark must be avoided, however, and for this a vapour concentration of 10% of LEL would be acceptable. On the other hand, for prolonged repair work without wearing a breathing mask the TLV must be achieved.

3.2 Tanks

Design for cleaning is important. Free solvent that can be removed by pumping or sucking out, in the case of water-immiscible solvents, by floating out as a top phase on water should be removed before attempting to evaporate solvent using steam or air blowing. If bund walls are high and tanks closely spaced, the ventilation on a still day in a tank bund may not disperse heavy vapour quickly and the atmosphere should be monitored so that personnel are not exposed to high vapour concentrations during degassing tanks. This is true particularly when steam is used, since this provides the latent heat for evaporating solvent. If air blowing is used the solvent surface tends to cool reducing the vapour generation.

It is possible to generate static electricity in a steam jet and if standard reinforced rubber steam hose is used, the metal jet or lance should be earthed to the tank as should any steam powered air mover.

3.3 Stills

The risk of generating a cloud of vapour is very much less in a still with a condenser since the steaming out of a still is similar to the operation of steam distillation whether direct steam or the boiling of water using the still's coils is employed. In this case, therefore, steam is much to be preferred as a medium for gas freeing.

4 Entry into vessels and sumps

A vessel that has held solvents should not be entered unless a support man is permanently stationed in a position to render assistance if needed. The support man in turn must not enter a tank without a further supporter outside or without wearing a lifeline.

The man entering the vessel should wear a lifeline and also breathing equipment unless the need for the latter can be eliminated conclusively. Frequent drills at rescuing an unconscious man from a tank should be carried out by a tank entry team. A portable breathing set with a second mask and checked to have a full air bottle is a desirable item to have at hand.

4.1 Unbreathable atmosphere

If inert gas is available on-site there is a risk that a tank's atmosphere may be depleted in oxygen and the atmosphere should be tested for this and for the presence of solvent vapour before permitting entry without an air supply.

Any vessel being certified for entry must be inspected for possible sources of ingress of solvent. Pipelines should be disconnected or spaded off. Valves should not be relied on to be 100% tight and no leakage, however small, is acceptable. Steam, air and water supply valves should be padlocked closed.

4.2 Reinspection

A fresh entry permit should be issued each day and no entry should take place before its issue. To ensure that there can be no misunderstanding about this, a copy of the entry permit should be posted at the tank entry point.

4.3 Alteration of conditions

A tank entry permit can only be valid if essential conditions do not change. When cleaning sludge from a tank it is possible to strike pockets of solvent occluded in the sludge. When this possibility exists a constant check on the atmosphere is required and an automatic monitor should be specified.

4.4 Test position

In checking for the presence of vapour it must be borne in mind that vapour is heavier than air and the sample point must be close to the tank bottom.

4.5 Mask air supply

If a portable compressor to supply breathing air is used, care must be taken to ensure that it draws its air supply from a source of clean air not contaminated with solvent vapour or with engine exhaust.

4.6 Manhole and sumps

Drain manholes, pumps and drainage interceptors should be treated as tanks from the point of view of entry certificates, lifelines, support man, etc.

5 Tank cleaning

Because of the nature of their operation, solvent recoverers need to clean tanks more frequently than is normal in chemical factories and material that needs to be removed is often difficult to handle. This makes large manholes at ground level very desirable for access. If, in addition, the tank bottom is sloped towards the manhole, cleaning or mopping out can often be achieved from outside the tank, which is desirable. Ventilation by injection of fresh air with an air mover into a manhole or large branch at the tank top is to be recommended since practically all solvent vapours are heavier than air.

5.1 Subliming solvents

It should not be forgotten that some solvents and chemicals (e.g. cyclohexane, *tert*-butanol, dioxane, cyclohexanol) can sublime into the tops and sides of a tank and when such materials have been stored particularly careful inspection is needed if tank entry or hot work on a tank is planned.

6 Hot work

If welding, burning or any other work creating a source of ignition is taking place in the danger area, it requires a Permit to Work. The responsible person issuing the permit should be aware that the separation distances laid down betwen storage and handling of flammable solvents and sources of ignition should be used as guidance and not as absolutes. For hot work, as for tank entry, permits need to be renewed each morning before work commences.

7 Pressure testing

After work which involves breaking joints on a solvent recovery unit, the plant should be pressure tested before being returned to service. This test need not be done to a pressure over that specified for the bursting disc since it is to detect gross leaks which cannot be corrected by pulling up joints while the plant is operating.

8 Routine inspections

8.1 Daily

Plant that is operational should be looked at daily or, if on shift work, at the start of each shift for leaks, failed pump seals, etc., and appropriate corrective action should be taken.

8.2 Bursting disc inspection

If a combination of bursting disc and safety valve is used on the still, a monthly check should be made on the bursting disc. If an overpressure is thought to have occurred or if the safety valve is thought to have blown, an immediate check should be made.

8.3 Instrument inspection

Instruments that have emergency safety functions should be tested by simulating the emergency. If this can be done as part of a normal batch cycle, a test each cycle is desirable but a weekly trip test is sufficient if the simulation requires a craftsman.

8.4 Corrosion inspection

As unfired pressure vessels, stills will need biennial inspections, but if it is believed that corrosion may have taken place a prompt check should be made. If a process whose moderate corrosion has been predicted on the basis of laboratory results is due to be done on the plant, corrosion test coupons should be installed in the plant and inspected regularly.

8.5 Tank vents

Tank vents need to be inspected regularly for blockage or failure of gauzes or flame traps due to corrosion. Much more frequent inspection is needed when liquids with certain properties are stored:

- liquids that sublime and need heated vents (e.g. *tert*-butanol);
- liquids that have subliming solids in solution (e.g. ammonium chloride);
- liquids that can evaporate leaving their inhibitors behind and then polymerize on the tank roof or vents, e.g. styrene, vinyltoluene;
- liquids containing volatile acids, e.g. hydrochloric acid.

Personal protection

1 Head and eyes

Splashes of solvent in the eyes are very painful and can lead to long-term damage. When breaking hoses, emptying drums, carrying out maintenance work, etc., goggles or a face shield should be worn.

Protection spectacles are very desirable when in the plant and storage areas. Hard hats are required to normal industrial standards.

2 Hands

For most solvents, PVC-coated gloves are satisfactory and comfortable to wear. However, DMF, THF and some other common solvents dissolve PVC quickly and, for them, butyl rubber is appropriate.

For laboratory use, polythene disposable gloves are needed for solvents that are very rapidly absorbed through the skin.

Barrier cream as a back up to the use of gloves is desirable.

3   Feet

When handling drums or heavy pieces of equipment, hard-toed boots or shoes should be compulsory. Footwear should be electrically conductive if static electricity is a hazard and should not be studded with nails that might cause a spark.

Industrial footwear with 'solvent-resistant' soles are frequently not to be recommended, but are needed for some solvents.

4   Body

When emptying drums an apron is useful for preventing spillage on overalls without creating the heat discomfort of heavy physical work in a PVC suit.

If toxic materials such as aniline are being handled, disposable paper overalls over normal cotton overalls are commonly used.

If a special risk (e.g. during plant cleaning) requires the wearing of PVC suits, they should be worn outside wellington boots and not tucked inside them.

5   Eating facilities

Suitable facilities for storing food and eating it clear of all possible solvent contamination are essential.

6   Clothing storage

Separate clean and dirty lockers should be provided for each operator.

7   Washing facilities

Showering facilities should be provided both for emergency decontamination and for routine cleanliness.

Paper disposable towels or hot-air drying are preferable to roller or other towels for drying hands.

First aid

1   First aid training

It is desirable that a high proportion of operatives are trained in first aid and specifically in the emergency treatment relevant to solvent hazards.

A list of trained first aiders should be permanently displayed and at least one should be available on site at all times.

First aid

2 Special antidotes

In addition to the standard equipment for problems within a first aider's competence, there should be ample supplies of any special antidotes for a chemical currently being handled.

3 Information to hospital

In the event of a patient being taken to hospital, suspected of being affected by a material being handled, any information on its effects and treatment should be communicated to the hospital and any special antidote should be supplied.

4 Clothing soaked with flammables

Particular care should be taken in handling a casualty whose clothing is soaked with flammable solvent. Heating in a first aid room should be flame proof. Smoking must be banned.

5 Safety showers

Close to vehicles unloading areas and process plant there should be frost-protected safety showers. These should be tested monthly.

6 Eyewash bottles

Near to all places where solvents are handled or processed eyewash bottles should be available and these should be checked monthly.

Fire emergency procedure

1 Fire fighting

Unless a solvent recovery plant is part of a large factory, it is unlikely that a site fire brigade being able to tackle a major fire will be a practicable proposition. Once the public fire brigade has been called, the staff of a solvent recovery plant should concentrate on shutting down their equipment to reduce, as far as possible, the spread of fire and cut off any flows of flammable solvents feeding it. If vehicles can be removed from the site without hazard to the driver this should be done (*see* page 140).

2 Small fires

Solvent recovery plant staff should be trained in the use of portable fire extinguishers and such extinguishers should be provided in easily accessible and visible positions near areas of high fire risk (e.g. laboratory, still, vehicle loading point). Forklift trucks should also carry a fire extinguisher.

3 Fire extinguishers

Dry powder extinguishers are the most effective for inexperienced fire fighters and are suitable for both chemical and electrical fires. They are, however, of limited use in a wind.

Carbon monoxide and Halon extinguishers are useful in a laboratory where delicate and expensive equipment may be damaged by foam or powder.

Alcohol-resistant foam is useful for small pool fires, particularly when they are contained in bunded areas.

A fire hose reel for washing away spillages and dealing with smouldering sources of re-ignition, e.g. wood, paper, is generally useful but should be fitted with a variable jet/spray nozzle and needs to be protected from freezing.

Extinguishers should be inspected annually by a competent person and the inspection date recorded.

4 Co-operation with fire brigade

In the design and layout of the plant, consideration must be given to the facilities the fire brigade may need, e.g. hydrants, static water, and the information they require to fight solvent fires.

A number of solvents, including most alcohols, cause standard foam to collapse and if such solvents are going to be handled stocks of alcohol-resistant foam may be required.

The local fire station should be kept informed of materials with unusual fire-fighting and toxic hazards that may be on-site.

5 Assembly points

In the event of a fire, all personnel not carrying out nominated duties should gather at their assembly point. This should be convenient to reach but in a safe area and not obstructing the access for emergency services. Written standing orders should lay down procedures for roll calls.

6 Emergency control point

This should be chosen in a safe area with good communications, including preferably a dedicated emergency telephone independent of the main switchboard. Information on the materials being stored on-site including their toxic and fire hazards should be readily available at the control point. Telephone numbers for all employees who need to be summoned should be prominently displayed.

7 Fire detection and warning

Only in exceptionally hazardous locations can an automatic fire alarm system be justified, but a manual system with break-glass buttons would be appropriate for all but the smallest installations and, since solvent fires can spread very rapidly, the manual system should be connected directly to the fire station. Standing orders should make it obligatory for the fire brigade to be called as soon as all but the most trivial fire is found in the solvent area.

10 Economic aspects of solvent recovery

In Chapter 8 the alternatives to solvent recovery as a means of disposing of used solvent were examined. All of these with the exception of using the calorific value (CV) of the solvent as a source of heat were likely to result in a charge to the process.

Use as a fuel

Examination of the CVs of the commonly used solvents will reveal that on a weight or a volume basis the hydrocarbons have a higher value than oxygenated solvents. Hydrocarbons are very widely used as fuels so there can be no case made that used hydrocarbons burnt as fuels represent a waste of natural resources. They are very sparingly water miscible and can therefore be burnt with a minimum of pre-treatment. They are also relatively cheap since the majority of fractions used as solvents are produced primarily as fuels or, on a very large scale, as industrial raw materials.

On the other hand, very few hydrocarbons used as solvents have a flash point higher than 32 °C (90 °F), which is the lowest flash point standard for industrial boiler fuels. This means that special safety precautions need to be taken in handling and firing them, although these do not mean that used solvents and standard commercial fuels cannot be burnt on the same boiler. The capital cost of an industrial steam boiler is of the same order as its annual fuel bill if it is operated on an 8000 h/yr basis, so to equip a plant for burning 'free' fuel can be a very attractive investment.

The only oxygenated solvent that may be cheap on a CV basis at its list price is methanol, which can be produced at very low cost in locations where natural gas has a low value. However, other used solvents may have a value below purchased fuel if recovery to a reusable standard is expensive and no market exists for a sub-standard material.

For use as boiler or furnace fuel, where temperatures do not reach those normally required in an incinerator, care is necessary in excluding all traces of chlorinated solvent from the fuel stream. Other solvents or solvent combinations may also be unacceptable since partial combustion can lead to unacceptable odours.

The fuel must also be low in ash if a modern oil- or gas-fired boiler is adapted to burn used solvent since even small accumulations in the boiler can cause blockages in the economiser tubes. Not only do such blockages call for frequent shut-downs for cleaning but they also cause uneven heating and possibly mechanical damage. Discharge of heavy

172 Economic aspects of solvent recovery

metals to the air is also unacceptable in many cases and the very fine ash can be entrained in the flue gases and cause problems in this respect.

The above consideration is based on the assumption that the used solvent is available as a liquid, but there will be circumstances in which the air leaving a process will contain solvent evaporated as part of the process. It is normal practice that the concentration of solvent will be less than its lower explosive limit (LEL) and figures between 50% and 25% are common if recapture on activated carbon is intended. The attraction of using this air with its solvent content in firing a boiler or air heater is obvious if suitable safeguards against explosion can be fitted. Table 10.1 shows the lower heat of combustion of a selection of solvents and their lower explosive limits. The gross heat release at a concentration of 50% LEL does not vary very greatly despite a range of 5 between the highest and lowest values of LEL, a twofold range in CV and almost threefold in molecular weight.

Table 10.1 Heat available from the combustion of solvents at a concentration of 50% LEL in air: under adiabatic conditions this will raise the air to about 1000 °F or 550 °C

Solvent	LEL (% v/v)	Lower calorific value (BTU/lb)	Heat release (BTU/100 ft^3 air, 50% LEL)
Methyl acetate	3.1	9200	2939
Acetone	2.6	12 250	2573
Hexane	1.2	19 246	2766
Ethanol	3.3	11 570	2446
n-Butanol	1.4	14 230	2053
Toluene	1.27	17 430	2836
MEK	1.8	13 480	2433
Methanol	6.0	8419	2251

It is worth observing that the two hydrocarbons included in Table 10.1 yield a high amount of heat but are both easy to recapture on activated carbon and are ready to use without further treatment apart from decanting the free water after desorption.

Liquid fuels can be stored easily. Low-cost tankage can therefore eliminate the problems which occur if used solvent arises when there is no requirement for it as a fuel. This is not so in the case of solvent-laden air, which can only be considered as a fuel in applications where a steady arising can be matched to a similar requirement for heat.

Destruction

The gap between useful concentrations of solvent in air and the very low levels of solvent permitted to be discharged (see Table 2.6) is very large. The valuable heat, in air which still needs treatment before it can be released, may be trivial and even insufficient to operate a catalytic fume incinerator without support fuel. On its own such a stream would also probably be an uneconomic application for a carbon bed adsorber and subsequent recovery. A factor that can have a major influence on the choice between destruction and recapture/recovery is whether the solvent stream consists solely of dilute effluent or whether the same solvent is also present on site as a concentrated stream, e.g.

a mother liquor. Under such circumstances one may be faced with only a marginal cost for the recovery of the small quantity of recaptured solvent.

For an existing operation that needs to be brought up to modern discharge standards, the retrofitting of equipment for destruction or recapture can present serious problems of layout. For recapture large plot areas are required for carbon bed adsorbers and such areas, close to manufacturing units, represent an often-overlooked capital cost in assessing the true economics. Air treatment by biological destruction also needs a large amount of space. The other possible method of destruction, incineration, either thermal or catalytic, does not occupy as big a site but when dealing with flammable solvents an incinerator can sterilize an even greater amount of valuable process land.

Solvent recovery

Even in instances where the opportunity to employ contaminated liquid solvents or solvent-laden air as a fuel exists, the difference in cost between standard fuels and solvents is likely to be the overriding influence in choosing solvent recovery. Except for hydrocarbons and methanol the cheapest solvents will have prices per BTU twice that of normal fuels and the expensive solvents such as pyridine, THF, DMAC, NMP and ACN may be up to ten or twenty times more costly on a heat basis. The chlorinated solvents cannot be burnt as fuels and on commercial incinerators the charges for destroying them are often about three times more than the price of buying the virgin solvent.

For the solvent user, therefore, the economic choice may not be whether to recover but which of the four recovery routes to take.

Sale to a merchant recoverer

To the producer of a used solvent this is an easy disposal route since apart from tank or drum storage, no facilities need to be provided or operated on his site. If as a means of disposal it has to be relied upon completely an understanding of the merchant recoverer's problems is necessary.

However costly a virgin solvent may be, there is no guarantee that a merchant recoverer can be found to buy, or be paid to uplift, a parcel that has been contaminated in use or by mistake. It is vital to the recoverer that a sale exists for the product arising from its recovery. To understand why a market may not exist one needs to consider the relationship between the merchant recoverer and the purchaser of recovered solvent.

Such a purchaser will only buy recovered solvent, in preference to virgin, if he can make an overall worthwhile saving and if he can depend on continuity of supply to an agreed specification. It makes little sense to buy small tonnages of recovered solvent, even at considerable discounts to virgin prices, since the overall cash savings will be unattractive. The merchant recoverer, needing to supply consistent quality and substantial quantity to attract his customer, must have more than one supplier of a solvent since no source of a contaminated solvent is guaranteed. It is clearly unreasonable to expect a guarantee of what is a waste product or at best a by-product stream.

The ideal situation for a recoverer is to be in receipt of a number of similar used solvent streams so that the withdrawal of any one will affect neither the quality nor the availability of his product too seriously. Large storage to act as a flywheel in the system is also a great advantage. To such an operation the occasional injection of a large parcel, arising perhaps from an accidental contamination, the breakdown of an in-house recovery plant or the disposal of a solvent inventory of an abandoned process, is welcome provided it is compatible with the rest of the stream. Because materials with seasonal sales need large storage (e.g. ethylene glycol for antifreeze or isopropanol for windscreen wash), they suit the merchant recoverer's operation particularly well and allow spot parcels to be accepted. The solvent user who is considering disposal of contaminated solvent on a regular basis should be sure that the recoverer has the ability to sell the solvent arisings as well as the technical resources to process it safely.

The buyers of recovered solvents are almost always able to use materials that do not meet virgin specifications, which are set to satisfy the most exacting requirements of the prime producers' customers. The balance between the availability of recovered solvents and the incentive to persuade buyers to use them typically results in a sales price of 60–80% of that of virgin material.

With this as the background to his business, it is clear that there will be solvents that have no potential down-market sale and therefore will not be of interest at any price to the merchant recoverer. Chloroform, for instance, is used in the pharmaceutical industry but is too toxic to be chosen by the potential customers of a recoverer. Certain contaminants that are difficult to remove down to an acceptable level will also disqualify a solvent to a recoverer. The presence of a thiol (mercaptan) in acetone is likely to make the mixture unacceptable.

On the other hand, low-boiling ketones, esters and aromatic hydrocarbons (except benzene), which can be incorporated into cellulose thinners and gun wash, find a ready outlet and represent uses for solvent mixtures that would be expensive and technically difficult to separate. They also typify applications for solvents that have a once-through use, of which some others are paint strippers, domestic adhesives and windscreen wash. In these the solvent is not recycled in use, with a make-up to replace losses, so there is no risk of small impurities building up in concentration. This permits a laxer specification on components which might be undesirable.

Such sales outlets do have the disadvantage that in most cases the solvent vapour is discharged directly to the atmosphere. The most immediate problem that this poses is that the smell of the solvent should be acceptable and consistent. If the foreign odour cannot be removed or masked the product will be unsaleable. The longer term matter of concern is that as the general public becomes more worried about the contribution of solvent-based products to volatile organic compound (VOC) pollution the sales of once-through solvent products will decline.

Having indicated the market limitations that confine a merchant recoverer, it is also apparent that this mode of operation has great advantages over both toll and in-house recovery. The merchant recoverer should collect several compatible streams and, by refining them as a blend, or consecutively without cleaning his refining equipment, can achieve much better plant utilization than the strict needs for segregation allow in toll processing.

It is also much easier to schedule production in merchant operation than it is in toll recovery where the timetable is influenced by the competing needs of a number of customers who have an interest in each keeping their solvent inventory, and hence their working capital, to a minimum. This tends to result in inefficiently short production campaigns.

Finally, the merchant recoverer will make products to a specification which balances the sales of a low-quality product with the available market outlets. In many cases this standard will be lower than that required for recycling a solvent to its original process, let alone getting back to a good-as-new purity. Since the recovery process will usually be fractional distillation a reduction in reflux ratio can be made if a higher impurity level is acceptable. This has the result, on a given column, of reducing the variable costs per unit of product and, usually more significant, increasing the throughput rate. The alternative to buying recovered solvent for the customer, however, is to purchase virgin material at the full market price so that lower processing costs will not normally alter the sales price.

There is no reason why a merchant recoverer should not also operate as a toll processor, although their operational philosophies tend to be different. The decisions which a merchant recoverer is entitled to take on materials that he owns are different from those of the processor handling goods belonging to someone else. In the one case yield can be sacrificed for throughput whereas in the other such an approach is unacceptable.

Toll processing

In the fine and speciality parts of the chemical industry, contract processing is common, enabling production to be commenced on a modest scale, coming on-stream quickly and without large capital investment. Toll recovery of solvents is one of the services that contract processors provide and, although not all will have the full range of chemical engineering unit operations available, it is usually possible in any instance to find the required equipment in appropriate materials of construction. It is not likely, however, that the equipment will be of the right size to be dedicated to a used solvent stream which, in any case, may be growing as a new product builds up its sales. It is likely, therefore, that a customer of a toll recoverer will be sharing refining equipment with other streams from other customers.

The frequency of campaigns should ideally be determined by the economics of the inputs of inventory and equipment. The results set out in Table 10.2 are based on the following assumptions:

- New solvent cost £300/Te.

- Process operating 300 days per year and consuming 24 Te of new or recovered solvent per day.

- Storage tanks in mild steel costing £20 per product tonne of capacity per year, including handling charges for contaminated and recovered solvent.

- Recovery plant throughput of 4 Te/h charged at £160/h.

- Time spent coming on stream, shutting down and cleaning out at the end of each campaign 70 h.

- Charge for working capital 15%.

As can be seen in Table 10.2, the variation in cost per tonne is very small between five and eight campaigns each year and in practice the cost of storage per tonne of capacity is likely to increase as the tank size is reduced, thus moving the minimum overall cost per tonne towards less frequent campaigns.

176 *Economic aspects of solvent recovery*

Table 10.2 Economics of campaign frequency

Campaigns per year	Solvent finance (k£)	Process cost (k£)	Storage cost (k£)	Cleaning cost (k£)	Total cost (k£)	Cost per Te (£)
1	324	288	288	11	911	126.5
2	162	288	144	22	616	85.6
3	108	288	96	34	526	73.1
4	81	288	72	45	486	67.5
5	65	288	58	56	467	64.9
6	54	288	48	67	457	63.5
7	46	288	41	78	453	62.9
8	41	288	36	90	455	63.2

The higher the purchase price of the solvent being recovered, the greater is the cost of financing the inventory and Table 10.3 shows the optimum frequency and length of campaigns for various solvent prices.

Table 10.3 Optimum frequency of toll recovery campaigns for various solvent costs

Solvent price (£/Te)	Optimum number of campaigns per year	Length of campaign (including cleaning) (h)
200	6.7	339
300	7.5	310
400	8.1	292
500	8.7	277
750	10.0	250
1000	11.0	234
1500	13.0	208

A further incentive towards a smaller inventory is the potential loss if a process has to be abandoned. It is likely that the solvent stock in this situation will have to be disposed of as a distress parcel with no chance of continuity which, from the analysis of the merchant recoverer's market position, is likely to result in a very low price or even a charge for removal.

Additional items would affect the cost of recovery without influencing the frequency of the campaigns:

1 Transport to and from the toll recoverer's premises is liable to be significant and its absence when in-house rather than toll recovery is considered may be a decisive factor. A cost of £20/Te of solvent processed might be typical and therefore a significant part of the total cost.

2 Disposal of residue which may be possible as a fuel but often involves incineration can well cost a similar sum to that for transport.

3 Yield of product will never be 100%. A loss of 2% would not be uncommon, depending primarily on the volatility of the solvent and the amount of residue present.

Using the optimum frequency of recovery campaigns and the items covered in (1) to (3) above, realistic total costs for solvents at a range of prices can be calculated (Table 10.4). These are not figures that can be used as accurate estimates in a practical situation, but they illustrate the calculations that may be made.

Table 10.4 Total costs of recovery for 7200 Te/yr

Solvent cost (£/Te)	Process and cleaning (k£/yr)	Finance cost (k£/yr)	Storage cost (k£/yr)	Transport, residue and 2% loss (k£/yr)	Total cost (£/Te)
200	363	32	43	317	104.9
300	372	43	39	331	109.0
400	378	53	36	346	112.9
500	386	62	33	360	116.8
750	400	81	29	396	125.8
1000	412	98	26	432	134.4
1500	433	125	22	504	150.6

It is clear from Table 10.4 that the overall cost of recovery does not, on this model, vary very much with the cost of the solvent being processed. Only finance costs and losses are affected by solvent price. It is obviously economic to be prepared to use more expensive recovery schemes if the solvent to be recycled has a high value, while the margin for a merchant recoverer on the refining of a solvent to be sold at 60% of £200 is clearly very small unless a cheaper refining scheme can be applied to it.

There are other considerations which may affect the choice of toll recovery in addition to economic factors. It is often preferable to use virgin solvent in the very early stages of a process so that the possible harmful effects of using solvent with a contaminant can be disregarded. This can, of course, give rise to the necessary inventory for recovery campaigns when the process teething troubles have been overcome.

The use of toll recovery facilities can have two other helpful aspects in the early stages of the production of a new chemical. Typically the sales, and therefore the rates of production, will increase over a 2–3 year period after start-up and the plant will be far from fully utilized initially. By using outside recovery on contract the low return on capital of under-utilized recovery equipment can be avoided. More important, the toll recoverer's technical staff will be available to cope with the problems of recovery at a time when the producer's team are likely to be fully extended.

With a suitable form of agreement, the know-how acquired by the toll processor can be made available for the design of in-house recovery plant if this is justified when full-scale operation is approached. Effectively this allows pilot-plant experience to be accumulated on a full-scale unit.

It must be stressed that the relationship between the producer of the recoverable used solvent and his chosen toll recoverer has to be one of trust and mutual confidence. As complete as possible information not only on the hazards involved in handling the pure solvent but also on the dangers of the impurities it may contain must be provided. This information must be constantly up-dated as changes and improvements are introduced into the manufacturing process. It should be realized that process improvements may not be beneficial to the solvent recoverer.

On his part, the recoverer must be prepared to have his premises, working methods,

178 *Economic aspects of solvent recovery*

records and laboratory testing inspected regularly. Inevitably the recoverer will acquire information that may be confidential and must be organized to safeguard it.

In-house recovery

Examination of Tables 10.2 and 10.4 will show that some of the costs that are involved with toll recovery can be avoided if the operation is done in-house. Assuming that a dedicated recovery plant is used there is no need for cleaning plant between campaigns or for the tank storage and inventory to bridge the gaps between campaigns. There is no need either for the transport to and fro to the recovery site.

For the basic case of £300/Te solvent using the same assumptions as for toll recovery, the comparative costs of in-house and toll operations are set out in Table 10.5.

Table 10.5 Comparison of toll and in-house recovery costs

	Toll		In-house	
	k£/yr	£/Te	k£/yr	£/Te
Process	288	40	468[a]	65[a]
Cleaning	84	12	Nil	Nil
Transport	144	20	Nil	Nil
Losses	43	6	43	6
Additional tankage	39	5	Nil	Nil
Residue disposal	144	20	144	20
Finance[b]	43	6	Nil	Nil
TOTAL	785	109	655	91

[a] Based on full utilization of plant; *see* Table 10.6.
[b] Additional finance for campaign running.

In evaluating the costs of in-house recovery, the make-up of the recovery plant hourly costs is clearly paramount:

- Utility costs. Assuming a product rate of 1 Te/h, to match the rate at which the used solvent is generated, and a reflux ratio of about 3:1 these will amount to about £10/h (compared with £40/h on the toll recovery example with four times this capacity).

- Labour costs. The labour cost of running a 1 Te/h plant and an otherwise similar 4 Te/h unit will be the same. Assuming a single operator on each shift and a half share of a graduate supervisor, the cost, inclusive of welfare, pension and other benefits, has been estimated at k£108/yr.

- The capital cost of the in-house unit has been assumed to be £1 million and that of the toll unit £2.3 million. It would seem reasonable to write the in-house plant off over 10 years as the expected life of the process it serves. The toll plant, being a general-purpose unit, has been written off over 15 years. Using an interest rate of 15% the finance charges are therefore k£176/yr for the in-house unit and k£325/yr for the toll.

- Maintenance: 4% of capital cost.

- Overheads: £10/h covering laboratory and administration costs for both toll and in-house recovery.
- Profit. The toll recoverer must allow for profit in his charges and this has been assumed to be 28% of the total charge.

From Table 10.3, the optimum time used on the toll recovery campaigns was $7.5 \times 310 = 2325$ h/yr. Assuming that the annual toll costs must be spread over 8000 h the breakdown of the toll costs on this stream is as shown in Table 10.6.

Table 10.6 Breakdown of toll recoverer's charges

Utilities	$2325 \times £40$	k£93
Labour	$\dfrac{2325}{8000} \times k£108$	31
Capital	$\dfrac{2325}{8000} \times k£325$	94
Maintenance	$\dfrac{2325}{8000} \times k£92$	27
Overheads	$2325 \times £10$	23
Profit	28% of k£288 + 84	104
TOTAL		k£372

The analysis of in-house recovery charges (Table 10.7) shows the importance of plant utilization in calculating the costs per tonne.

Table 10.7 Effect of incomplete occupation on in-house recovery costs

	Utilization		
	100%	50%	30%
Utilities (k£)	72	36	22
Labour (k£)	108	81[a]	81[a]
Capital (k£)	176	176	176
Maintenance (k£)	40	20	12
Overheads (k£)	72	30	22
TOTAL (k£)	468	349	315
Cost per Te (£)	65	97	144

[a] On the assumption that the plant will be manned on a three-shift basis until it approaches its rated capacity.

There may be reasons why the apparent economic superiority of in-house recovery when the full design capacity of the plant is reached is not attractive enough to justify its use:

- If distillation is the recovery process, as it is in most cases, the generation of flammable solvent vapours on a site may pose unacceptable safety problems, especially if the site is small or in a built-up area.

- On such a site the control of odours may prove difficult.

- Distillation processes use substantial amounts of heat and cooling which may outstrip existing facilities at an existing site. This may increase the capital investment of a new solvent recovery project.

- Distillation columns tend to be very prominent features on a factory site and may be unacceptable to the planning departments or residents of the surrounding area.

There are counter-balancing non-economic arguments for favouring on-site refining:

- Reducing the carriage of solvents on roads, which is perceived by the public to be very hazardous.

- Avoiding transport interruptions from strikes, weather and other causes outside the control of the used solvent producer. These could bring the process to a halt in a short period if the arrival of new solvent or the removal of used were interrupted.

- Reducing the risk that confidential know-how will be obtained by outsiders whom the producer cannot control.

- Maintaining the solvent user's control of the solvent at all times. This may be essential for processes carried out under FDA regulations.

- Avoiding in some cases the trans-frontier shipment of material classified as a hazardous waste.

Whatever the relative costs of in-house and toll recovery, they are likely to be modest compared with the costs and profits of the solvent-using process as a whole. Therefore, in examining the pros and cons of in-house recovery, the overall security of the system is likely to outweigh the narrower cost considerations.

The fact that toll recovery on a campaign basis involves considerable tank storage is a significant factor in its favour. If in-house recovery is chosen it is advisable to have a suitable large rentable tank or an incinerator with adequate spare capacity available at short notice to provide security against breakdowns. This may, indeed, be an essential feature of an in-house strategy.

Since high recovery rates (better than 90%) are now frequently being achieved, stocks of used solvent that may accumulate if recovery facilities are shut down, while the process continues to run, may take a long time to use up. The option of incineration may not, therefore, be so expensive as the rental of a large tank with a slowly reducing contents.

Return to supplier

The increasing concern for the impact of solvents on the environment has led to major manufacturers taking a growing interest in their disposal. This has been most marked in the case of the chlorinated solvents, particularly 1,1,1-trichloroethane and trichloroethylene.

Because the scale of their use by individual users is small, the segregation of degreasing solvents for toll recovery is impractical. Also for reasons of scale and because they are used in industries where distillation expertise does not normally exist, in-house recovery is seldom attractive. Because of their high chlorine content, they are also expensive to incinerate.

Since parcels of used degreasing solvents tend to be small, they are not attractive for collection by market recoverers but manufacturers and distributors are making deliveries to users and can bring back used solvent, which because of losses in use is always a smaller volume than the amount purchased. Credits or disposal charges depend on the quantity and solvent content of the material collected. In some countries this pattern of operation is required by law as part of their action against pollution.

A similar service is available from suppliers of hydrocarbon solvents in safety cans where the transport logisitics are the overriding economic factor in safe and environmentally acceptable disposal.

Because the impurities in both the above-mentioned groups of used solvents are primarily oil and grease, it is possible to pass the used solvent, after filtration, evaporation and water removal, through the supplier's plant to make the final product indistinguishable from new.

The serious environmental problems arising from letting chlorofluorocarbons (CFCs) evaporate has led to their manufacturers offering a recovery and destruction service for used materials. Because many of them have very low boiling points it is often beyond the technical capability of merchant recoverers to handle CFCs whereas their manufacturers have the equipment to deal with them.

The vital requirements for such recovery by the original makers of the solvents is to keep the solvents carefully segregated so that impurities that cannot be eliminated satisfactorily in the reprocessing do not enter the system. A small amount of cellulose gun wash, for instance, in used trichloroethane would make its recovery impossible to a standard that could be considered 'good as new.'

Solvents that present safety problems in recovery and which have been involved in accidents are also ones that their manufacturers are likely to take back after use. Clearly it is important that such solvents do not get a 'bad name' as being dangerous to recover and ones that form peroxides (e.g. THF) and other unstable derivatives can be handled more safely by their manufacturers, with their large technical resources, than they can be by most of their users.

Another special case is the recovery of pyridine, which is one of the most expensive solvents in general use. Because unlike the great majority of solvents it is chemically reactive, it can be separated from solvent mixtures readily and therefore its producers can purify it by methods other than fractionation, which too often cannot make an absolutely clean separation.

Apart from the safety and environmental reasons for manufacturers becoming involved in the recycling of their products, it should be recognized that parcels of slightly off-specification recovered solvents being sold at prices below the virgin solvent price can have an effect on the market much greater than their quantity justifies.

11 Future of solvent recovery

The future of solvent recovery will depend upon the changes in the pattern of solvent use industrially and domestically. Understanding of the harm that their final decomposition products do to the environment is based on the generation of ozone and other photochemical oxidants at low levels of the atmosphere and destruction of the ozone layer at very high levels.

Ozone generation

Where the generation of ozone becomes an acute local problem, due to a combination of high sunlight, atmospheric conditions and concentrated industrial activity, significant improvements can be achieved by reformulation of products using organic solvents that are not photochemically reactive. Los Angeles Rule 66 can be considered as a model of this approach and classifies solvents in three groups of activity:

- Group 1   Olefins, ketones, ethers, esters, aldehydes.
- Group 2 C_8 and heavier aromatic hydrocarbons except ethylbenzene.
- Group 3   Ethylbenzene, toluene, trichloroethylene and ketones with a branched hydrocarbon structure such as MIBK.

The relatively inactive paraffinic and naphthenic solvents would therefore seem to be attractive for reformulations in which activity should be reduced. However, they will not reduce the overall generation of ozone and may indeed increase it if their lower solvent power increases the amount of solvent required for a particular duty. The importance of solvents in atmospheric pollution at a low level should not be underestimated. The discharge of used solvents into the environment is the source of a surprisingly high proportion of the man-made part of the air pollution known as volatile organic compounds (VOC) which is the precursor of photochemical oxidants such as ozone.

In 1990, the contribution of used solvents from industrial and domestic sources to Western European VOCs was approximately equal to that of petrol distribution and use, each source being about 40% of the 10 million ton total.

In all the heavily industrialized countries of the world, legislation is in force or pending to reduce the contribution both from cars and from solvents. The impact of public opinion and legislation on the field of solvent recovery will be significant, affecting both the capture for reuse of solvents in industry and the reformulation of domestic products.

184 *Future of solvent recovery*

The whole production of solvents world-wide is eventually discharged into the environment though a proportion is incinerated in such a way that VOCs are not formed. While it is reasonable to assume that much of the industrial solvent usage (and loss) can be eliminated by improved equipment, this is not a practicable approach to the solvent included in the multitude of small packages for the domestic market.

Table 11.1 gives an approximate breakdown of the solvent losses in Western Europe.

Table 11.1 Breakdown of solvent losses (%) in Western Europe

	Industrial	Domestic
Surface coating	20	20
Metal coating	9	1
Household products	—	8
Pharmaceuticals	6	—
Adhesives	4	3
Dry cleaning	—	4
Miscellaneous	15	5
Total	54	41

Small-scale use of solvents where consumption is on a domestic scale have been included under the 'domestic' heading.

At present, most of the output of merchant solvent recoverers ends up in the domestic market where pressures to reformulate away from solvent-based products are likely to be severe. Since a merchant recoverer must have, to survive, outlets for the solvents he recovers, a reconsideration of his role is necessary.

At the same time, the industrial solvent user will be improving the recapture of losses as the oil industry is already planning to do throughout the gasoline distribution system. The pharmaceutical and paper coating industries aim to feed their process less than 10% of newly purchased solvent and 90%+ of recovered and recycled material, and such targets are achievable. Although very high rates of recapture can be achieved, it is also important that the material recaptured can be reprocessed for reuse at similar high yields. To do this requires consideration of the recovery process at an early stage of any new development. There is seldom a unique solvent for any system and it is natural for chemists to use, in the development stage, a solvent or solvent mixture with which they are familiar. If consideration of recovery is left until the process has become fixed, perhaps for example in the pharmaceutical industry with FDA approval, it will be a time-consuming and expensive matter to alter the solvents for improved recoverability. Consultation on solvent choice with those who understand the problems of recovery should, therefore, be made at an early stage in process development. Such a high recycle rate has generated a new problem for the on-site processors of used solvent. Impurities that were purged from the system as part of the losses can build up to unacceptable levels when the amount of the purge is greatly reduced. This can be particularly difficult to cope with if very volatile impurities (e.g. formaldehyde) are adsorbed on an activated charcoal bed from both process and distillation column vents and continuously returned to the process. While a purge to an on-site incinerator is a possible solution on a large plant, the provision of on-site incineration for a small operation is often impractical. Since there are great difficulties in transporting very volatile liquids to a merchant incinerator, chemical destruction may be necessary.

Even impurities that are easy to handle when they have been removed from the solvent stream may be difficult to separate at high concentrations. This may be particularly difficult on a continuous distillation unit which lacks sufficient columns to devote one to removing an impurity intermediate in volatility between heavy and light products.

Economic considerations would often call for the working up of purge streams of this sort to be done by a merchant recoverer, assuming he could find an outlet for a less-than-pure recovered solvent. Indeed, outside the pharmaceutical industry they might be returned to the original user. The regulation of the pharmaceutical industry seldom allows solvent to be sent for recovery off-site and subsequent return to pharmaceutical production.

Ozone destruction

While the low-level production of ozone is only partly man-made so that the acute problems currently associated with it can be reduced by taking action in parts of the world affected by it, the high-level destruction of ozone is largely due to the activities of man and creates world-wide problems. Two solvents are responsible for sizeable damage to the ozone layer, 1,1,1-trichloroethane and 1,1,2-trichlorotrifluoroethane (CFC 113). The latter has an ozone depletion potential (ODP) of 0.80 but its use is less than a quarter of that of 1,1,1-trichloroethane (ODP 0.15) and the damage contributed by each is of the same order of magnitude. Carbon tetrachloride also has a high ODP (1.04), but its use except as a raw material for chlorofluorocarbon production is negligible. Of the other chlorohydrocarbon industrial solvents, only MDC has an effect on the ozone layer and it is a small one.

The two solutions to the solvent depletion of ozone are to reformulate using other solvents or to achieve such high recapture rates that the discharge to atmosphere is almost eliminated. A major justification behind the near fivefold rise in world-wide production of 1,1,1-trichloroethane in the period 1970–90 was that the loss of solvent in metal cleaning is about 20% less than for the competitive solvents trichloroethylene and perchloroethylene. Since metal cleaning represents about three-quarters of the 1,1,1-trichloroethane market world wide, this is the area above all where lower losses are needed. One of the major changes that could reduce losses is much more widespread recycling of dirty solvent either in-house or by commercial recoverers, particularly among smaller users. It is, perhaps, instructive to note that solvent recovery in the UK was started primarily to recover tax levied on mineral spirits. When this tax was abolished the recovery of mineral spirits ceased very quickly.

Although CFC 113 is used more in the electronics industry than in the metal-working industry, it too could be recovered very much more effectively by using better equipment and work practices. It has a much higher volatility (b.p. 47.6 °C) than 1,1,1-trichloroethane and is used more frequently in a number of blends, which means that a high standard of segregation is needed if sent to an outside recoverer. However, its very high price compared with 1,1,1-trichloroethane makes its recovery financially attractive even if water removal and reblending are needed.

Oil prices

Organic solvents are in the main derived from petroleum. The exceptions are small amounts formed as byproducts in agriculture and coke ovens. They can be produced by relatively simple distillation (e.g. mineral spirits) or by processes involving several stages of synthesis (e.g. THF, NMP). Their manufacturing costs can, therefore, be very sensitive to the cost of the naphtha fraction of crude oil from which they are derived. Alternatively, because of the value added through various stages, the cost may be almost divorced from the petroleum market.

A breakdown of the types of solvent consumed in Western Europe (Table 11.2) shows approximately the current division into chemical types.

Table 11.2 Breakdown of solvent consumption in Western Europe

	Wt%
Aliphatic hydrocarbons	28
Aromatic hydrocarbons	20
Halogenated hydrocarbons	18
Alcohols	14
Ketones	10
Esters	7
Glycol ethers	3

The market trend, strongly influenced by environmental pressures, is likely to be away from aromatic hydrocarbons in formulations for domestic use. This will probably also apply to chlorinated hydrocarbons. This is not necessarily true of their industrial use, but further improvements in recapturing and recycling will reduce their production and sales.

A trend towards aliphatic hydrocarbons—low in toxicity, photochemical activity and price and easy to dispose of by incineration—seems probable. Aliphatic hydrocarbons are difficult to restore by recovery to a 'good as new' state. Their low, mild odour is spoilt by the smell of cracked resins or even by the cracking of n-alkanes ($>C_9$) at their atmospheric pressure boiling points. For C_7 and higher alkanes the large numbers of isomers present in crude oil results in the industrially available fractions having both light and heavy components that can be lost in handling or repeated redistillation. Neither of these effects may be important for in-house recovery, but the merchant recoverer has great difficulty in making products that are consistent and marketable.

The value of used aliphatic hydrocarbons before recovery may be little different to that of purchased fuel oil or gas, and it is possible exists that in some circumstances it will be more economic to burn the used solvent for its heat value than to try to recover it to a high standard. The possibility exists that an oil company supplying a hydrocarbon solvent and having the expertise in burning 'difficult' fuels may be able to offer a service and an attractive price for such a disposal.

Environmental regulation

While the regulations to protect the environment will have a great impact on solvent use and solvent users, the effect on solvent recoverers will be equally marked. There is little technical difficulty in designing equipment to collect and clean the vents from a process using a single solvent or an unchanging group of solvents. The merchant recoverer handles a wide range of solvents and will often be approached to add to this range. Adsorption equipment to cover such a situation on a segregated basis will usually be impractical and it is probable that merchant recoverers will need to have on-site vapour incineration to meet the requirements of the regulatory authorities. This will prove to be a difficult requirement to meet on some small sites where adequate distances from flammable liquid storage and process plant to sources of ignition must be maintained.

The capital requirements for a solvent recovery operation which have in the past been small in comparison with most chemical enterprises seem likely to rise and the value of existing units complete with the necessary permissions to store and handle used solvents will reflect this change. Since such valuable permissions are usually reviewed in the aftermath of an accident, the need for ever greater safety precautions against fires and process exotherms will be justified by the value of the asset to be protected.

Such precautions will depend on a high calibre of staff. A greater awareness of the need for solvent recovery to have a high status within the solvent-using industries is important to attract and hold engineers and chemists of the right ability.

12 Significance of physical properties of solvents

The solvent recoverer setting out to process a solvent mixture needs to have available to him a wide range of information about the material to be handled. This should include the boiling points of the pure components of the mixture and the departures from ideal behaviour when these components are in mixtures. He also has to know the fire and toxic hazards that may be involved and the storage and handling properties of the recovered products he may make.

In Appendix 1, the relevant information, as far as it is available, has been listed for a range of the most commonly used industrial solvents along with a brief commentary on the practical problems that are associated with their handling and recovery. In case the application of, and background to, this information may not be immediately clear, reference should first be made to the appropriate section of this chapter.

Name

It may seen almost frivolous to stress that correct and agreed names should be used for the solvents handled but, particularly in the case of a commercial solvent recoverer handling used solvents from his suppliers, this is an area of real importance.

Such solvents may be known by familiar initials in the prime user's works where IPA can be used for isopropyl acetate because there is no risk of confusion with isopropanol since the latter is not used on the premises. The confusion can, of course, too easily take place at the recoverer's works.

Benzene can be confused (in dealings within Europe) with benzin, a very different product but nearly impossible to separate from a mixture of the two once an accidental contamination has taken place.

Trike, used indiscriminately for 1,1,1-trichloroethane and trichloroethylene by factory floor operatives in different firms, is another pitfall for the recoverer.

Boiling point

The great majority of solvent recovery involves distillation which, of course, involves evaporation and condensation. For both of these processes, a minimum temperature difference of 15–20 °C between the fluid being worked on and the heat transfer medium is necessary. This is particularly true if a plant is not specifically designed for a job and therefore has normally sized heat exchangers.

Although a difference in boiling points between the components of a mixture may lead one to an indication of an easy or difficult separation by fractionation, this is only the crudest of guidelines and should not be relied upon.

Freezing point

Several of the solvents in Appendix 1 have freezing points above the temperatures that may be reached in plants built out of doors and it is risky to try to handle them without traced lines and, particularly, heated tank vents. Air-cooled heat exchangers can be seriously damaged by stresses caused by the difference in temperature between a cold blocked tube and the rest of the bundle at process temperature.

Liquid expansion coefficient

Apart from the obvious use of this to calculate the density of a liquid at temperatures other than the standard 20 °C (at which the specific gravity is measured), it can be very important to calculate liquid expansion when batch distilling relatively high-boiling materials. Assuming that the still is charged cold, enough ullage must be left in the kettle to allow heating to boiling point without the charge expanding up the column.

As an example, acetone will expand by 5.5% over 41 °C when heated from ambient temperature, say 15 °C, to its boiling point of 56 °C. However, a high-boiling solvent such as nitrobenzene will expand by 20% before it starts to boil at atmospheric pressure.

Specific gravity

Most storage tanks are built to store safely liquids of density 1.0. If a plant were to be built specifically for the recovery of chlorinated solvents, which might include perchloroethylene (s.g. 1.63), the tankage would be built so that it could be safely filled at this density. However, normally the nominal water tonnage of a storage tank should not be exceeded.

A recoverer should also be very conscious of the large differences between concentrations being quoted as w/w, or w/v or v/v when densities may vary from 1.63 as quoted above to 0.63 for pentane.

Most pumps for general-purpose usage are supplied with motors of sufficient horse power so that they do not overload when handling water. This is insufficient if pumping material of much higher density and it may be necessary to throttle back pumps particularly when pumping against a low head, e.g. transferring from a full to an empty tank. When new pumps are being specified, the highest density likely to be handled should be quoted.

Flash point

The figures quoted throughout are for the closed cup method, which corresponds to the vapour just above the liquid surface in , say, a storage tank being submitted to a source of ignition. It corresponds to the lower explosive limit (LEL) except in the case of chlorinated hydrocarbons.

A hydrocarbon fraction has a flash point (in °C) related to its initial boiling point (IBP), also in °C, by the empirical equation

$$\text{flash point} = (\text{IBP} \times 0.73) - 72.6 \tag{12.1}$$

Flammable limits in air

The upper explosive limit (UEL) of a solvent is the concentration of solvent in air that is just too rich to explode. It varies over a very wide range from about 6% to 10% for hydrocarbons up to 36.5% for methanol.

The lower explosive limit (LEL) is normally between 1% and 3% by volume in air. It is generally safe, if a flash point cannot be measured, to estimate it as the temperature corresponding to 1% v/v in air.

Between the LEL and UEL solvent–air vapours can explode if the third essential ingredient of an explosion, a source of ignition, is present. This can be a naked flame, a mechanical spark, lightning or non-flameproof electrical equipment. In the case of hydrocarbons and other solvents of low electrical conductivity, it can also be an electrostatic spark. Static electricity can be generated whenever new surface is created and if this happens at a greater rate than the charge can be dissipated, a hazardous situation arises. Spraying solvent or pumping a two-phase mixture of solvent and air or solvent and water are common ways in which new surface can be created. Pumping velocities of less than 1 m/s are advisable, if two-phase flow is possible, for solvents between their LEL and UEL.

An additional precaution is to blanket tanks of such solvents with nitrogen or other inert gas containing less than 8% oxygen, since under these circumstances an explosion cannot take place. Normal practice is to use nitrogen with about 3% oxygen to allow for air mixing with the inert gas.

As Tables 12.1 and 9.2 (p. 145) show, many solvents will normally be within their flammable limits if stored at ambient temperature and therefore present a constant explosion risk. Even when they are below their LEL it is always possible, when processing, that warm material may be run into a storage tank.

Table 12.1 Flammable limits for non-hydrocarbons

Solvent	LEL (%)	Flash point (°C)	UEL (%)	UEL temperature (°C)	B.p. (°C)
Ethanol	3.3	13	19.0	41	78
n-Propanol	2.1	25	13.5	53	97
n-Butanol	1.4	35	11.2	67	118
MEK	2.6	−6	12.8	28	80
Ethyl acetate	2.2	−4	11.5	23	77

It should also be noted that benzene, which has a melting point of about 5 °C, will generate an explosive vapour when it is solid.

Antoine and Cox constants

In order to calculate vapour pressures and concentrations at temperatures other than a solvent's atmospheric boiling point, two correlations are commonly used.

Antoine equation

$$\log p = A - \frac{B}{C+T} \tag{12.2}$$

where p is the vapour pressure in mmHg, T the temperature in °C and A, B and C are constants for each pure solvent covering a range up to the solvent's atmospheric boiling point.

Cox equation

This is, in general, the less accurate of the two correlations. A review of the values of C in the Antoine equation shows that they mostly fall in the range 210–250. The Cox equation

$$\log p = A - \frac{B}{T+230} \tag{12.3}$$

allows A and B to be calculated from two known points on the vapour pressure–temperature relationship. It also allows the relative volatility, α, of an ideal binary mixture to be estimated quickly throughout the temperature span of a distillation column:

$$\alpha = p_1/p_2$$
$$\log \alpha = \log p_1 - \log p_2$$
$$= (A_1 - A_2) - \frac{B_1 - B_2}{T+230} \tag{12.4}$$

Both equations 12.1 and 12.2 can be used to calculate the temperatures at which the vapour of a solvent falls within the explosive limits or poses an acute danger to health.

Odour threshold

This is an extremely subjective measurement and cannot be relied upon as a measure of vapour concentration. Not only does the ability to detect odour depend on the individual, but a person who has once been exposed to a solvent vapour is likely to have a much higher threshold for the rest of the day.

However, if odour is detected, this represents a minimum concentration and, for many of the more toxic solvents, the odour threshold is above the threshold limit value.

It is also important for environmental reasons to know which solvents (e.g. ethyl acetate) are so easy to detect by nose that they represent a problem to neighbours.

IDLH

The immediate danger to life and health (IDLH) value represents a maximum vapour concentration from which a person can escape within half an hour without irreversible health damage or effects that would impair the ability to escape. Such information is clearly important in rescues and emergencies and should be compared with the saturated vapour concentration and the lower explosive limit. Since a spark would cause an explosion if it occurred in an atmosphere that was flammable, such an atmosphere should never be entered even in an emergency. It should, of course, be noted that the saturated vapour concentration quoted is at 21 °C and that at a temperature only 10 °C lower the saturated vapour concentration would be almost halved for a solvent with a boiling point of about 100 °C.

Threshold limit value (TLV)

The TLV represents the concentration in air at which most healthy people can work without harm. It is expressed in ppm (by volume per million parts of air). This is usually a time-weighted average (TLV–TWA) figure and for many solvents a short-term exposure limit (TLV–STEL) is allowable for exposure over a 15 min period four times per day.

Saturated vapour concentration (SVC)

This is a convenient yardstick for both toxic and fire hazards at 21 °C. In general, if the saturated vapour concentration generated by a solvent is less than five times its TLV, it is unlikely, although not impossible, that a toxic atmosphere will result from its use.

If the saturated vapour concentration is less than the IDLH, then in an emergency it will not be suicidal to rescue without an air mask a workmate overcome by fumes in a tank or an enclosed work room, subject to explosion risk.

If it is less than the LEL the danger of explosion of vapour in a tank is very low, although this does not mean that a pool of liquid will not burn once it is set alight.

Finally if it lies between the LEL and UEL or above the UEL, the danger of explosion exists.

To calculate the saturated vapour concentration of a solvent at a given temperature, work out the vapour pressure using the Antoine or Cox equation. The SVC is then

$p/(760-p) \times 10^6$ ppm

In the reference literature, TLV is sometimes quoted in mg/m^3. To convert from this to ppm at 20 °C:

ppm = (mg/m^3) × 24.04/molecular weight

UEL and LEL are usually quoted in % by volume and can be converted into ppm by multiplying by 10^4.

Vapour density

The density of solvent vapour relative to air gives a measure of the ease with which vapour can be dispersed. All solvents are denser than air, although methanol is only slightly so. To remove vapour, therefore, low-level is more effective than high-level extraction. Vapour freeing of tanks should be done whenever possible by drawing vapour from a low-level manhole while fresh air should be forced into the top of the tank.

The danger of vapour flowing down ditches or gathering in bunded areas should not be overlooked.

Miscibility with water

Although it is often stated in reference books that hydrocarbon and chlorinated solvents are 'insoluble' in water, this is never strictly true. All solvents are water miscible to some extent. Since one of the most frequent problems for the recoverer is to remove water from solvents to a tight specification and, increasingly, to remove solvents from aqueous effluent streams, accurate data in this area are very important.

Oxygen demand

The biodegradability of solvents varies very widely and the correspondence between laboratory testing and works-scale operation is not very good. The figures included in Appendix 1 refer to the biological oxygen demand (BOD), which in this case is measured over a 5 day test, and the chemical oxygen demand (COD), done under severe laboratory conditions and representing the highest oxygen demand that a solvent is likely to require practically. The theoretical oxygen demand (ThOD) is useful, as the maximum load that can be put on the environment, if experimental results are not available. All results are quoted in terms of weight of oxygen per unit weight of solvent.

Autoignition temperature

Some solvents have such low autoignition temperatures that contact with hot oil or even high-pressure steam pipelines will cause them to catch fire. They may also ignite on hot electric motors or light fittings if these are not correctly specified. In planning a general-purpose solvent recovery unit, consideration should be given to the sorts of solvents that may be handled in the future. Choice of electrical and other equipment which would handle diethyl ether and carbon disulphide safely may not be economic but a limitation of this sort needs to be appreciated by the equipment users.

Hildebrand solubility parameter

In choosing a solvent for a polymer, it is important to consider recovery problems but it is also necessary, of course, to pick an effective solvent. The solubility parameter is a measurement of the force by which molecules attract one another. If the fields of force surrounding two different molecules are not the same they will probably not mix. Every resin has a solubility parameter range and tends to be soluble in a solvent that lies in that range.

The solubility parameter at 20 °C can be calculated as:

$$\delta = (Ld)^{1/2} \tag{12.5}$$

where

δ = solubility parameter;
L = latent heat in cal/g;
d = liquid density in g/cm^3.

Solubility parameters of blended solvents can be obtained by a weighted average of their components.

The solubility parameter is linked to the Kauri butanol number (KB), an empirical laboratory measurement of solvent power:

$$KB = 50\delta - 345 \tag{12.6}$$

Dipole moment and dielectric constant

Both of these properties are of assistance when choosing a solvent or assembling a short list of possible solvents for a duty. The dielectric constant is a good indication of the solubility of inorganic salts in a solvent.

Evaporation time

This also is important, particularly when formulating paints, printing inks, adhesives and polishes. The rates quoted are all ratios of evaporation time to that of the very volatile diethyl ether. In many reference works butyl acetate is used as the standard. On the diethyl ether scale butyl acetate is 11.8.

Loss per transfer

For both environmental and economic reasons it has become important to identify losses of solvent in storage and handling operations. For solvents boiling above about 80 °C losses to the atmosphere when handling are not of economic importance if the solvent is cold, but the loss of saturated vapour when making a transfer of, say, n-pentane without linking the vents of supplying and receiving vessel can be more than 0.25% of the liquid transferred. This represents not only a loss of solvent but also a contribution to VOC.

Latent heat

Short-cut methods for calculating the trays and reflux in fractionating tend to rely on the molar latent heats of the components in the system being the same.

For quick scanning of a problem no serious error is introduced if, in a binary system, they are within 20% of each other but, for instance, the latent heats of isopropanol and n-hexane are 38% different. In such a case VLE data need to be replotted on the basis of 'corrected' molecular weights (and hence mole fractions) to give equal molar latent heats.

Heat of fusion

Apart from its obvious use to calculate the amount of heat that will be required for thawing out solvent that has frozen, the heat of fusion can yield useful information as to how pure a solvent is and what amount of impurity will guard against a freeze-up.

The relationship between the concentration of an impurity or a solute and the freezing point depression is

$$\ln(1-x) = \frac{L\theta}{RT_m^2} \tag{12.7}$$

where
  x = mole fraction of solute;
  L = molal heat of fusion of solvent;
  θ = reduction of freezing point in K;
  R = universal gas constant (1.987);
  T_m = freezing point of pure solvent in K.

When the value of x is below 0.10 it is accurate enough for all engineering purposes to reduce equation 12.7 to

$$x = \frac{L\theta}{RT_m^2} \tag{12.7a}$$

Hence to reduce benzene to a freezing point of $-1\,°C$ it is necessary to add an impurity of 0.098 mole fraction.

Specific heat

It is very common in the operation of a small distillation column for the reflux to be returned to the top of the column at a temperature well below its boiling point. This means that the measured reflux can be a good deal less than the true value since the cold reflux condenses vapour at its point of return to the column. Thus for isopropanol:

Liquid reflux flow as measured	10 kmol/h
Reflux temperature	20 °C
Column top temperature	82 °C
Specific heat of reflux	36.6 cal/mol/°C
Latent heat of reflux	9540 cal/mol

Heat needed to raise reflux to boiling point:

$10 \times (82 - 20) \times 36.6 \text{ kcal/h} = 22\,692 \text{ kcal/h}$

which is the equivalent of the latent heat of

$\dfrac{22\,692}{9540} = 2.38 \text{ kmol/h}$

and provides almost a quarter more reflux than that 'set' by a liquid flow control.

Heat of combustion

The cost of incineration of a solvent depends in part on whether additional fuel must be purchased to achieve incineration temperatures high enough to destroy the solvent completely. Under incineration conditions, all the water present will be in a vapour form so the quoted heat of combustion figures correspond to the net or lower calorific value.

Relative volatility

In considering the possibility of a solvent mixture for recovery, the first technique to look at is almost always fractional distillation. Reference to Chapter 6 will show that a relative volatility (α) of less than 1.5 will make recovery by this method unlikely if a high-purity recovered solvent is demanded. When for a binary mixture α is more than 10, the separation by distillation is very easy.

An overestimate of the value of α for a binary separation can result in a significant underestimate of the minimum number of stages required if the relative volatility is low while the effect is within the normal chemical engineering margin of safety for values of α over 4 (Figure 12.1).

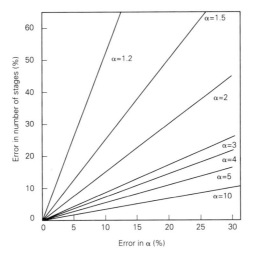

Fig. 12.1 Error in estimation of stage requirements by using the Fenske equation with incorrect value of α

A 13 volume collection entitled *Vapor–Liquid Equilibrium Data Collection* has been published by Dechema and can be consulted through the Dechema Data Bank. In Appendix 1, the sources quoted for relative volatility and Van Laar constants refer to volume and page number in this collection.

198 *Significance of physical properties of solvents*

Azeotropes

Even with very high figures for α over most of the composition range, separation by fractionation can be prevented by the presence of an azeotrope. A very comprehensive list of azeotropes is available in Horsley's *Azeotropic Data* published by the American Chemical Society.

Vapour–liquid equilibrium (VLE) diagrams

Diagrams of VLE data are very useful because they give at a glance an impression of the type of system under consideration and what problems it will present to separation by distillation. They show if there is an azeotrope in a binary system and, approximately, the values of γ and α. They also show in the shape of a maximum, or a horizontal section, a reason to suspect that the liquid will split into two phases.

The members of homologous series (e.g. benzene, toluene, xylene) tend to form ideal mixtures in which the value of γ for each component is close to 1.0 throughout the concentration range. In solvent recovery, unlike solvent production, it is rare to come across a mixture of homologues since each component of a mixture is wanted for its specific characteristics. It is therefore usual to have to process very non-ideal solutions.

Figure A36, Appendix 2, shows a typical relationship between γ and composition for the binary mixture *n*-hexane (1)–ethyl acetate (2). It will be seen that for either component the value of γ is not significantly different from ideality ($\gamma = 1.0$) at mole fractions of more than 0.5. At low concentrations which correspond to the top and bottom of a continuous fractionating column separating the binary mixture, the less concentrated component can behave in a very non-ideal manner.

The vapour pressure of the components will be

$$p_1 = \gamma_1 x_1 P_1 \tag{12.8}$$

and

$$p_2 = \gamma_2 x_2 P_2 \tag{12.9}$$

$$\alpha = \left(\frac{\frac{p_1}{p_1+p_2}}{\frac{p_2}{p_1+p_2}} \right) \frac{x_2}{x_1} \tag{12.10}$$

$$= \frac{\gamma_1 P_1}{\gamma_2 P_2} = \alpha^* \left(\frac{\gamma_1}{\gamma_2} \right) \tag{12.11}$$

where α^* is the ideal mixture value of α.

At the column top, where the more volatile component (1) is nearly pure, $\gamma_1 = 1.00$ and equation 12.11 reduces to

$$\alpha = \frac{\alpha^*}{\gamma_2} \tag{12.12}$$

γ is usually greater than unity except in solutions of molecules of very different size.

Among solvents this is usually only found when one of the components is a chlorinated hydrocarbon.

Since γ_2 is usually >1, the actual relative volatility is often much less than the ideal relative volatility at the column top. Indeed, if γ_2 is greater than α^* there will be an azeotrope at the column top (Figure A.35, Appendix 2) preventing a pure component 1 being produced by fractional distillation.

Within the range of normal industrial solvent fractionation γ is usually little affected by temperature and pressure, and it has already been seen that α^* changes with pressure substantially (page 89). It is possible that by reducing the system pressure an azeotrope will no longer be present because at this lower pressure $\alpha^* > \gamma_2^\infty$.

At the column bottom, equation 12.9 reduces to

$$\alpha = \alpha^* \gamma_1 \tag{12.13}$$

because at this end of the column component 2 is very concentrated and $\gamma_2 = 1.0$. Here, therefore, it is much easier to strip component 1 from component 2 than an ideal solution would indicate. This explains how effective a normal batch distillation, which has only a single stripping plate, the reboiler, can be in many binary separations.

Activity coefficients are not only useful in distillation applications. For screening solvents for water miscibility or for steam or air stripping from water, high values of γ in aqueous solutions indicate which solvents are hydrophobic and therefore easy to separate from water and which are hydrophilic.

If it is necessary to calculate a vapour–liquid equilibrium diagram for a binary mixture, one must have the Antoine or Cox constants and values of γ^∞ for both components. The activity coefficients are used in the form $\ln \gamma_1^\infty$, which is the constant A_{12} in the van Laar equation, and $\ln \gamma_2^\infty = A_{21}$.

Most of the values for A_{12} and A_{21} are taken from the Dechema collection mentioned previously and are at atmospheric boiling point. Examples of the method for doing the calculation are given in Appendix 2.

APPENDIX 1

Physical properties, activity coefficients and recovery notes for selected solvents

n-PENTANE

1. **Names**
                                                   n-Pentane
                                                   Below 40 °C Petroleum ether

2. **Physical properties**

Molecular weight	72
Empirical formula	C_5H_{12}
Boiling point (°C)	36.1
Freezing point (°C)	-129
Specific gravity (20/4 °C)	0.626
Liquid expansion coefficient (per °C)	0.00095
Surface tension (at 20 °C in dyn/cm)	16
Absolute viscosity (at 25 °C in cP)	0.235

3. **Fire and health hazards**

Flash point (closed cup) (°C)	-40
Autoignition temperature (°C)	260
Lower explosive limit (ppm)	15000
Upper explosive limit (ppm)	78000
IDLH (ppm)	5000
Odour threshold (ppm)	10
TWA–TLV (ppm)	600
Saturated vapour concentration (70 °F in ppm)	720000
Vapour density (relative to air)	2.5

4. **Solvent properties**

Hildebrand solubility parameter	7.0
Dipole moment (D)	0
Dielectric constant (20 °C)	1.844
Evaporation time (diethyl ether = 1.0)	1.0

5. **Aqueous effluent characteristics**

Solubility in water (at 25 °C in % w/w)	0.0038
Solubility of water in (at 25 °C in % w/w)	0.012
Biological oxygen demand (w/w)	
Chemical oxygen demand (w/w)	
Theoretical oxygen demand (w/w)	3.56

6. **Handling details**

Hazchem number	1265
Hazard class	3 Y
Vapour pressure (at 21 °C in mmHg)	442
Loss per transfer (% of liquid transferred)	0.27

7. **Vapour pressure constants (mmHg, log base 10)**

Antoine equation	$A = 6.87632$
	$B = 1075.780$
	$C = 233.205$
Cox chart	$A = 6.82847$
	$B = 1050.1$

8. **Thermal information**

Latent heat (cal/mol)	6120
Specific heat (cal/mol/°C)	40.32
Heat of combustion (cal/mol)	774072

Azeotropes and activity coefficients of component X: n-pentane

Component Y	Azeotrope °C	Azeotrope % w/w X	α	A_{XY} $=\ln \gamma_X^\infty$	A_{YX} $=\ln \gamma_Y^\infty$	Source[11] Vol.	Page
Hydrocarbons							
n-Hexane		None	3	−0.07	−0.11	6(a)	123
n-Heptane		None	7	−0.34	−0.15	6(a)	127
Cyclohexane		None	4	0.18	0.11	6(a)	119
Benzene		None	7	0.60	0.50	6(a)	118
Toluene		None	20	0.43	1.31	6(a)	160
Ethylbenzene		None					
Xylenes		None			0.43		
Alcohols							
Methanol	31	8	8[a]	2.40	3.43	2(a)	198
Ethanol	34	95	20[a]	1.87	2.53	2(c)	375
n-Propanol							
Isopropanol	35	94					
n-Butanol		None	>10	1.44	2.31	2(b)	169
Isobutanol							
sec-Butanol							
Cyclohexanol							
Ethylene glycol		None		5.39			
Methyl Cellosolve							
Ethyl Cellosolve							
Butyl Cellosolve							
Chlorinated hydrocarbons							
Methylene dichloride	31	51	2[a]	1.06	0.85	6(a)	100
Chloroform							
Carbon tetrachloride							
1,2-Dichloroethane							
1,1,1-Trichloroethane							
Trichloroethylene							
Perchloroethylene							
Monochlorobenzene							
Ketones							
Acetone	32	80	5[a]	1.39	1.74	3+4	187
Methyl ethyl ketone							
MIBK							
Cyclohexanone							
N-Methylpyrrolidone				2.44			
Ethers							
Diethyl ether	33	40					
Diisopropyl ether							
1,4-Dioxane							
Tetrahydrofuran		None					
Esters							
Methyl acetate	34	78					
Ethyl acetate							
Butyl acetate							
Miscellaneous							
Dimethylformamide		None					
Dimethyl sulphoxide		None					
Pyridine		None					
Acetonitrile	35	90	>20	2.07	3.27	6(a)	101
Furfuraldehyde							
Water	35	99		11.6			

[a] The relative volatility between the less volatile component of the binary mixture and the azeotrope.

Recovery notes: *n*-pentane

Pentane is used as an industrial solvent both as a nearly pure product and as a large proportion of the low-boiling petroleum solvents which have the generic name petroleum ethers. This name indicates their original use as cheap substitutes for diethyl ether that were also safer for the cleaning of delicate domestic fabrics in the late 19th century.

Their use these days is largely because of their high volatility, which allows the evaporation of solvents at low temperatures from unstable products.

The principle problem that pentane presents is in condensing, since its atmospheric boiling point at 36.1 °C (further reduced to 34.6 °C if water is present) is uncomfortably close to cooling tower water in hot and humid conditions.

The loss on handling at 21 °C (70 °F) for tanks vented to the atmosphere is 0.27% and losses from the diurnal breathing of freely vented overground tanks are unacceptable on anything but a very short-term operation.

Another hazard that pentane, along with other paraffinic hydrocarbons, can present is derived from their high thermal expansion coefficients in relation to their densities. A long section of overground pipework filled with pentane, if it is isolated between two valves, can, in hot and sunny weather, develop a high enough internal pressure to crack cast iron pump cases or burst flexible hoses. This is especially true if the pentane has been chilled in process and is therefore cold when shut in the pipe. Drums can develop a high pressure and should have at least a 4% ullage, with more in hot climates.

Care must also be taken when proposing to pump pentane that a pump with a low NPSH is specified. Vacuum should not be considered for sucking pentane out of drums. Air-operated double diaphragm and other reciprocating pumps tend to gasify low-boiling liquids as they accelerate them at the start of each stroke and are therefore not very suitable.

Neoprene and PVC gloves and aprons are suitable as protection and splashes in the eye are sore, although they do no lasting harm, so goggles should be worn when handling pentane.

n-HEXANE

1 Names
          n-Hexane
          62/68 Hexane

2 Physical properties

Molecular weight	86
Empirical formula	C_6H_{14}
Boiling point (°C)	69
Freezing point (°C)	−95
Specific gravity (20/4 °C)	0.659
Liquid expansion coefficient (per °C)	0.0013
Surface tension (at 20 °C in dyn/cm)	18.4
Absolute viscosity (at 25 °C in cP)	0.31

3 Fire and health hazards

Flash point (closed cup) (°C)	−22
Autoignition temperature (°C)	225
Lower explosive limit (ppm)	12 000
Upper explosive limit (ppm)	75 000
IDLH (ppm)	5 000
Odour threshold (ppm)	10
TWA–TLV (ppm)	50
Saturated vapour concentration (70 °F in ppm)	200 000
Vapour density (relative to air)	2.99

4 Solvent properties

Hildebrand solubility parameter	7.3
Dipole moment (D)	0
Dielectric constant (20 °C)	1.9
Evaporation time (diethyl ether = 1.0)	1.3

5 Aqueous effluent characteristics

Solubility in water (at 25 °C in % w/w)	0.000 95
Solubility of water in (at 25 °C in % w/w)	0.011
Biological oxygen demand (w/w)	0.04
Chemical oxygen demand (w/w)	2.21
Theoretical oxygen demand (w/w)	3.52

6 Handling details

Hazchem number	1208
Hazard class	3 YE
Vapour pressure (at 21 °C in mmHg)	127
Loss per transfer (% of liquid transferred)	0.09

7 Vapour pressure constants (mmHg, log base 10)

Antoine equation	$A = 6.910\,58$
	$B = 1189.64$
	$C = 226.28$
Cox chart	$A = 6.9386$
	$B = 1212.1$

8 Thermal information

Latent heat (cal/mol)	6880
Specific heat (cal/mol/°C)	42.0
Heat of combustion (cal/mol)	993 042

Azeotropes and activity coefficients of component X: *n*-hexane

Component Y	Azeotrope °C	Azeotrope % w/w X	α	$A_{XY} = \ln \gamma_X^\infty$	$A_{YX} = \ln \gamma_Y^\infty$	Source[11] Vol.	Page
Hydrocarbons							
n-Pentane		None	3	−0.11	−0.07	6(a)	123
n-Heptane		None	2.5	−0.10	−0.08	6(a)	604
Cyclohexane		None	1.7	0.07	0.06	6(a)	273
Benzene	68	95	1.6[a]	0.44	0.37	6(a)	535
Toluene		None	4	0.33	0.31	6(a)	593
Ethylbenzene		None	10				
Xylenes		None	10	0.17	0.40	6(a)	607
Alcohols							
Methanol	50	72	>10[a]	2.60	3.01	2(a)	253
Ethanol	59	79	5[a]	1.90	2.41	2(a)	453
n-Propanol	66	96	12[a]	1.88	1.54	2(a)	585
Isopropanol	63	77	5[a]	1.55	2.10	2(b)	99
n-Butanol	68	97	>10[a]	1.07	2.44	2(b)	202
Isobutanol	68	98	11[a]	1.12	2.25	2(d)	370
sec-Butanol	67	92	8[a]	1.28	1.96	2(b)	250
Cyclohexanol		None					
Ethylene glycol		None					
Methyl Cellosolve		None					
Ethyl Cellosolve	66	95	>10[a]	1.54	1.70	2(b)	295
Butyl Cellosolve		None					
Chlorinated hydrocarbons							
Methylene dichloride		None					
Chloroform	60	16	1.7[a]	0.57	0.32	6(a)	430
Carbon tetrachloride		None	1.5	0.24	0.17	6(a)	403
1,2-Dichloroethane		None					
1,1,1-Trichloroethane	67	71	<1.5[a]	0.34	0.29	6(a)	473
Trichloroethylene		None	4	0.43	0.38	6(a)	462
Perchloroethylene		None	10	0.41	0.34	6(a)	453
Monochlorobenzene		None	11	0.46	0.60	6(a)	529
Ketones							
Acetone	50	41	4[a]	1.64	1.47	3+4	225
Methyl ethyl ketone	64	71	3[a]	1.03	1.18	3+4	301
MIBK	Possible						
Cyclohexanone				2.60			
N-Methylpyrrolidone							
Ethers							
Diethyl ether		None	3				
Diisopropyl ether	67	47					
1,4-Dioxane	60	98	5[a]	1.23	1.14	3+4	471
Tetrahydrofuran	63	50					
Esters							
Methyl acetate	52	39					
Ethyl acetate	65	62	2[a]	0.89	0.98	5	514
Butyl acetate							
Miscellaneous							
Dimethylformamide		None	20	2.60	2.55	6(c)	332
Dimethyl sulphoxide		None		4.32			
Pyridine		None					
Acetonitrile	57	72					
Furfuraldehyde							
Water	62	94		13.1	7.54		

[a] The relative volatility between the less volatile component of the binary mixture and the azeotrope.

Recovery notes: *n*-hexane

Of the azeotropes that *n*-hexane forms, those with methanol, ethanol, acetonitrile and water form two phases. It is also not fully miscible with ethylene glycol, furfural, NMP and DMF (Figure 6.4).

Although it is possible to purchase nearly pure *n*-hexane, a lot of hexane used commercially is a mixture with its isomers (methylpentanes and dimethylbutanes). Hence in use the solvent mixture may change in composition as it is recycled, either losing the least volatile component because all solvent is not stripped out of the product, or the most volatile component because it is preferentially lost to the atmosphere in handling and condensing.

n-Hexane has been identified as being toxic by inhalation, causing nerve damage, and many people will only notice its smell at levels high enough to cause possible health damage.

Hexane can be recovered from air very satisfactorily using steam-regenerated activated carbon beds and, because of its low water miscibility, needs only decanting to prepare it for reuse in most cases.

Since hexane has a very mild odour, it is difficult to recover it with a smell as good as virgin material. As a result, recovery is usually done in-house and seldom by merchant recoverers, who have difficulty in finding customers.

n-HEPTANE

1 Names   *n*-Heptane

2 Physical properties

Molecular weight	100
Empirical formula	C_7H_{16}
Boiling point (°C)	98.4
Freezing point (°C)	−90.6
Specific gravity (20/4 °C)	0.684
Liquid expansion coefficient (per °C)	0.0009
Surface tension (at 20 °C in dyn/cm)	19.3
Absolute viscosity (at 25 °C in cP)	0.41

3 Fire and health hazards

Flash point (closed cup) (°C)	−4
Autoignition temperature (°C)	230
Lower explosive limit (ppm)	10 000
Upper explosive limit (ppm)	70 000
IDLH (ppm)	4250
Odour threshold (ppm)	220
TWA–TLV (ppm)	400
Saturated vapour concentration (70 °F in ppm)	51 800
Vapour density (relative to air)	3.5

4 Solvent properties

Hildebrand solubility parameter	7.5
Dipole moment (D)	0.0
Dielectric constant (20 °C)	1.924
Evaporation time (diethyl ether = 1.0)	2.2

5 Aqueous effluent characteristics

Solubility in water (at 25 °C in % w/w)	0.005
Solubility of water in (at 25 °C in % w/w)	0.005
Biological oxygen demand (w/w)	1.92
Chemical oxygen demand (w/w)	0.06
Theoretical oxygen demand (w/w)	3.52

6 Handling details

Hazchem number	1206
Hazard class	3 YE
Vapour pressure (at 21 °C in mmHg)	37.4
Loss per transfer (% of liquid transferred)	0.03

7 Vapour pressure constants (mmHg, log base 10)

Antoine equation	$A = 6.89386$
	$B = 1264.370$
	$C = 216.640$
Cox chart	$A = 7.04265$
	$B = 1365.1$

8 Thermal information

Latent heat (cal/mol)	7645
Specific heat (cal/mol/°C)	50.7
Heat of combustion (cal/mol)	1 076 000

Azeotropes and activity coefficients of component X: n-heptane

Component Y	Azeotrope °C	Azeotrope % w/w X	α	A_{XY} =ln γ_X^∞	A_{YX} =ln γ_Y^∞	Source[11] Vol.	Page
Hydrocarbons							
n-Pentane		None	7	−0.15	−0.34	6(a)	127
n-Hexane		None	2.5	−0.08	−0.10	6(a)	604
Cyclohexane		None	2	0.07	0.04	6(a)	300
Benzene		None	2	0.52	0.25	6(b)	123
Toluene		None	1.8	0.26	0.26	6(b)	169
Ethylbenzene		None	3	0.44	0.36	6(c)	491
Xylenes		None	3	0.19	0.22	6(b)	186
Alcohols							
Methanol	59	49	>10[a]	2.79	2.26	2(a)	274
Ethanol	72	52	10[a]	2.35	2.43	2(a)	488
n-Propanol	85	65	2.5[a]	1.63	1.84	2(a)	595
Isopropanol	76	50	5[a]	1.71	2.42	2(b)	113
n-Butanol	94	82	3[a]	0.85	2.53	2(b)	219
Isobutanol	91	73	5[a]	1.50	2.21	2(d)	378
sec-Butanol	88	63	2.5[a]	1.34	1.89	2(d)	281
Cyclohexanol		None					
Ethylene glycol	98	97					
Methyl Cellosolve	92	77					
Ethyl Cellosolve	97	86					
Butyl Cellosolve							
Chlorinated hydrocarbons							
Methylene dichloride		None					
Chloroform		None	3.5	0.29	0.36	6(b)	77
Carbon tetrachloride		None	2.2	0.12	0.05	6(b)	72
1,2-Dichloroethane	81	24	2.5[a]	1.22	0.84	6(c)	444
1,1,1-Trichloroethane							
Trichloroethylene		None					
Perchloroethylene							
Monochlorobenzene		None	3	0.61	0.51	6(b)	119
Ketones							
Acetone	56	10	8[a]	1.70	1.17	3+4	242
Methyl ethyl ketone	77	30	5[a]	1.16	1.32	3+4	311
MIBK	98	87			0.64		
Cyclohexanone							
N-Methylpyrrolidone				2.79	2.89		
Ethers							
Diethyl ether							
Diisopropyl ether		None	2/3	0.05	1.43	3+4	559
1,4-Dioxane	92	56	2[a]	1.30	0.97	3+4	478
Tetrahydrofuran							
Esters							
Methyl acetate	56	4					
Ethyl acetate	77	5					
Butyl acetate		None	3/1.5	0.50	0.99	5	591
Miscellaneous							
Dimethylformamide	97	95	20[a]	1.82	2.19	6(b)	98
Dimethyl sulphoxide							
Pyridine	96	75	2.8[a]	0.64	1.54	6(b)	113
Acetonitrile	69	54	10[a]	2.95	2.25	6(b)	84
Furfuraldehyde		None	>10	2.19	2.40	3+4	50
Water	79	87	>20[a]	14.5			

[a]The relative volatility between the less volatile component of the binary mixture and the azeotrope.

Recovery notes: *n*-heptane

Because *n*-heptane must be extremely pure for use as the standard zero in the testing of motor fuels for motor octane number, it is available for use as a pure solvent. In practice, such a grade is seldom used except as a laboratory reagent. Narrow boiling range petroleum cuts primarily consisting of *n*- and isoalkanes have solvent properties similar to *n*-heptane and are very much less expensive. Repeated recovery may result in the more volatile compounds not being stripped completely from solutions but the overall solvent and volatility properties of recovered solvent will not be significantly different from those of virgin material. Unlike toluene, which has a similar volatility, heptane is not a good motor fuel and the risk of loss by pilfering is much less.

As the length of the carbon chain grows, alkanes become less stable at high temperatures but there is little risk of cracking of heptane taking place under solvent recovery conditions. Its very low water miscibility means that heptane can be recovered very satisfactorily using steam-regenerated activated carbon beds and the water phase after recovery has a very low BOD.

The fire hazards of heptane are high. A low autoignition temperature makes the use of hot oil heating systems, which normally operate in the range 270–290 °C, undesirable.

The atmosphere above the liquid surface of heptane in a storage tank is explosive under all likely ambient temperatures. All alkanes have a very low electrical conductivity and therefore a long relaxation time for the dissipation of static electricity, and this combination means that inert gas blanketing for storage is desirable.

The health hazards are low. The TLV is above the odour threshold and there have been no reported problems akin to those indicated for hexane in nerve damage. Although in past times the narrow cuts available as heptane concentrates contained high toluene contents, this is no longer a potential hazard.

CYCLOHEXANE

1 Names

Cyclohexane
Hexamethylene
Benzene hydride

2 Physical properties

Molecular weight	84
Empirical formula	C_6H_{12}
Boiling point (°C)	80.7
Freezing point (°C)	6.5
Specific gravity (20/4 °C)	0.778
Liquid expansion coefficient (per °C)	0.00012
Surface tension (at 20 °C in dyn/cm)	24.98
Absolute viscosity (at 25 °C in cP)	0.980

3 Fire and health hazards

Flash point (closed cup) (°C)	−17
Autoignition temperature (°C)	260
Lower explosive limit (ppm)	13 000
Upper explosive limit (ppm)	84 000
IDLH (ppm)	10 000
Odour threshold (ppm)	25
TWA–TLV (ppm)	300
Saturated vapour concentration (70 °F in ppm)	155 700
Vapour density (relative to air)	2.9

4 Solvent properties

Hildebrand solubility parameter	8.2
Dipole moment (D)	0.3
Dielectric constant (20 °C)	2.02
Evaporation time (diethyl ether = 1.0)	1.5

5 Aqueous effluent characteristics

Solubility in water (at 25 °C in % w/w)	0.0055
Solubility of water in (at 25 °C in % w/w)	0.01
Biological oxygen demand (w/w)	0
Chemical oxygen demand (w/w)	
Theoretical oxygen demand (w/w)	3.43

6 Handling details

Hazchem number	1145
Hazard class	3 YE
Vapour pressure (at 21 °C in mmHg)	78.8
Loss per transfer (% of liquid transferred)	0.046

7 Vapour pressure constants (mmHg, log base 10)

Antoine equation	$A = 6.85146$
	$B = 1206.470$
	$C = 223.136$
Cox chart	$A = 7.04736$
	$B = 1295.8$

8 Thermal information

Latent heat (cal/mol)	7140
Specific heat (cal/mol/°C)	37.0
Heat of combustion (cal/mol)	871 920
Heat of fusion (cal/mol)	623

Azeotropes and activity coefficients of component X: cyclohexane

Component Y	Azeotrope °C	Azeotrope % w/w X	α	A_{XY} $=\ln \gamma_X^\infty$	A_{YX} $=\ln \gamma_Y^\infty$	Source[11] Vol.	Page
Hydrocarbons							
n-Pentane		None	4	0.11	0.18	6(a)	119
n-Hexane		None	1.7	0.06	0.07	6(a)	273
n-Heptane		None	2	0.04	0.07	6(a)	300
Benzene	78	50	1.5	0.35	0.35	6(a)	210
Toluene		None	3	0.20	0.41	6(a)	284
Ethylbenzene		None	5	0.20	0.55	6(a)	310
Xylenes		None	5	0.30	0.44	6(a)	315
Alcohols							
Methanol	54	62	9[a]	2.29	2.95	2(a)	239
Ethanol	65	69	3[a]	2.27	1.80	2(a)	430
n-Propanol	74	80	5[a]	1.32	2.30	2(a)	579
Isopropanol	69	68	3[a]	1.39	2.39	2(b)	84
n-Butanol	80	96	9[a]	1.13	2.27	2(b)	188
Isobutanol	78	86	5[a]	1.16	2.00	2(b)	288
sec-Butanol	76	82					
Cyclohexanol		None	>20	0.64	2.73	2(d)	516
Ethylene glycol		None					
Methyl Cellosolve	80	92	10[a]	1.68	2.90	2(b)	128
Ethyl Cellosolve		None					
Butyl Cellosolve							
Chlorinated hydrocarbons							
Methylene dichloride		None					
Chloroform		None					
Carbon tetrachloride		None	<1.5	1.07	0.97	6(a)	158
1,2-Dichloroethane	74	50	2[a]	1.07	0.97	6(a)	158
1,1,1-Trichloroethane							
Trichloroethylene	80	83	1.5[a]	0.25	0.42	6(a)	155
Perchloroethylene							
Monochlorobenzene		None	5	0.76	0.29	6(a)	202
Ketones							
Acetone	53	33	5[a]	1.46	1.60	3+4	213
Methyl ethyl ketone	72	60	1.5[a]	0.89	1.26	3+4	296
MIBK		None	4	0.39	1.07	3+4	354
Cyclohexanone		None	>10	0.81	1.00	3+4	337
N-Methylpyrrolidone		None		2.10			
Ethers							
Diethyl ether							
Diisopropyl ether		None	1.5	0.04	0.01	3+4	555
1,4-Dioxane	80	75	3[a]	0.95	1.10	3+4	468
Tetrahydrofuran	60	3		0.50	0.49		
Esters							
Methyl acetate	55	20	5[a]	1.28	1.25	5	393
Ethyl acetate	73	46		0.83	0.98		
Butyl acetate		None	5	0.38	0.76	5	585
Miscellaneous							
Dimethylformamide		None	>20	2.04	2.67	6(c)	200
Dimethyl sulphoxide							
Pyridine		None	4	1.42	1.71	6(a)	171
Acetonitrile	62	67					
Furfuraldehyde		None	>10	1.99	2.33	3+4	45
Water	70	92					

[a]The relative volatility between the less volatile component of the binary mixture and the azeotrope.

Recovery notes: cyclohexane

Cyclohexane is stable at its boiling point and so is suitable as an entrainer for azeotropic distillation. It is produced in large quantity as a raw material for nylon manufacture, and therefore costs little more than benzene, from which it is derived.

Apart from its low flash point, the only serious handling problem it presents is a freezing point that is inconveniently high for outdoor operations in the winter. Since it would be hard to justify the installation and maintenance of a lagged and traced pipework system to guard against blockages on a few days in each year, consideration should be given to adding an impurity which would reduce the freezing point of the mixture to a safe level. Methylcyclopentane (b.p. 72 °C, freezing point -142 °C) would be the ideal choice and various narrow-range aliphatic fractions (62/68, SBP2, etc.) may also be suitable, depending on the application for which cyclohexane is chosen.

Although cyclohexane is not a very good motor fuel, a risk of theft for this purpose does exist.

BENZENE

1 Names

Benzene
Benzole
Benzol
Not benzine or benzin

2 Physical properties

Molecular weight	78
Empirical formula	C_6H_6
Boiling point (°C)	80
Freezing point (°C)	5.5
Specific gravity (20/4 °C)	0.879
Liquid expansion coefficient (per °C)	0.00138
Surface tension (at 20 °C in dyn/cm)	29
Absolute viscosity (at 25 °C in cP)	0.65

3 Fire and health hazards

Flash point (closed cup) (°C)	−11
Autoignition temperature (°C)	562
Lower explosive limit (ppm)	13 000
Upper explosive limit (ppm)	71 000
IDLH (ppm)	2 000
Odour threshold (ppm)	4
TWA–TLV (ppm)	1
Saturated vapour concentration (70 °F in ppm)	117 000
Vapour density (relative to air)	2.7

4 Solvent properties

Hildebrand solubility parameter	9.2
Dipole moment (D)	0
Dielectric constant (20 °C)	2.28
Evaporation time (diethyl ether = 1.0)	2.6

5 Aqueous effluent characteristics

Solubility in water (at 25 °C in % w/w)	0.18
Solubility of water in (at 25 °C in % w/w)	0.063
Biological oxygen demand (w/w)	0.12
Chemical oxygen demand (w/w)	1.4
Theoretical oxygen demand (w/w)	3.08

6 Handling details

Hazchem number	1114
Hazard class	3 WE
Vapour pressure (at 21 °C in mmHg)	79.4
Loss per transfer (% of liquid transferred)	0.04

7 Vapour pressure constants (mmHg, log base 10)

Antoine equation	$A = 6.87987$
	$B = 1196.760$
	$C = 219.161$
Cox chart	$A = 7.04500$
	$B = 1290.9$

8 Thermal information

Latent heat (cal/mol)	7332
Specific heat (cal/mol/°C)	31.8
Heat of combustion (cal/mol)	7564444
Heat of fusion (cal/mol)	2375

Appendix 1 215

Azeotropes and activity coefficients of component X: benzene

Component Y	Azeotrope °C	Azeotrope % w/w X	α	A_{XY} $=\ln \gamma_X^\infty$	A_{YX} $=\ln \gamma_Y^\infty$	Source[11] Vol.	Page
Hydrocarbons							
n-Pentane		None	7	0.50	0.60	6(a)	118
n-Hexane	68	5	1.6[a]	0.37	0.44	6(a)	535
n-Heptane		None	2	0.25	0.52	6(b)	123
Cyclohexane	77	50	<1.5[a]	0.35	0.35	6(a)	210
Toluene		None	2.5	0.30	0.03	7	283
Ethylbenzene		None	5	−0.14	0.18	7	306
Xylenes		None	5	−0.07	−0.04	7	310
Alcohols							
Methanol	57	61	10[a]	1.96	2.28	2(a)	205
Ethanol	68	68	2.5[a]	1.40	1.80	2(a)	399
n-Propanol	77	83	4[a]	1.27	2.07	2(a)	556
Isopropanol	72	67	3[a]	1.32	1.94	2(b)	67
n-Butanol		None	7	0.79	1.45	2(b)	179
Isobutanol	79	92	5[a]	0.87	1.67	2(b)	287
sec-Butanol	79	85	5[a]	0.84	0.90	2(b)	248
Cyclohexanol							
Ethylene glycol		None		3.46			
Methyl Cellosolve		None	7	0.83	1.80	2(b)	127
Ethyl Cellosolve		None					
Butyl Cellosolve							
Chlorinated hydrocarbons							
Methylene dichloride		None					
Chloroform		None	1.5/2	−0.45	−0.48	7	67
Carbon tetrachloride		None	<1.5	0.10	0.07	7	11
1,2-Dichloroethane	80	82	1.5[a]	0.02	0.01	7	139
1,1,1-Trichloroethane		None	<1.5	0.03	0.03	7	121
Trichloroethylene		None	<1.5	0.04	0.01	7	116
Perchloroethylene		None	4	0.25	0.23	7	112
Monochlorobenzene		None	5	0.03	0.04	7	243
Ketones							
Acetone		None	2/3	0.33	0.47	3+4	197
Methyl ethyl ketone	78	55	<1.5[a]	0.16	0.24	3+4	284
MIBK		None	5	0.03	0.07	3+4	351
Cyclohexanone							
N-Methylpyrrolidone		None		0.33			
Ethers							
Diethyl ether		None	5	−0.06	−0.32	3+4	516
Diisopropyl ether		None	<1.5	0.16	0.20	3+4	553
1,4-Dioxane	82	88	1.8[a]	0.14	0.10	3+4	465
Tetrahydrofuran							
Esters							
Methyl acetate		None	2	0.31	0.22	5	375
Ethyl acetate	77	6	<1.5[a]	0.08	0.12	5	496
Butyl acetate		None	4.5	−0.09	−0.08	5	583
Miscellaneous							
Dimethylformamide		None	12	0.36	0.25	7	183
Dimethyl sulphoxide		None	10	0.99	1.18	7	169
Pyridine		None	3	0.24	0.21	7	221
Acetonitrile	73	66	2[a]	0.92	1.13	7	123
Furfuraldehyde		None	>10	0.49	0.59	3+4	44
Water	69	91		7.46	5.42		

[a]The relative volatility between the less volatile component of the binary mixture and the azeotrope.

Recovery notes: benzene

Because of benzene's relatively high melting point, pipelines traced with steam or electric heating and similar precautions against freezing may need to be taken in handling it. The toxicity of benzene makes it difficult to break pipelines safely to clear blockages. so high standards of plant design are necessary to avoid such work. In cold weather, hoses should be drained when not in use.

Few people can detect by smell a level of benzene vapour several times the TLV, so very great care is needed in handling it and tests both of the air in the work area and of the urine and blood of operatives who may be exposed to it are necessary.

Benzene is most useful as an azeotropic entrainer for drying ethanol, IPA and ACN, but in all these cases there are alternatives and it will be seldom that the problems of handling benzene do not cause one of them to be preferred.

Benzene can be stored in any of the usual metals used for plant construction but attacks natural rubber, butyl rubber and neoprene. Viton elastomers can be used for gaskets and diaphragms.

Benzene vapour is heavier than air. Ventilation in plant and laboratory should be designed with this in mind.

Benzene is a possible component of motor fuel and is therefore tempting to pilfer, but since the concentration of benzene in commercial motor fuels is low such theft can be readily measured.

TOLUENE

1. **Names**

   Toluene
   Toluol
   Methylbenzene

2. **Physical properties**

Molecular weight	92
Empirical formula	C_7H_8
Boiling point (°C)	110.6
Freezing point (°C)	−95
Specific gravity (20/4 °C)	0.867
Liquid expansion coefficient (per °C)	0.0011
Surface tension (at 20 °C in dyn/cm)	28.5
Absolute viscosity (at 25 °C in cP)	0.59

3. **Fire and health hazards**

Flash point (closed cup) (°C)	4
Autoignition temperature (°C)	480
Lower explosive limit (ppm)	12 700
Upper explosive limit (ppm)	70 000
IDLH (ppm)	2000
Odour threshold (ppm)	0.2
TWA–TLV (ppm)	100
Saturated vapour concentration (70 °F in ppm)	31 000
Vapour density (relative to air)	3.2

4. **Solvent properties**

Hildebrand solubility parameter	8.9
Dipole moment (D)	0.4
Dielectric constant (20 °C)	2.38
Evaporation time (diethyl ether = 1.0)	4.6

5. **Aqueous effluent characteristics**

Solubility in water (at 25 °C in % w/w)	0.052
Solubility of water in (at 25 °C in % w/w)	0.033
Biological oxygen demand (w/w)	1.19
Chemical oxygen demand (w/w)	1.41
Theoretical oxygen demand (w/w)	3.13

6. **Handling details**

Hazchem number	1294
Hazard class	3 YE
Vapour pressure (at 21 °C in mmHg)	23.2
Loss per transfer (% of liquid transferred)	0.013

7. **Vapour pressure constants (mmHg, log base 10)**

Antoine equation	$A = 8.53516$
	$B = 1950.940$
	$C = 237.147$
Cox chart	$A = 7.12773$
	$B = 1448.2$

8. **Thermal information**

Latent heat (cal/mol)	7985
Specific heat (cal/mol/°C)	41.0
Heat of combustion (cal/mol)	891 112

Azeotropes and activity coefficients of component X: toluene

Component Y	Azeotrope °C	Azeotrope % w/w X	α	$A_{XY} = \ln \gamma_X^\infty$	$A_{YX} = \ln \gamma_Y^\infty$	Source[11] Vol.	Page
Hydrocarbons							
n-Pentane	None		20	1.31	0.43	6(a)	160
n-Hexane	None		4	0.31	0.34	6(a)	593
n-Heptane	None		1.8	0.26	0.26	6(b)	169
Cyclohexane	None		2.9	0.41	0.20	6(a)	284
Benzene	None		2.5	0.03	0.30	7	283
Ethylbenzene	None		2.8	0.02	0.08	7	443
Xylenes	None		2.8	−0.06	−0.80	7	444
Alcohols							
Methanol	64	31	10[a]	2.04	1.95	2(a)	269
Ethanol	77	32	10[a]	1.57	1.65	2(a)	470
n-Propanol	93	51	3[a]	1.19	1.46	2(a)	592
Isopropanol	81	31	6[a]	1.34	1.14	2(b)	108
n-Butanol	105	72	3[a]	0.83	1.26	2(b)	207
Isobutanol	101	55	2[a]	0.98	1.31	2(b)	289
sec-Butanol	95	45	2[a]	0.82	1.19	2(d)	277
Cyclohexanol	None						
Ethylene glycol	110	98					
Methyl Cellosolve	106	75					
Ethyl Cellosolve	109	89					
Butyl Cellosolve	None		8	0.31	0.58	2(d)	560
Chlorinated hydrocarbons							
Methylene dichloride	None						
Chloroform	None		4	−0.24	−0.10	7	352
Carbon tetrachloride	None		2.5	0.23	0.01	7	334
1,2-Dichloroethane	None		2	0.05	1.05	7	382
1,1,1-Trichloroethane	None		2				
Trichloroethylene	None		2	−0.21	−0.08	7	370
Perchloroethylene	None						
Monochlorobenzene	None		2	0.00	−0.03	7	416
Ketones							
Acetone	None			0.44	0.60	7	236
Methyl ethyl ketone	None		3	0.45	0.30	3+4	308
MIBK	111	97	1.5[a]	0.12	0.16	3+4	356
Cyclohexanone	None		4	−0.09	0.06	3+4	339
N-Methylpyrrolidone	None			0.51			
Ethers							
Diethyl ether	None		13	0.21			
Diisopropyl ether	None		4	0.14	0.14	3+4	558
1,4-Dioxane	102	20	<1.5[a]	0.18	0.23	3+4	477
Tetrahydrofuran	None						
Esters							
Methyl acetate	None						
Ethyl acetate	None		3	0.21	0.19	5	520
Butyl acetate	None		1.5	−0.04	−0.87	5	587
Miscellaneous							
Dimethylformamide	None		4	0.31	0.96	7	390
Dimethyl sulphoxide	None		10	1.31	1.91	7	386
Pyridine	110	78	1.5[a]	0.30	0.51	7	406
Acetonitrile	81	20	4[a]	1.24	1.21	7	373
Furfuraldehyde	None		5/10	1.28	0.66	3+4	48
Water	84	82	20	8.13	8.11		

[a] The relative volatility between the less volatile component of the binary mixture and the azeotrope.

Recovery notes: toluene

Toluene is not miscible with water and it is not satisfactorily removed from an air stream by water scrubbing but it can be removed with a high-boiling hydrocarbon absorbing oil.

Removal using carbon bed adsorbers is very satisfactory since toluene is stable and does not need drying when the bed is regenerated using steam, unless the recovered toluene is to be used for an unusual purpose, e.g. urethane paint.

Toluene's very low solubility in water means that its removal from water does not present a problem unless the system contains a water-miscible solvent that increases its solubility. Steam stripping from water is effective and easy.

Mild steel and non-ferrous metals are not affected by toluene and it does not go off-colour in contact with mild steel. Neoprene, natural rubber and butyl rubber all swell and deteriorate in the presence of toluene.

Unless, by repeated overexposure, an individual has lost his sense of smell for toluene, its odour is a sufficient safeguard. Skin and, particularly, eye contact may be harmful.

The explosive limits of toluene correspond to its vapour pressure at 4 °C and 37 °C. The vapour phase in a toluene storage tank is always potentially explosive and inert gas blanketing is very desirable. Particular care against the generation of static electricity by pumping two phase mixtures (e.g. water or air and toluene) should be shown and bonding of pipelines, hoses and receptacles (e.g. drums) is necessary. Plastic containers should never be used.

Pumping speeds of less than 1 m/s and the use of an anti-static additive, if it is not harmful in the process, are useful safety measures.

Toluene is very stable at any temperature typical of a solvent recovery process so there is no risk of it cracking or decomposing if it is held at its boiling point.

Of all solvents used industrially, toluene is the most attractive to steal for use as a motor fuel. It has a high octane rating and an adequate volatility particularly when blended 50:50 with commercial motor fuel.

Because commercial motor fuel already contains substantial concentrations of toluene, theft is difficult to prove or even detect with absolute confidence.

Since industrially used toluene is very unlikely to have had any motor fuel tax paid on it, its theft for this purpose is likely to be a crime against both the owner and the tax authorities. It is treated very seriously by the latter.

ETHYLBENZENE

1 **Names**   Ethylbenzene
                                             Phenylethane

2 **Physical properties**

   Molecular weight   106
   Empirical formula C_8H_{10}
   Boiling point (°C)   136
   Freezing point (°C)   −95
   Specific gravity (20/4 °C)   0.867
   Liquid expansion coefficient (per °C)   0.0009
   Surface tension (at 20 °C in dyn/cm)   29.2
   Absolute viscosity (at 25 °C in cP)   0.72

3 **Fire and health hazards**

   Flash point (closed cup) (°C)   21
   Autoignition temperature (°C)   460
   Lower explosive limit (ppm)   10 000
   Upper explosive limit (ppm)   67 000
   IDLH (ppm)   2000
   Odour threshold (ppm)   140
   TWA–TLV (ppm)   100
   Saturated vapour concentration (70 °F in ppm) 9960
   Vapour density (relative to air)   3.7

4 **Solvent properties**

   Hildebrand solubility parameter   8.3
   Dipole moment (D)   0.6
   Dielectric constant (25 °C)   2.41
   Evaporation time (diethyl ether = 1.0)   10.9

5 **Aqueous effluent characteristics**

   Solubility in water (at 20 °C in % w/w)   0.020
   Solubility of water in (at 25 °C in % w/w) 0.033
   Biological oxygen demand (w/w)   0.028
   Chemical oxygen demand (w/w)   –
   Theoretical oxygen demand (w/w)   3.17

6 **Handling details**

   Hazchem number   1175
   Hazard class   3 YE
   Vapour pressure (at 21 °C in mmHg)   8.0
   Loss per transfer (% of liquid transferred) 0.005

7 **Vapour pressure constants (mmHg, log base 10)**

   Antoine equation $A = 6.96580$
                                             $B = 1429.550$
                                             $C = 213.767$
   Cox chart $A = 7.19691$
                                             $B = 1579.7$

8 **Thermal information**

   Latent heat (cal/mol)   8480
   Specific heat (cal/mol/°C)   43.4
   Heat of combustion (cal/mol)   1 046 962

Azeotropes and activity coefficients of component X: ethylbenzene

Component Y	Azeotrope °C	Azeotrope % w/w X	α	$A_{XY} = \ln \gamma_X^\infty$	$A_{YX} = \ln \gamma_Y^\infty$	Source[11] Vol.	Page
Hydrocarbons							
n-Pentane	None	None					
n-Hexane	None	None	10				
n-Heptane	None	None	3	0.36	0.44	6(c)	491
Cyclohexane	None	None	5	0.55	0.20	6(a)	310
Benzene	None	None	5	0.18	−0.14	7	306
Toluene	None	None	2.8	0.08	0.02	7	443
Xylenes	None	None					
Alcohols							
Methanol	None	None	>20	2.33	1.85	2(c)	245
Ethanol	None	None	20	1.94	1.44	2(a)	499
n-Propanol	97	9	8[a]	1.33	1.24	2(a)	601
Isopropanol	None	None		1.67	0.71		
n-Butanol	115	33	4[a]	0.99	0.99	2(b)	228
Isobutanol	107	20					
sec-Butanol	None	None					
Cyclohexanol							
Ethylene glycol	133	86					
Methyl Cellosolve	117	46	3[a]	1.41	1.77	2(b)	132
Ethyl Cellosolve	126	57	2[a]	0.88	1.09	2(b)	299
Butyl Cellosolve	None	None					
Chlorinated hydrocarbons							
Methylene dichloride							
Chloroform							
Carbon tetrachloride	None	None	5	−0.03	−0.29	7	464
1,2-Dichloroethane	None	None	5	−0.12	−0.68	7	466
1,1,1-Trichloroethane							
Trichloroethylene							
Perchloroethylene							
Monochlorobenzene	None	None	<1.5	0.01	−0.01	7	469
Ketones							
Acetone							
Methyl ethyl ketone	None	None	8	0.78	0.32	3+4	316
MIBK	None	None					
Cyclohexanone							
N-Methylpyrrolidone				0.53			
Ethers							
Diethyl ether							
Diisopropyl ether	None	None	10	0.29	0.15	3+4	563
1,4-Dioxane	None	None					
Tetrahydrofuran							
Esters							
Methyl acetate							
Ethyl acetate	None	None	8	0.25	0.06	5	538
Butyl acetate	None	None					
Miscellaneous							
Dimethylformamide	134	85	2[a]				
Dimethyl sulphoxide							
Pyridine	None	None					
Acetonitrile	None	None	10	1.59	0.94	7	465
Furfuraldehyde	132	95	3[a]	0.93	1.09	3+4	51
Water	92	67					

[a] The relative volatility between the less volatile component of the binary mixture and the azeotrope.

Recovery notes: ethylbenzene

Ethylbenzene is not considered to be photochemically active under Rule 66, and can be used in place of xylene mixtures in formulations. Apart from this and its flash point, which brings it under UK regulations as a highly flammable petroleum spirit, it is nearly identical with xylene in recovery and use.

Xylenes

1 Names

Xylenes
Xylol
Mixed xylenes
Dimethylbenzenes
C8 aromatics

2 Physical properties

Molecular weight	106
Empirical formula	C_8H_{10}
Boiling point (°C)	138
Freezing point (°C)	
Specific gravity (20/4 °C)	0.871
Liquid expansion coefficient (per °C)	0.001
Surface tension (at 20 °C in dyn/cm)	28.7
Absolute viscosity (at 25 °C in cP)	0.7

3 Fire and health hazards

Flash point (closed cup) (°C)	23*
Autoignition temperature (°C)	480
Lower explosive limit (ppm)	11 400
Upper explosive limit (ppm)	70 000
IDLH (ppm)	10 000
Odour threshold (ppm)	1.1
TWA–TLV (ppm)	100
Saturated vapour concentration (70 °F in ppm)	9180
Vapour density (relative to air)	3.7

4 Solvent properties

Hildebrand solubility parameter	8.8
Dipole moment (D)	0.3
Dielectric constant (20 °C)	2.4
Evaporation time (diethyl ether = 1.0)	11.5

5 Aqueous effluent characteristics

Solubility in water (at 25 °C in % w/w)	0.02
Solubility of water in (at 25 °C in % w/w)	0.05
Biological oxygen demand (w/w)	0.1
Chemical oxygen demand (w/w)	2.56
Theoretical oxygen demand (w/w)	3.17

6 Handling details

Hazchem number	1307
Hazard class	3 Y
Vapour pressure (at 21 °C in mmHg)	7.0
Loss per transfer (% of liquid transferred)	0.005

7 Vapour pressure constants (mmHg, log base 10)

Antoine equation	$A = 6.990\,53$
	$B = 1453.43$
	$C = 215.310$
Cox chart	$A = 7.208\,07$
	$B = 1601.1$

8 Thermal information

Latent heat (cal/mol)	8692
Specific heat (cal/mol/°C)	42.4
Heat of combustion (cal/mol)	1 033 924

*See recovery notes.

Azeotropes and activity coefficients of component X: xylenes

Component Y	Azeotrope °C	Azeotrope % w/w X	α	$A_{XY} = \ln \gamma_X^\infty$	$A_{YX} = \ln \gamma_Y^\infty$	Source[11] Vol.	Page
Hydrocarbons							
n-Pentane	None			0.43			
n-Hexane	None		10	0.40	0.18	6(a)	608
n-Heptane	None		3	0.22	0.19	6(b)	186
Cyclohexane	None		5	0.44	0.30	6(a)	315
Benzene	None		5	−0.04	−0.07	7	310
Toluene	None		2.5	−0.80	−0.06	7	444
Ethylbenzene	None		<1.5				
Alcohols							
Methanol	None		>20	2.40	1.80	2(c)	246
Ethanol	None		20	1.85	1.54	2(a)	500
n-Propanol	97	7	20[a]	1.39	1.26	2(c)	575
Isopropanol	None						
n-Butanol	115	27	4[a]	1.00	1.11	2(b)	229
Isobutanol	108	12	5[a]	1.03	0.97	2(b)	292
sec-Butanol	None		5	0.97	1.04	2(d)	282
Cyclohexanol	140	90					
Ethylene glycol	135	93					
Methyl Cellosolve	120	45	3[a]	1.12	1.34	2(b)	133
Ethyl Cellosolve	128	50	2[a]	0.75	1.06	2(b)	301
Butyl Cellosolve	144	4					
Chlorinated hydrocarbons							
Methylene dichloride				−0.23			
Chloroform							
Carbon tetrachloride	None		10	0.14	−0.27	7	480
1,2-Dichloroethane	None		8	0.23	0.07	7	490
1,1,1-Trichloroethane							
Trichloroethylene							
Perchloroethylene							
Monochlorobenzene	None		<1.5	−0.09	−0.13	7	508
Ketones							
Acetone				0.74			
Methyl ethyl ketone				0.53			
MIBK							
Cyclohexanone	None						
N-Methylpyrrolidone							
Ethers							
Diethyl ether				0.09			
Diisopropyl ether							
1,4-Dioxane							
Tetrahydrofuran							
Esters							
Methyl acetate	None						
Ethyl acetate	None		9	0.46	0.47	5	541
Butyl acetate	None						
Miscellaneous							
Dimethylformamide	136	88	2[a]	0.99	0.93	7	481
Dimethyl sulphoxide							
Pyridine	None		2/3	0.29	0.33	7	482
Acetonitrile	None		5/10	1.71	0.80	7	499
Furfuraldehyde	139	90	3[a]	1.02	1.05	3+4	52
Water	93	63					

[a] The relative volatility between the less volatile component of the binary mixture and the azeotrope.

Recovery notes: xylenes

Xylenes for solvent purposes consist of a mixture of three dimethylbenzene isomers, *ortho-*, *meta-* and *para-*xyelene, and ethylbenzene. The physical properties quoted are for a typical mixture and the only property that is significantly altered by the ratio of the isomers is the flash point of the mixture. This can be significant in the UK and other countries where legislation primarily aimed at the safe storage of petrol regulates the storage and handling of hydrocarbons with flash points of less than 73 °F (22.8 °C) by the Abel method.

o-Xylene, which is often removed from a mixed xylene stream for use as a raw material for making phthalic anhydride, is the least volatile of the isomers and is needed in the mixture to keep the flash point high. Pure *o*-xylene has a flash point of 30 °C.

Ethylbenzene, separated from the other isomers by super-fractionation for use in making styrene or as a Rule 66 solvent, has a flash point of 21 °C.

p-Xylene, extracted from mixed xylenes as a source of terephthalic acid, also has a low flash point of 25 °C.

m-Xylene, which only has a small requirement as a chemical raw material, tends to be the most concentrated component of mixed xylenes at 40–50% and has a flash point of 27 °C.

However, small traces of toluene in mixed xylenes cause the mixture to have a flash point lower than the weighted average of its main components and the reproducibility of the test method means that a flash point of at least 24 °C is normally required for satisfactory operation. In recovery operations, the effect of *n*-butanol (pure flash point 35 °C) and ethyl Cellosolve (pure flash point 40 °C), but which both form azeotropes with the C_8 isomers, can cause 'recovered xylene' to have a low flash point.

The C_8 aromatics are stable at their boiling point and need no inhibitors in use or recovery.

The xylenes are very good motor fuels from the knock-rating point of view and therefore represent a theft risk, although excessive concentrations of material boiling in the 140 °C range will cause bad startability in cold weather.

All normal materials of construction and protective clothing are suitable for use with xylenes except for natural, neoprene and butyl rubbers. PVC gloves have a limited life.

METHANOL

1 Names

Methanol
Methyl alcohol
Wood alcohol
Carbinol
Not methylated spirits

2 Physical properties

Molecular weight	32
Empirical formula	CH_4O
Boiling point (°C)	64
Freezing point (°C)	−98
Specific gravity (20/4 °C)	0.792
Liquid expansion coefficient (per °C)	0.0012
Surface tension (at 20 °C in dyn/cm)	22.6
Absolute viscosity (at 25 °C in cP)	0.6

3 Fire and health hazards

Flash point (closed cup) (°C)	15
Autoignition temperature (°C)	470
Lower explosive limit (ppm)	60 000
Upper explosive limit (ppm)	365 000
IDLH (ppm)	25 000
Odour threshold (ppm)	100
TWA–TLV (ppm)	200
Saturated vapour concentration (70 °F in ppm)	130 000
Vapour density (relative to air)	1.11

4 Solvent properties

Hildebrand solubility parameter	14.5
Dipole moment (D)	1.7
Dielectric constant (20 °C)	32.6
Evaporation time (diethyl ether = 1.0)	6.3

5 Aqueous effluent characteristics

Solubility in water (at 25 °C in % w/w)	Total
Solubility of water in (at 25 °C in % w/w)	Total
Biological oxygen demand (w/w)	1.12
Chemical oxygen demand (w/w)	1.4
Theoretical oxygen demand (w/w)	1.5

6 Handling details

Hazchem number	1.230
Hazard class	2 PE
Vapour pressure (at 21 °C in mmHg)	103
Loss per transfer (% of liquid transferred)	0.02

7 Vapour pressure constants (mmHg, log base 10)

Antoine equation	$A = 8.08097$
	$B = 1582.271$
	$C = 239.726$
Cox chart	$A = 8.23606$
	$B = 1579.9$

8 Thermal information

Latent heat (cal/mol)	8426
Specific heat (cal/mol/°C)	19.5
Heat of combustion (cal/mol)	149 851

Azeotropes and activity coefficients of component X: methanol

Component Y	Azeotrope °C	Azeotrope % w/w X	α	A_{XY} =$\ln \gamma_X^\infty$	A_{YX} =$\ln \gamma_Y^\infty$	Source[11] Vol.	Page
Hydrocarbons							
n-Pentane	31	8	8[a]	3.43	2.40	2(a)	198
n-Hexane	50	26	>10[a]	3.01	2.60	2(a)	253
n-Heptane	59	51	>10[a]	2.26	2.79	2(a)	274
Cyclohexane	54	38	>10[a]	2.95	2.29	2(a)	239
Benzene	56	38	10[a]	2.28	1.96	2(a)	205
Toluene	63	69	>10[a]	1.95	2.04	2(a)	269
Ethylbenzene		None	20	2.09	2.23	2(c)	245
Xylenes		None	20				
Alcohols							
Ethanol		None	1.7[b]	0.04	0.02	2(a)	60
n-Propanol		None	5	0.06	0.24	2(a)	122
Isopropanol		None	2	−0.12	−0.10	2(a)	123
n-Butanol		None	10	0.15	0.38	2(a)	169
Isobutanol		None	6	0.21	0.22	2(a)	171
sec-Butanol		None	10	−0.38	−0.06	2(c)	128
Cyclohexanol		None	>10				
Ethylene glycol		None	>20	0.07	0.61	2(a)	62
Methyl Cellosolve		None	>15			2(c)	98
Ethyl Cellosolve		None	>15				
Butyl Cellosolve		None	>15				
Chlorinated hydrocarbons							
Methylene dichloride	38	7	9[a]	2.37	0.93	2(a)	24
Chloroform	53	13	3[a]	2.07	0.99	2(a)	18
Carbon tetrachloride	56	21	5[a]	2.33	1.91	2(a)	1
1,2-Dichloroethane	60	33	8[a]	2.36	1.53	2(a)	47
1,1,1-Trichloroethane	56	22		1.93	1.70		
Trichloroethylene	59	38	7[a]	1.97	2.08	2(a)	40
Perchloroethylene	64	59	7	2.82	2.47	2(a)	37
Monochlorobenzene		None	>20	1.69	2.16	2(a)	204
Ketones							
Acetone	56	12	1.5/2[a]	0.60	0.61	2(a)	76
Methyl ethyl ketone	64	70	2.5[a]	0.72	0.72	2(a)	133
MIBK		None	6	0.78	1.20	2(a)	248
Cyclohexanone		None	>10				
N-Methylpyrrolidone		None	>10	−0.64			
Ethers							
Diethyl ether		None	7	1.52	1.12	2(a)	170
Diisopropyl ether	59	24	2[a]	1.18	1.42	2(a)	261
1,4-Dioxane	78	91	5[a]	0.51	0.78	2(a)	148
Tetrahydrofuran	61	31	2[a]	0.90	0.76	2(a)	140
Esters							
Methyl acetate	54	19	3.5[a]	1.00	1.06	2(a)	92
Ethyl acetate	62	46	2.5[a]	1.62	1.02	2(a)	154
Butyl acetate		None	18	1.05	1.21	2(c)	216
Miscellaneous							
Dimethylformamide		None	20	−0.21	−0.62	2(a)	114
Dimethyl sulphoxide		None	>20	−1.09	−1.40	2(c)	62
Pyridine		None	5	0.04	0.07	2(a)	180
Acetonitrile	63	19	3[a]	1.09	0.88	2(a)	43
Furfuraldehyde		None	>20	0.00	0.00	2(c)	140
Water		None	5	0.80	0.66	1	48

[a]The relative volatility between the less volatile component of the binary mixture and the azeotrope.
[b]$\text{Log}_{10}\alpha = 0.22 - 0.11x + 0.30x^2 - 0.14x^3$, where x is the mole fraction of methanol.

228 *Appendix 1*

Recovery notes: methanol

VLE, flash point and solubility data for methanol are given in Figures A.1 and A.2 and Table A.1.

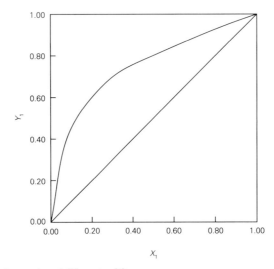

Fig. A.1 VLE diagram for methanol (1)–water (2)

Methanol is stable at its boiling point and, apart from having a comparatively high latent heat, does not present any difficulties in either batch or continuous distillation. It is moderately easy to strip from water, certainly to the point at which an effluent water would have an acceptable BOD. The mixture of less than 25% of methanol in water is not flammable.

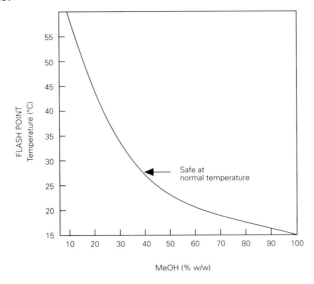

Fig. A.2 Flash point vs. water content for methanol

Table A.1 Solubilities of hydrocarbons in methanol in g per 100 ml at various temperatures

Hydrocarbon	Temperature (°C)							
	0	10	20	30	40	50	60	70
Pentane	62.0	81	Misc.	Misc.	Misc.	Misc.	Misc.	Misc.
Hexane	32.4	37.0	42.7	49.5	60.4	83	Misc.	Misc.
3-Methylpentane	38.9	45.0	53.0	65	91	Misc.	Misc.	Misc.
2,2-Dimethylbutane	59	80	Misc.	Misc.	Misc.	Misc.	Misc.	Misc.
2,3-Dimethylbutane	49.5	59.3	76	170	Misc.	Misc.	Misc.	Misc.
Heptane	18.1	20.0	22.5	25.4	28.7	32.7	37.8	45.0
Octane	12.2	13.6	15.2	16.7	18.4	20.6	23.0	26.0
3-Methylheptane	15.4	17.0	19.0	21.2	24.2	27.4	31.4	36.5
2,2,4-Trimethylpentane	24.9	27.9	31.4	35.3	40.2	46.0	56.0	76
Nonane	8.4	9.5	10.5	11.6	12.9	14.2	15.5	17.0
2,2,5-Trimethylhexane	16.2	17.9	20.0	22.1	24.7	28.0	31.6	36.0
Decane	6.2	6.8	7.4	8.1	8.9	9.8	10.9	12.0
Cyclopentane	68	86	140	Misc.	Misc.	Misc.	Misc.	Misc.
Methylcyclopentane	38.0	41.5	50.0	59.5	74	110	Misc.	Misc.
Cyclohexane	—	—	34.4	38.4	43.5	50.3	60	74
Methylcyclohexane	26.9	29.8	33.2	37.2	42.2	48.8	57.5	70.9

Methanol does not have an offensive odour and because of its low vapour density it is easily dispersed in air.

Removal of hydrocarbons from methanol cannot be done easily by fractionation because of the existence of many azeotropes, but the partition of methanol between water and hydrocarbons is very strongly in favour of the former, although methanol separated by this route will seldom have a good water miscibility test. Because of its small molecular diameter methanol should not be dried by molecular sieves, but silica gel, calcium oxide and anhydrous potassium carbonate are effective.

Protection against eye splashes and against absorption via cuts and other breaks in the skin is required. Methanol burns with a non-luminous flame and in fire fighting great care must be taken not to be trapped by an unnoticed fire. Normal foam is not effective in fighting methanol fires because it causes the foam to collapse. If methanol is stored on a site, supplies of a special foam are needed.

A mixture of methanol and ethanol forms a useful test mixture for fractionating columns provided that freedom from water can be guaranteed.

ETHANOL

1. **Names**
   Ethanol*
   Ethyl alcohol
   Grain alcohol

2. **Physical properties**

Molecular weight	46
Empirical formula	C_2H_6O
Boiling point (°C)	78
Freezing point (°C)	-114
Specific gravity (20/4 °C)	0.789
Liquid expansion coefficient (per °C)	0.0011
Surface tension (at 20 °C in dyn/cm)	22.3
Absolute viscosity (at 25 °C in cP)	1.08

3. **Fire and health hazards**

Flash point (closed cup) (°C)	13
Autoignition temperature (°C)	423
Lower explosive limit (ppm)	33 000
Upper explosive limit (ppm)	190 000
IDLH (ppm)	>20 000
Odour threshold (ppm)	4400†
TWA–TLV (ppm)	1000
Saturated vapour concentration (70 °F in ppm)	60 000
Vapour density (relative to air)	1.59

4. **Solvent properties**

Hildebrand solubility parameter	13.4
Dipole moment (D)	1.7
Dielectric constant (20 °C)	24.3
Evaporation time (diethyl ether = 1.0)	6.3

5. **Aqueous effluent characteristics**

Solubility in water (at 25 °C in % w/w)	Total
Solubility of water in (at 25 °C in % w/w)	Total
Biological oxygen demand (w/w)	1.25
Chemical oxygen demand (w/w)	2.0
Theoretical oxygen demand (w/w)	2.09

6. **Handling details**

Hazchem number	1170
Hazard class	2 SE
Vapour pressure (at 21 °C in mmHg)	45.7
Loss per transfer (% of liquid transferred)	0.014

7. **Vapour pressure constants (mmHg, log base 10)**

Antoine equation	$A = 8.11220$
	$B = 1592.864$
	$C = 226.184$
Cox chart	$A = 8.24183$
	$B = 1651.2$

8. **Thermal information**

Latent heat (cal/mol)	9200
Specific heat (cal/mol/°C)	27.14
Heat of combustion (cal/mol)	295 550

*Much ethanol in industrial use is in the form of industrial methylated spirits (IMS) which has been denatured to make it unpleasant to drink.

†IMS usually has a much lower odour threshold because of the denaturants it contains.

Azeotropes and activity coefficients of component X: ethanol

Component Y	Azeotrope °C	Azeotrope % w/w X	α	A_{XY} $=\ln \gamma_X^\infty$	A_{YX} $=\ln \gamma_Y^\infty$	Source[11] Vol.	Page
Hydrocarbons							
n-Pentane	34	5	>20[a]	2.53	1.87	2(c)	375
n-Hexane	59	21	5[a]	2.41	1.90	2(c)	453
n-Heptane	98	72	10[a]	2.43	2.35	2(a)	488
Cyclohexane	65	31	3[a]	1.80	2.27	2(a)	430
Benzene	68	32	2.5[a]	1.80	1.40	2(a)	399
Toluene	77	63	10[a]	1.65	1.57	2(a)	470
Ethylbenzene	None		20	1.44	1.94	2(a)	499
Xylenes	None		20	1.54	1.85	2(a)	500
Alcohols							
Methanol	None		1.7	0.02	0.04	2(a)	60
n-Propanol	None		2.4	0.23	0.21	2(a)	336
Isopropanol	None		<1.5	−0.01	−0.05	2(a)	341
n-Butanol	None		4	0.01	0.03	2(a)	365
Isobutanol	None		3	0.06	0.01	2(a)	376
sec-Butanol	None		2.4	0.18	0.05	2(a)	366
Cyclohexanol	None		20	0.16	1.06	2(c)	421
Ethylene glycol	None		>20	0.72	1.88	2(c)	297
Methyl Cellosolve	None						
Ethyl Cellosolve	None					2(c)	350
Butyl Cellosolve	None						
Chlorinated hydrocarbons							
Methylene dichloride	40	2	9[a]	3.35	0.46	2(a)	293
Chloroform	59	7	3[a]	1.48	0.69	2(a)	285
Carbon tetrachloride	65	16	3[a]	2.05	1.62	2(a)	280
1,2-Dichloroethane	70	37	2.5[a]	1.60	1.28	2(a)	299
1,1,1-Trichloroethane							
Trichloroethylene	71	28	3[a]	2.02	1.57	2(a)	295
Perchloroethylene	77	63	15[a]	1.79	1.80	2(c)	285
Monochlorobenzene	None		20	1.53	1.75	2(a)	397
Ketones							
Acetone	None		2.8	0.58	0.62	2(a)	320
Methyl ethyl ketone	74	39	<1.5[a]	0.57	0.68	2(a)	342
MIBK	None		5	1.00	0.74	2(c)	423
Cyclohexanone							
N-Methylpyrrolidone				−0.41			
Ethers							
Diethyl ether	None		15	1.25	0.91	2(a)	374
Diisopropyl ether	64	17	3[a]	1.35	1.39	2(a)	458
1,4-Dioxane	78	91	3[a]	0.65	0.81	2(a)	348
Tetrahydrofuran	66	10	2[a]	0.66	0.43	2(c)	328
Esters							
Methyl acetate	57	3	2.8[a]	0.62	0.62	2(a)	330
Ethyl acetate	72	31	1.8[a]	0.85	0.78	2(a)	352
Butyl acetate	None		10	0.71	1.10	2(c)	426
Miscellaneous							
Dimethylformamide				−0.51	−0.32	2(c)	371
Dimethyl sulphoxide							
Pyridine	None		3	0.05	−0.05	2(c)	355
Acetonitrile	73	56	1.5[a]	1.39	0.69	2(a)	298
Furfuraldehyde	None		>20	1.37	1.66	2(a)	385
Water	78	96	4[a]	1.77	0.96	1	181

[a]The relative volatility between the less volatile component of the binary mixture and the azeotrope.

Recovery notes: ethanol

Ethanol is stable at its boiling point. It is easy to strip from water, but forms a water azeotrope that prevents it being dried by ordinary fractionation. The relative volatility of the azeotrope from dry ethanol (i.e. above the azeotrope) is very low so that it is difficult to make very dry ethanol by redistilling, say, 1% water content material.

Ethanol can be dried by a number of routes:

- Azeotropic distillation by forming a ternary azeotrope with an added entrainer (Table 7.4)
- Extractive distillation. Suitable entrainers include MEG, glycerine and ethyl Cellosolve
- Pervaporation. The ethanol–water system is very well tried for this duty
- Molecular sieves. Particularly suitable if very low water content is needed but methanol, present as a denaturant, may also be adsorbed
- Distillation in the presence of a salt (Figure A.3). The relative volatility of ethanol

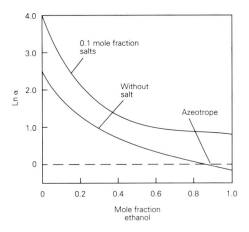

Fig. A.3 Effect of salt in solution on ethanol–water system

from water is radically altered by the presence of salts in the distillation column. Sodium and potassium acetate, or a mixture of the two, can raise the relative volatility of the ethanol–water by a factor of 2 over the whole composition range. This eliminates the azeotrope and takes the value of α up to about 50 at the base of the column. It is, however, a difficult process to operate as the melting point of the mixed salts is about 250 °C. A large number of plants using this technology were built in the 1930s

- Extraction from water using a solvent.

A number of solvents including high-boiling alcohols and ketones and hydrocarbons will extract ethanol from water with enough selectivity that the ethanol stripped from the solvent will be drier than the ethanol–water azeotrope. Losses of solvent to the water phase are unacceptable, so the effluent water needs to be stripped. If the recovered ethanol is required at a water content of 1–2% this is a method worth considering.

Alumina is not suitable for drying ethanol, as some decomposition may occur during bed regeneration, to yield ethylene.

It should be noted that the presence of 4% methanol in the ethanol, the most common of the denaturants, can interfere with water removal and other processing of IMS. There are very many different ways of denaturing, and if methanol should be undesirable it is likely that an alternative can be found.

Customs and Excise control of ethanol will usually require that 'methylating,' or restoring the denaturant to IMS if it should be necessary after recovery, is done under licence and the granting of licences is strictly conditional on having the right equipment and security.

Ethanol is one of the least toxic of all solvents, although careful supervision of operators is necessary to guard against alcohol abuse.

n-PROPANOL

1 Names
    n-Propanol
    Propan-1-ol
    n-Propyl alcohol

2 Physical properties

Molecular weight	60
Empirical formula	C_7H_8O
Boiling point (°C)	97
Freezing point (°C)	−126
Specific gravity (20/4 °C)	0.804
Liquid expansion coefficient (per °C)	0.000 96
Surface tension (at 20 °C in dyn/cm)	23.7
Absolute viscosity (at 25 °C in cP)	1.722

3 Fire and health hazards

Flash point (closed cup) (°C)	25
Autoignition temperature (°C)	440
Lower explosive limit (ppm)	21 000
Upper explosive limit (ppm)	135 000
IDLH (ppm)	4000
Odour threshold (ppm)	30
TWA–TLV (ppm)	200
Saturated vapour concentration (70 °F in ppm)	18 000
Vapour density (relative to air)	2.08

4 Solvent properties

Hildebrand solubility parameter	11.9
Dipole moment (D)	1.7
Dielectric constant (25°C)	20.1
Evaporation time (diethyl ether = 1.0)	9.0

5 Aqueous effluent characteristics

Solubility in water (at 25 °C in % w/w)	Total
Solubility of water in (at 25 °C in % w/w)	Total
Biological oxygen demand (w/w)	1.5
Chemical oxygen demand (w/w)	–
Theoretical oxygen demand (w/w)	2.40

6 Handling details

Hazchem number	1274
Hazard class	2 SE
Vapour pressure (at 21 °C in mmHg)	13.4
Loss per transfer (% of liquid transferred)	0.001

7 Vapour pressure constants (mmHg, log base 10)

Antoine equation	$A = 8.37895$
	$B = 1788.020$
	$C = 227.438$
Cox chart	$A = 8.25022$
	$B = 1755.8$

8 Thermal information

Latent heat (cal/mol)	9780
Specific heat (cal/mol/°C)	34.2
Heat of combustion (cal/mol)	437 760

Azeotropes and activity coefficients of component X: *n*-propanol

Component Y	Azeotrope °C	Azeotrope % w/w X	α	$A_{XY}=\ln \gamma_X^\infty$	$A_{YX}=\ln \gamma_Y^\infty$	Source[11] Vol.	Source[11] Page
Hydrocarbons							
n-Pentane	None						
n-Hexane	66	4	12[a]	1.54	1.88	2(a)	585
n-Heptane	88	36	2.5[a]	1.84	1.63	2(a)	595
Cyclohexane	74	20	5[a]	3.00	1.32	2(a)	579
Benzene	77	17	4[a]	2.07	1.27	2(a)	556
Toluene	93	49	3[a]	1.46	1.19	2(a)	592
Ethylbenzene	97	91	8[a]	1.24	1.33	2(a)	601
Xylenes	97	93	20[a]	1.26	1.39	2(c)	575
Alcohols							
Methanol	None		5	0.24	0.06	2(a)	122
Ethanol	None		2.4	0.21	0.23	2(a)	336
Isopropanol	None		2	−0.06	0.13	2(a)	531
n-Butanol	None		2.3	−0.03	0.02	2(a)	539
Isobutanol	None		1.5	0.37	0.19	2(a)	540
sec-Butanol							
Cyclohexanol							
Ethylene glycol							
Methyl Cellosolve	None		2.5	0.50	0.23	2(c)	490
Ethyl Cellosolve							
Butyl Cellosolve							
Chlorinated hydrocarbons							
Methylene dichloride	None						
Chloroform	None						
Carbon tetrachloride	73	8	5[a]	2.00	1.14	2(a)	514
1,2-Dichloroethane	81	19	2.8[a]	1.30	1.06	2(a)	520
1,1,1-Trichloroethane	73	7		1.78	2.17		
Trichloroethylene	82	17	2.5[a]	1.70	0.97	2(a)	518
Perchloroethylene	94	48	5[a]	1.73	1.57	2(a)	517
Monochlorobenzene	97	83	5[a]	1.13	1.26	2(a)	552
Ketones							
Acetone							
Methyl ethyl ketone	None		2.2	0.49	0.48	2(c)	496
MIBK	94	35					
Cyclohexanone							
N-Methylpyrrolidone							
Ethers							
Diethyl ether							
Diisopropyl ether	None		5	1.25	1.27	2(a)	586
1,4-Dioxane	95	55	1.5[a]	0.55	0.59	2(a)	533
Tetrahydrofuran	None		3	0.32	0.19	2(c)	497
Esters							
Methyl acetate	None		6	1.00	1.22	2(a)	530
Ethyl acetate	None		2.8	0.65	0.52	2(a)	536
Butyl acetate	94	40					
Miscellaneous							
Dimethylformamide							
Dimethyl sulphoxide							
Pyridine	None		1.5	−0.13	−0.24	2(c)	512
Acetonitrile	81	28					
Furfuraldehyde							
Water	87	71		2.74	1.16	1	286

[a]The relative volatility between the less volatile component of the binary mixture and the azeotrope.

Recovery notes: *n*-propanol

n-Propanol is the highest boiling monohydric alcohol which is miscible with water in all proportions. Comparison with its homologues shows that it has considerable hydrophobicity; $\log P$ values are as follows:

Methanol	−0.64
Ethanol	−0.31
Isopropanol	+0.05
n-Propanol	+0.25
tert-Butanol	+0.37
sec-Butanol	+0.61
Isobutanol	+0.76
n-Butanol	+0.88

Methanol and ethanol partition between water and hydrocarbons strongly in favour of the former. Normal and isobutanol make two phases in contact with water over most of the concentration range. However, *n*-propanol occupies an intermediate position.

While it is fully miscible with water in all proportions, the partition coefficient in ternary mixtures of *n*-propanol with water and a possible entrainer such as cyclohexane or benzene is favourable to the hydrocarbon phase (Table A.2). The lower values of partition coefficient are for those systems in which it is more likely to be economically advantageous to recycle part of the potential reflux to a phase separation with the feed.

Table A.2 Partition coefficients (% w/w) between water and entrainer for ternary azeotropes of alcohol, water and entrainer (all at 20/25 °C except those marked)

| | Entrainer | | | |
Alcohol	DIPE	Benzene	Cyclohexane	Overall average
Ethanol	3.4	4.1	25.6	9.8
Isopropanol	2.4	0.7	2.4	1.6
n-Propanol	1.5	0.8 (45 °C)	1.7 (35 °C)	1.3
sec-Butanol	—	0.6	—	0.4

The alcohol to be dehydrated may have any concentration of water in it. If it has been recovered by water washing or carbon bed adsorption from air or by steam stripping from water, the alcohol to be treated will usually be of azeotropic composition or wetter. The high water content of the *n*-propanol–water azeotrope (29% w/w) make it attractive to use an entrainer that does double duty as a liquid–liquid extraction agent.

It is also worth considering multi-stage counter-current extraction rather then merely a single stage of mixing and separation, in which case it is important to ascertain whether or not a system contains a solutrope. This corresponds to an azeotrope in distillation and presents a barrier to advance in a liquid–liquid extraction. It can be spotted from a change of slope from positive to negative on the tie lines.

The alcohols isopropanol, *tert*-butanol and *n*-propanol all display this phenomenon with some potential azeotropic entrainers (Figure A.4). The barrier of about 17% *n*-propanol in water represents only a modest improvement over what can be achieved by

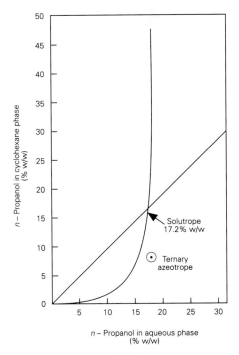

Fig. A.4 *n*-Propanol–water–*n*-hexane solutrope at 35 °C

a single stage and would indicate that, using cyclohexane as an entrainer, a multi-stage contacting column is probably not justifiable.

Potential entrainers for the *n*-propanol–water system are shown in Table A.3.

Table A.3 Ternary azeotropes containing *n*-propanol and water which form two liquid phases on condensing

Entrainer	Entrainer (% w/w)	*n*-Propanol (% w/w)	Water (% w/w)
Benzene[a]	82.3	10.1	7.6
Cyclohexane[a]	81.5	10.0	8.5
Diisobutene	59.1	31.6	9.3
Diisopropyl ether	92.4	2	5.6
Trichloroethylene	81	12	7
Carbon tetrachloride	84	11	5
n-Propyl acetate	59.5	19.5	21
(also reported as)	73	10	17

[a] Form solutropes.

Although propyl acetate carries the most water in the ternary azeotrope, the organic phase on an entrainer-free basis is very wet (Figure A.5) compared with that for cyclohexane (Figure A.6) and *n*-propyl acetate is not wholly stable in the presence of water.

The optimum choice of the entrainer, taking account of toxicity, stability and freedom of problems with peroxides, is probably cyclohexane.

n-Propanol presents no problem in being dried by pervaporation, but the water separated from the organic phase may contain too much *n*-propanol to allow for its

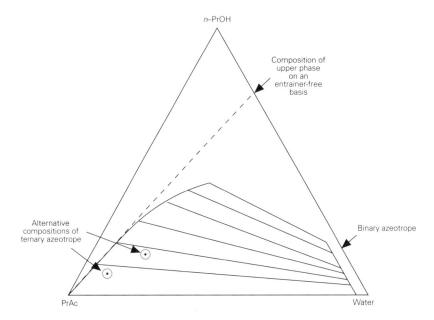

Fig. A.5 Ternary solubility diagram for *n*-propanol–water–*n*-propyl acetate (% w/w, 30 °C)

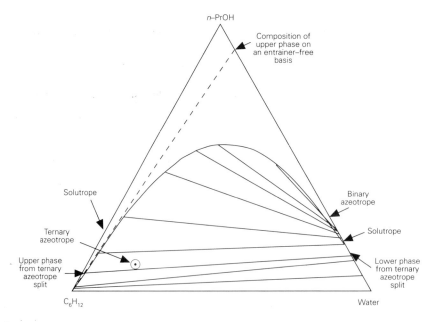

Fig. A.6 Ternary solubility diagram for *n*-propanol–water–cyclohexane (% w/w, 35 °C)

disposal without stripping. As Figure A.7 shows, it is easy to strip *n*-propanol from water and a simple steam stripping without a fractionating column will yield a distillate close to the azeotrope in composition.

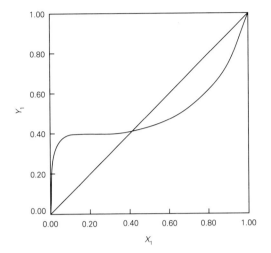

Fig. A.7 VLE diagram for *n*-propanol (1)–water (2) at 760 mmHg

240 *Appendix 1*

ISOPROPANOL

1. **Names**
   Isopropanol
   Propan-2-ol
   Isopropyl alcohol
   IPA*

2. **Physical properties**

Molecular weight	60
Empirical formula	C_3H_8O
Boiling point (°C)	82
Freezing point (°C)	−88
Specific gravity (20/4 °C)	0.786
Liquid expansion coefficient (per °C)	0.00105
Surface tension (at 20 °C in dyn/cm)	22
Absolute viscosity (at 25 °C in cP)	2.0

3. **Fire and health hazards**

Flash point (closed cup) (°C)	12
Autoignition temperature (°C)	456
Lower explosive limit (ppm)	20000
Upper explosive limit (ppm)	120000
IDLH (ppm)	20000
Odour threshold (ppm)	90
TWA–TLV (ppm)	400
Saturated vapour concentration (70 °F in ppm)	46000
Vapour density (relative to air)	2.07

4. **Solvent properties**

Hildebrand solubility parameter	11.5
Dipole moment (D)	1.66
Dielectric constant (20 °C)	18.3
Evaporation time (diethyl ether = 1.0)	7.4

5. **Aqueous effluent characteristics**

Solubility in water (at 25 °C in % w/w)	Total
Solubility of water in (at 25 °C in % w/w)	Total
Biological oxygen demand (w/w)	1.59
Chemical oxygen demand (w/w)	2.3
Theoretical oxygen demand (w/w)	2.40

6. **Handling details**

Hazchem number	1.219
Hazard class	2 SE
Vapour pressure (at 21 °C in mmHg)	35.1
Loss per transfer (% of liquid transferred)	0.015

7. **Vapour pressure constants (mmHg, log base 10)**

Antoine equation	$A = 8.87829$
	$B = 2010.33$
	$C = 252.636$
Cox chart	$A = 8.24362$
	$B = 1673.2$

8. **Thermal information**

Latent heat (cal/mol)	9540
Specific heat (cal/mol/°C)	36.6
Heat of combustion (cal/mol)	432060

*If the possibility of confusion with isopropyl acetate exists, this should not be used.

Azeotropes and activity coefficients of component X: isopropanol

Component Y	Azeotrope °C	Azeotrope % w/w X	α	A_{XY} $=\ln \gamma_X^\infty$	A_{YX} $=\ln \gamma_Y^\infty$	Source[11] Vol.	Page
Hydrocarbons							
n-Pentane	35	6					
n-Hexane	63	23	5[a]	2.10	1.55	2(b)	99
n-Heptane	76	51	5[a]	2.42	1.71	2(b)	113
Cyclohexane	69	32	3[a]	2.39	1.39	2(b)	84
Benzene	72	33	3[a]	1.94	1.32	2(b)	67
Toluene	81	69	6[a]	1.34	1.14	2(b)	108
Ethylbenzene	None			0.71	1.67		
Xylenes	None						
Alcohols							
Methanol	None		2	−0.10	−0.12	2(a)	123
Ethanol	None		<1.5	−0.05	−0.01	2(a)	341
n-Propanol	None		2	0.13	−0.06	2(a)	531
n-Butanol	None		2	−0.08	0.17		
Isobutanol	None		2	0.08	0.08	2(b)	63
sec-Butanol	None			0.07	0.08	2(b)	62
Cyclohexanol	None						
Ethylene glycol	None						
Methyl Cellosolve							
Ethyl Cellosolve							
Butyl Cellosolve							
Chlorinated hydrocarbons							
Methylene dichloride	None						
Chloroform	61	4	3[a]				
Carbon tetrachloride	69	18	3[a]	1.89	1.16	2(b)	40
1,2-Dichloroethane	75	43					
1,1,1-Trichloroethane							
Trichloroethylene	75	30		1.15	1.39		
Perchloroethylene	82	70		1.75	1.68	2(b)	42
Monochlorobenzene	None			1.23	1.56		
Ketones							
Acetone	None		3	0.84	0.89	2(b)	44
Methyl ethyl ketone	78	32	1.5[a]	0.45	0.38	2(b)	54
MIBK	None		4	0.38	0.54	2(b)	90
Cyclohexanone							
N-Methylpyrrolidone				−0.28			
Ethers							
Diethyl ether			6/7				
Diisopropyl ether	66	15	3[a]	1.43	1.06	2(b)	101
1,4-Dioxane	None		2/3	0.25	0.58	2(b)	56
Tetrahydrofuran	None		2	0.33	0.33	2(b)	55
Esters							
Methyl acetate	None		5	0.77	0.93	2(b)	50
Ethyl acetate	75	25	1.5[a]	0.89	0.87	2(b)	59
Butyl acetate	80	52					
Miscellaneous							
Dimethylformamide	None			0.12	0.16		
Dimethyl sulphoxide				0.11	−0.07		
Pyridine							
Acetonitrile	75	48					
Furfuraldehyde							
Water	80	88	5[a]	2.47	1.09	1	330

[a] The relative volatility between the less volatile component of the binary mixture and the azeotrope.

Recovery notes: isopropanol

Isopropanol (IPA) behaves in a very similar way to ethanol, except that the relative volatility of its water azeotrope from dry IPA is sufficient to allow very dry material to be made from a feed with a water content below that of the azeotrope (Figure A.8; compare with Figure A.7 for *n*-propanol).

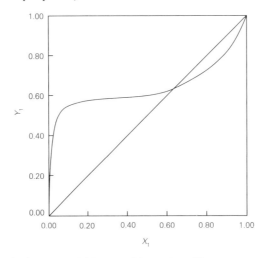

Fig. A.8 VLE diagram for isopropanol (1)–water (2) at 760 mmHg

The methods of drying IPA are identical with those for ethanol. IPA has a low toxicity and does not have an unpleasant smell. It contains neither inhibitors nor denaturants.

In France, its recovery is under the control of the Excise authorities but there seems to be no serious risk of pilfering for potable use.

IPA, along with other low-boiling alcohols, causes standard fire-fighting foam to collapse and special alcohol-fire foam in bulk and in portable extinguishers should be available where IPA is stored or handled.

n-BUTANOL

1 **Names**

    n-Butanol
    Butyl alcohol
    Butan-1-ol

2 **Physical properties**

Molecular weight	74
Empirical formula	$C_4H_{10}O$
Boiling point (°C)	118
Freezing point (°C)	-89
Specific gravity (20/4 °C)	0.810
Liquid expansion coefficient (per °C)	0.00094
Surface tension (at 20 °C in dyn/cm)	24.6
Absolute viscosity (at 25 °C in cP)	3.0

3 **Fire and health hazards**

Flash point (closed cup) (°C)	35
Autoignition temperature (°C)	367
Lower explosive limit (ppm)	14 000
Upper explosive limit (ppm)	112 000
IDLH (ppm)	8000
Odour threshold (ppm)	2.5
TWA–TLV (ppm)	50
Saturated vapour concentration (70 °F in ppm)	6300
Vapour density (relative to air)	2.55

4 **Solvent properties**

Hildebrand solubility parameter	11.4
Dipole moment (D)	1.66
Dielectric constant (20 °C)	17.1
Evaporation time (diethyl ether = 1.0)	20.5

5 **Aqueous effluent characteristics**

Solubility in water (at 25 °C in % w/w)	7.3
Solubility of water in (at 25 °C in % w/w)	20.4
Biological oxygen demand (w/w)	1.15
Chemical oxygen demand (w/w)	1.72
Theoretical oxygen demand (w/w)	2.21

6 **Handling details**

Hazchem number	1120
Hazard class	3 Y
Vapour pressure (at 21 °C in mmHg)	4.8
Loss per transfer (% of liquid transferred)	

7 **Vapour pressure constants (mmHg, log base 10)**

Antoine equation	$A = 7.83800$
	$B = 1558.19$
	$C = 196.881$
Cox chart	$A = 8.25925$
	$B = 1871.7$

8 **Thermal information**

Latent heat (cal/mol)	10 434
Specific heat (cal/mol/°C)	41.0
Heat of combustion (cal/mol)	585 044

Azeotropes and activity coefficients of component X: *n*-butanol

Component Y	Azeotrope °C	Azeotrope % w/w X	α	$A_{XY} = \ln \gamma_X^\infty$	$A_{YX} = \ln \gamma_Y^\infty$	Source[11] Vol.	Source[11] Page
Hydrocarbons							
n-Pentane	None		>10	2.31	1.44	2(b)	169
n-Hexane	68	3	10[a]	2.44	1.07	2(b)	202
n-Heptane	94	18	3[a]	2.53	0.85	2(b)	219
Cyclohexane	80	4	9[a]	2.27	1.13	2(b)	188
Benzene	None		7	1.45	0.79	2(b)	179
Toluene	105	28	3[a]	1.25	0.83	2(b)	207
Ethylbenzene	115	67	4[a]	0.99	0.99	2(b)	228
Xylenes	115	73	4[a]	1.11	1.00	2(b)	229
Alcohols							
Methanol	None		10	0.38	0.15	2(a)	169
Ethanol	None		4	0.03	0.01	2(a)	365
n-Propanol	None		2.3	0.02	−0.02	2(a)	539
Isopropanol	None		2	0.17	−0.08	2(d)	55
Isobutanol	None		1.5	0.00	−0.08	2(b)	161
sec-Butanol	None		2	−0.11	−0.04	2(b)	154
Cyclohexanol	None		5	−0.07	0.02	2(b)	193
Ethylene glycol	None		>20	1.89	0.94	2(d)	6
Methyl Cellosolve	None						
Ethyl Cellosolve	None						
Butyl Cellosolve	None		4			2(d)	200
Chlorinated hydrocarbons							
Methylene dichloride	None						
Chloroform	None		10	0.98	0.38	2(b)	136
Carbon tetrachloride	77	3	10[a]	2.12	0.89	2(b)	135
1,2-Dichloroethane	None		5/10	1.20	0.84	2(b)	137
1,1,1-Trichloroethane							
Trichloroethylene	87	3					
Perchloroethylene	109	30	2[a]	1.14	1.33	2(d)	154
Monochlorobenzene	115	56	3[a]	0.93	0.89	2(b)	174
Ketones							
Acetone	None		10	0.49	0.21	2(b)	140
Methyl ethyl ketone	None		5	0.60	0.28	2(b)	144
MIBK	114	30	<1.5[a]	1.07	0.53	2(d)	196
Cyclohexanone							
N-Methylpyrrolidone							
Ethers							
Diethyl ether							
Diisopropyl ether	None		10	1.22	0.85	2(b)	203
1,4-Dioxane	None		2	0.41	0.19	2(b)	147
Tetrahydrofuran	None		6/7	0.15	0.21	2(b)	146
Esters							
Methyl acetate							
Ethyl acetate	None		7	0.82	0.59	2(b)	148
Butyl acetate	116	63	1.5[a]	0.43	0.81	2(b)	197
Miscellaneous							
Dimethylformamide				−0.24	−0.32		
Dimethyl sulphoxide	None		6	0.04	−1.18	2(d)	163
Pyridine	119	70	<1.5[a]	−0.12	−0.52	2(b)	166
Acetonitrile	None		6	1.35	1.29	2(d)	156
Furfuraldehyde							
Water	93	58		3.96	1.14	1	406

[a]The relative volatility between the less volatile component of the binary mixture and the azeotrope.

Recovery notes: *n*-butanol

n-Butanol is stable at its boiling point and requires no denaturants or inhibitors. It does attack aluminium when it is hot but all other normal materials of construction are suitable for use in conjunction with *n*-butanol.

It is relatively non-toxic in that its IDLH is above its vapour concentration at 20 °C and it has a strong enough odour to alert most people to its presence at concentrations well below its TLV.

n-Butanol has somewhat unusual qualities as far as its water solubility is concerned in that it is less soluble in hot than cold water (Table A.4). Because water and butanol are not soluble in all proportions, unlike the lower alcohols, they can be separated by distillation without the use of an azeotropic entrainer (Figure 7.1).

Table A.4 *n*-Butanol–water solubilities at different temperatures

Temperature (°C)	BuOH in water (% w/w)	Water in BuOH (% w/w)
10	8.7	19.4
20	7.6	19.7
30	7.0	20.1
40	6.6	21.5
50	6.3	22.4

Feed should be added to the system at the correct position according to its composition:

Butanol column	Water	0–20% w/w
Water column	Water	92–100% w/w
Decanter	Water	20–92% w/w

The VLE diagram for the butanol–water system and the high values of γ^∞ for both butanol in water and water in butanol show that the fractionating approach to the azeotrope is very easy from both directions so that the columns required for the continuous separation need only a few plates.

Because the split of the condensate from the tops of both columns is, if anything, improved by being done at a high temperature, no cooler is needed between the condenser and the decanter. The density difference between the butanol-rich phase at 0.85 and the water-rich phase at 0.99 is large enough to allow operations with a modest-sized decanter.

Certain impurities, particularly the lower alcohols, have a very marked adverse effect on the phase split and, since such materials will concentrate at the column tops with no means of escape from the system, it is important that butanol for drying does not contain them.

For small quantities of wet butanol to be dried, batchwise, the choice lies between sending water with about 7% of butanol to waste, storing it for recycling or using an entrainer in a similar way to that used for drying ethanol or isopropanol. Reference to the table of butanol's azeotropes shows that a number of possible entrainers do not form binary azeotropes with it and, of these, diisopropyl ether is probably the best choice since both benzene and chloroform introduce toxicity hazards. Hence if there is not sufficient water available requiring stripping off butanol to make recycling attractive, a conventional azeotropic distillation can be used.

ISOBUTANOL

1. **Names**

    Isobutanol
    2-methylpropan-1-ol
    Isobutyl alcohol
    IBA

2. **Physical properties**

Molecular weight	74
Empirical formula	$C_4H_{10}O$
Boiling point (°C)	108
Freezing point (°C)	−108
Specific gravity (20/4 °C)	0.807
Liquid expansion coefficient (per °C)	0.00095
Surface tension (at 20 °C in dyn/cm)	22.8
Absolute viscosity (at 25 °C in cP)	3.96

3. **Fire and health hazards**

Flash point (closed cup) (°C)	25
Autoignition temperature (°C)	430
Lower explosive limit (ppm)	16000
Upper explosive limit (ppm)	109000
IDLH (ppm)	8000
Odour threshold (ppm)	1
TWA–TLV (ppm)	50
Saturated vapour concentration (70 °F in ppm)	11500
Vapour density (relative to air)	2.56

4. **Solvent properties**

Hildebrand solubility parameter	10.7
Dipole moment (D)	1.7
Dielectric constant (20 °C)	17.7
Evaporation time (diethyl ether = 1.0)	17.3

5. **Aqueous effluent characteristics**

Solubility in water (at 25 °C in % w/w)	8.7
Solubility of water in (at 25 °C in % w/w)	15.0
Biological oxygen demand (w/w)	1.62
Chemical oxygen demand (w/w)	2.6
Theoretical oxygen demand (w/w)	2.59

6. **Handling details**

Hazchem number	1120
Hazard class	3 Y
Vapour pressure (at 21 °C in mmHg)	8.6
Loss per transfer (% of liquid transferred)	0.0043

7. **Vapour pressure constants (mmHg, log base 10)**

Antoine equation	$A = 8.53516$
	$B = 1950.940$
	$C = 237.147$
Cox chart	$A = 8.25506$
	$B = 1816.5$

8. **Thermal information**

Latent heat (cal/mol)	10220
Specific heat (cal/mol/°C)	53.0
Heat of combustion (cal/mol)	584600

Azeotropes and activity coefficients of component X: isobutanol

Component Y	Azeotrope °C	Azeotrope % w/w X	α	A_{XY} =ln γ_X^∞	A_{YX} =ln γ_Y^∞	Source[11] Vol.	Page
Hydrocarbons							
n-Pentane	None						
n-Hexane	68	2	11[a]	2.25	1.12	2(d)	370
n-Heptane	91	27	5[a]	2.21	1.50	2(d)	378
Cyclohexane	78	14	5[a]	2.00	1.16	2(b)	288
Benzene	79	8	5[a]	1.67	0.87	2(b)	287
Toluene	101	45	2[a]	1.31	0.98	2(b)	289
Ethylbenzene	107	80					
Xylenes	108	88	5[a]	0.97	1.03	2(b)	292
Alcohols							
Methanol	None		6	0.22	0.21	2(a)	171
Ethanol	None		3	0.01	0.06	2(a)	376
n-Propanol	None		1.5	0.19	0.37	2(a)	540
Isopropanol	None		2	0.08	0.08	2(b)	63
n-Butanol	None		1.5	−0.08	0.00	2(b)	161
sec-Butanol	None		1.5	−0.23	−0.20	2(b)	243
Cyclohexanol							
Ethylene glycol							
Methyl Cellosolve	None						
Ethyl Cellosolve	None						
Butyl Cellosolve							
Chlorinated hydrocarbons							
Methylene dichloride	None						
Chloroform	None						
Carbon tetrachloride	76	5	5[a]	1.83	0.90	2(d)	342
1,2-Dichloroethane	84	16	5[a]	1.57	0.83	2(b)	272
1,1,1-Trichloroethane							
Trichloroethylene	85	9	3.5[a]	1.35	0.91	2(d)	345
Perchloroethylene	103	40	2.5[a]	1.30	1.22	2(d)	344
Monochlorobenzene	107	63	3[a]	1.02	0.93	2(d)	357
Ketones							
Acetone	None		10	0.43	0.34	2(d)	348
Methyl ethyl ketone	None						
MIBK	108	91					
Cyclohexanone							
N-Methylpyrrolidone							
Ethers							
Diethyl ether							
Diisopropyl ether							
1,4-Dioxane	101	4	<1.5[a]	0.35	0.18	2(b)	279
Tetrahydrofuran							
Esters							
Methyl acetate	None						
Ethyl acetate	None						
Butyl acetate	None						
Miscellaneous							
Dimethylformamide							
Dimethylsulphoxide	None		10	−0.33	−1.41	2(b)	275
Pyridine	None		1.7	−0.47	−0.21	2(d)	350
Acetonitrile	None		5	0.98	0.44	2(d)	346
Furfuraldehyde							
Water	90	67		3.71	1.35	1	439

[a] The relative volatility between the less volatile component of the binary mixture and the azeotrope.

Recovery notes: isobutanol

Comparison of the VLE diagram of isobutanol with that of *n*-butanol (see Figure A.10) shows that it is less easy to strip water azeotrope from dry isobutanol, but in most other respects they are similar and reference should be made to the notes on *n*-butanol recovery (*see sec*-butanol).

sec-BUTANOL

1 Names

sec-Butanol
Butan-2-ol
2-Hydroxybutane
Methyl ethyl carbinol

2 Physical properties

Molecular weight	74
Empirical formula	$C_4H_{10}O$
Boiling point (°C)	99
Freezing point (°C)	-115
Specific gravity (20/4 °C)	0.807
Liquid expansion coefficient (per °C)	0.00091
Surface tension (at 20 °C in dyn/cm)	23.0
Absolute viscosity (at 25 °C in cP)	3.7

3 Fire and health hazards

Flash point (closed cup) (°C)	21
Autoignition temperature (°C)	405
Lower explosive limit (ppm)	17 000
Upper explosive limit (ppm)	98 000
IDLH (ppm)	10 000
Odour threshold (ppm)	75
TWA–TLV (ppm)	100
Saturated vapour concentration (70 °F in ppm)	17 600
Vapour density (relative to air)	2.56

4 Solvent properties

Hildebrand solubility parameter	10.8
Dipole moment (D)	1.7
Dielectric constant (20 °C)	
Evaporation time (diethyl ether = 1.0)	13.0

5 Aqueous effluent characteristics

Solubility in water (at 25 °C in % w/w)	19.8
Solubility of water in (at 25 °C in % w/w)	65.1
Biological oxygen demand (w/w)	1.87
Chemical oxygen demand (w/w)	2.5
Theoretical oxygen demand (w/w)	2.59

6 Handling details

Hazchem number	1120
Hazard class	3 Y
Vapour pressure (at 21 °C in mmHg)	13.2
Loss per transfer (% of liquid transferred)	

7 Vapour pressure constants (mmHg, log base 10)

Antoine equation	$A = 7.47429$
	$B = 1314.188$
	$C = 186.500$
Cox chart	$A = 8.25102$
	$B = 1766.8$

8 Thermal information

Latent heat (cal/mol)	9916
Specific heat (cal/mol/°C)	40.00
Heat of combustion (cal/mol)	636 400

Azeotropes and activity coefficients of component X: *sec*-butanol

Component Y	Azeotrope °C	Azeotrope % w/w X	α	A_{XY} $=\ln \gamma_X^\infty$	A_{YX} $=\ln \gamma_Y^\infty$	Source[11] Vol.	Page
Hydrocarbons							
n-Pentane	None						
n-Hexane	67	8	8[a]	1.96	1.28	2(b)	250
n-Heptane	88	37	2.5[a]	1.89	1.34	2(d)	281
Cyclohexane	76	18					
Benzene	79	15	5[a]	0.90	0.84	2(b)	248
Toluene	95	55	2[a]	1.19	0.82	2(b)	277
Ethylbenzene			5				
Xylenes	None		5	1.04	0.97	2(d)	282
Alcohols							
Methanol	None		10	−0.06	−0.38	2(c)	128
Ethanol	None		2.4	0.05	0.18	2(a)	366
n-Propanol	None						
Isopropanol	None			0.08	0.07	2(b)	62
n-Butanol	None		2	−0.04	−0.11	2(b)	154
Isobutanol	None		1.5	−0.20	−0.23	2(b)	243
Cyclohexanol							
Ethylene glycol							
Methyl Cellosolve							
Ethyl Cellosolve							
Butyl Cellosolve							
Chlorinated hydrocarbons							
Methylene dichloride							
Chloroform							
Carbon tetrachloride	74	8		2.72			
1,2-Dichloroethane	88	12		0.96	0.78		
1,1,1-Trichloroethane							
Trichloroethylene	84	15					
Perchloroethylene	97	57	3.5[a]	1.20	1.38	2(d)	240
Monochlorobenzene			5	1.04	0.98	2(d)	258
Ketones							
Acetone							
Methyl ethyl ketone	None		2.2	0.33	0.31	2(d)	249
MIBK							
Cyclohexanone							
N-Methylpyrrolidone							
Ethers							
Diethyl ether							
Diisopropyl ether							
1,4-Dioxane	99	40					
Tetrahydrofuran							
Esters							
Methyl acetate							
Ethyl acetate							
Butyl acetate	None						
Miscellaneous							
Dimethylformamide							
Dimethyl sulphoxide							
Pyridine	89	32	<1.5[a]	−0.12	−0.43	2(d)	255
Acetonitrile	81	14	2.5[a]	0.99	0.81	2(d)	241
Furfuraldehyde							
Water	87	73		3.56	1.10	1	419

[a] The relative volatility between the less volatile component of the binary mixture and the azeotrope.

Recovery notes: *sec*-butanol

sec-Butanol is stable at its atmospheric pressure boiling point and needs no inhibitors. It has a flash point low enough to mean that it is within its explosive range at workshop temperatures and often at storage temperature, so that inert gas blanketing is advisable.

Although *sec*-butanol has a higher TLV than *n*-butanol, it is not so easy to detect by smell and odour is not to be relied upon as a warning of toxic concentrations.

As can be seen in Figure A.9, the *sec*-butanol azeotrope with water is single phase and

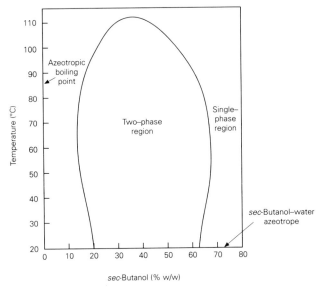

Fig. A.9 Solubility of water in *sec*-butanol

therefore the techniques for drying *n*-butanol and isobutanol do not work for it. While it is therefore more like ethanol and isopropanol, which require the addition of an entrainer to dry them by distillation, *sec*-butanol is not very hydrophilic and liquid–liquid extraction can be linked to distillation in water removal processes.

Table A.5 shows that the amount of water needing to be removed from the water azeotrope is comparatively large.

Table A.5 Water content of single-phase alcohol azeotropes

Alcohol	Water content of azeotrope (% w/w)	Water to be removed per kg of dry solvent (kg)
Ethanol	4.0	0.042
Isopropanol	12.6	0.144
n-Propanol	28.3	0.395
sec-Butanol	26.8	0.366

Table A.6 shows some possible entrainers drying *sec*-butanol. If no recovery of solvent from the aqueous phase from this process is made, then the loss of *sec*-butanol as a percentage of that fed to the system is as follows:

butyl acetate 1.95%
diisobutene 3.8%
benzene 2.1%

Table A.6 Possible entrainers for drying *sec*-butanol

Compound	Phase	Entrainer (% w/w)	Water (% w/w)	sec-Butanol (% w/w)	Vol.% of phases
Butyl acetate	Azeotrope	52	20	28	100
	Top phase	62	6	32	86
	Bottom phase	1	94	5	14
Diisobutene	Azeotrope	70	11	19	100
	Top phase	79	1	20	92
	Bottom phase	0.5	91	9.5	8
Benzene	Azeotrope	86	6	8	100
	Top phase	91	0.4	8.6	94
	Bottom phase	1	94	5.3	6

Since, as examination of Figure A.10 will show, it is very easy to strip *sec*-butanol from water, it will usually be worth doing a recovery and a liquid–liquid extraction, using some of the entrainer, on the binary azeotrope feed (Figure A.11).

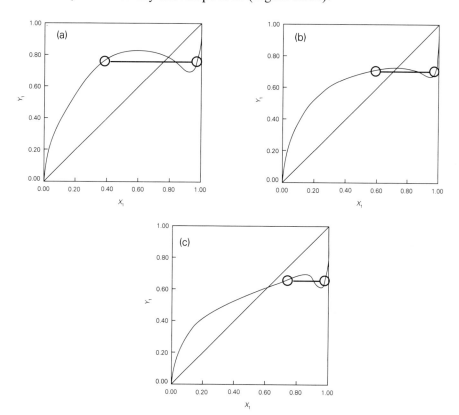

Fig. A.10 Comparison of VLE diagrams of (a) water (1)–*n*-butanol (2), (b) water (1)–isobutanol (2) and (c) water (1)–*sec*-butanol (2)

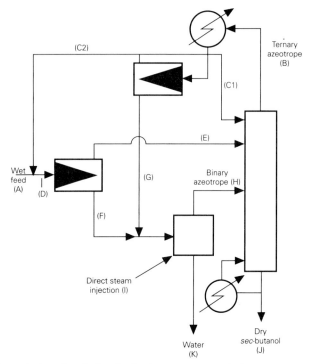

Fig. A.11 Single-column drying of *sec*-butanol using benzene as an entrainer

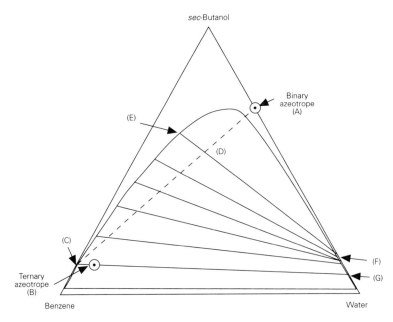

Fig. A.12 Ternary solubility diagram for *sec*-butanol–benzene–water (% w/w, 30 °C)

If the plant feed is assumed to have the composition of the *sec*-butanol–water binary azeotrope and the chosen entrainer is benzene the composition of the various flows in Figure A.11 can be identified on the triangular diagram (Figure A.12). It is assumed that the feed (A) is mixed with half its weight of the upper phase from the overheads separator (C2), the remainder of this being returned to the column as reflux. It can be seen from Table A.7 that only 8.56 kg of water per 100 kg of feed needs to be removed by azeotropic distillation by combining distillation with extraction. If azeotropic distillation alone were used all the water (27 kg per 100 kg of feed) would have to be removed by this method with a comparatively high cost in both fuel and plant occupation.

Table A.7 Size and composition of streams in Figure A.11

Stream	Entrainer (% w/w)	*sec*-Butanol (% w/w)	Water (% w/w)	Total (% w/w)
A	0	73	27	100
B	124.5	12.2	0.8	14
C1	79.0	7.5	0.35	86.85
C2	45.5	4.3	0.2	50
D	45.5	77.3	27.2	150
E	45.5	75.1	7.6	128.2
F	Trace	2.2	19.6	21.8
G	Trace	0.4	7.2	7.6
H	Trace	2.6	0.96	3.56
I	0	0	2.0	2.0
J	0	73.0	0	73.0
K	0	0	29.0	29.0

Cyclohexane is another possible choice as an entrainer employing a ternary azeotrope.

For *sec*-butanol that needs drying but only contains a small concentration of water, it is worth considering the use of diisopropyl ether, which does not form an azeotrope with *sec*-butanol and can therefore be used to take water very selectively as an overhead. It unfortunately has a low carrying capacity for water. The butanols tend to promote foaming when boiling at low concentration in water.

Cyclohexanol

1 Names

Cyclohexanol
Cyclohexyl alcohol
Hexalin
Adronal

2 Physical properties

Molecular weight	100
Empirical formula	$C_6H_{12}O$
Boiling point (°C)	161
Freezing point (°C)	23.6
Specific gravity (20/4 °C)	0.947
Liquid expansion coefficient (per °C)	0.000 77
Surface tension (at 20 °C in dyn/cm)	32
Absolute viscosity (at 25 °C in cP)	54.5

3 Fire and health hazards

Flash point (closed cup) (°C)	68
Autoignition temperature (°C)	300
Lower explosive limit (ppm)	20 000
Upper explosive limit (ppm)	
IDLH (ppm)	3500
Odour threshold (ppm)	0.15
TWA–TLV (ppm)	50
Saturated vapour concentration (70 °F in ppm)	1000
Vapour density (relative to air)	3.45

4 Solvent properties

Hildebrand solubility parameter	11.4
Dipole moment (D)	1.7
Dielectric constant (20 °C)	15.0
Evaporation time (diethyl ether = 1.0)	403

5 Aqueous effluent characteristics

Solubility in water (at 25 °C in % w/w)	4.3
Solubility of water in (at 25 °C in % w/w)	11.8
Biological oxygen demand (w/w)	0.08
Chemical oxygen demand (w/w)	2.2
Theoretical oxygen demand (w/w)	2.83

6 Handling details

Hazchem number	
Hazard class	
Vapour pressure (at 21 °C in mmHg)	1.14
Loss per transfer (% of liquid transferred)	

7 Vapour pressure constants (mmHg, log base 10)

Antoine equation	$A = 8.35237$
	$B = 2258.560$
	$C = 251.624$
Cox chart	$A = 8.27876$
	$B = 2110.6$

8 Thermal information

Latent heat (cal/mol)	10 900
Specific heat (cal/mol/°C)	50.0
Heat of combustion (cal/mol)	891 000
Heat of fusion (cal/mol)	419

Azeotropes and activity coefficients of component X: cyclohexanol

Component Y	Azeotrope °C	Azeotrope % w/w X	α	A_{XY} $=\ln \gamma_X^\infty$	A_{YX} $=\ln \gamma_Y^\infty$	Source[11] Vol.	Page
Hydrocarbons							
n-Pentane	None						
n-Hexane	None						
n-Heptane	None						
Cyclohexane	None		>20	2.73	0.64	2(d)	516
Benzene	None						
Toluene	None						
Ethylbenzene	None						
Xylenes	140	10					
Alcohols							
Methanol							
Ethanol	None		20	1.06	0.16	2(c)	421
n-Propanol							
Isopropanol							
n-Butanol	None		5	0.02	−0.07	2(b)	193
Isobutanol							
sec-Butanol							
Ethylene glycol	None		10	1.39	0.95	2(d)	14
Methyl Cellosolve							
Ethyl Cellosolve							
Butyl Cellosolve	None						
Chlorinated hydrocarbons							
Methylene dichloride							
Chloroform							
Carbon tetrachloride							
1,2-Dichloroethane							
1,1,1-Trichloroethane							
Trichloroethylene							
Perchloroethylene							
Monochlorobenzene	None		3	0.95	0.46	2(b)	393
Ketones							
Acetone	None		>20	1.54	1.34	2(d)	510
Methyl ethyl ketone							
MIBK							
Cyclohexanone	None		<1.5	0.11	0.12	2(b)	395
N-Methylpyrrolidone							
Ethers							
Diethyl ether							
Diisopropyl ether							
1,4-Dioxane							
Tetrahydrofuran							
Esters							
Methyl acetate							
Ethyl acetate	None		>20	1.42	0.47	2(d)	511
Butyl acetate							
Miscellaneous							
Dimethylformamide							
Dimethyl sulphoxide							
Pyridine							
Acetonitrile							
Furfuraldehyde	157	95					
Water	98	70	>20[a]	2.76	1.52	1	514

[a] The relative volatility between the less volatile component of the binary mixture and the azeotrope.

Recovery notes: cyclohexanol

Commercial cyclohexanol contains a small percentage (less than 1%) of cyclohexanone and its freezing point is somewhat depressed from the reagent grade. Although cyclohexanone is slightly more volatile, the relative volatility is so small that losses of the ketone will not cause the freezing point to rise significantly. Even so, lagged and traced pipelines and tank vents are required for storing cyclohexanol.

As the VLE diagram shows (Figure A.13), the water azeotrope is very easily stripped both from water and from cyclohexanol. The loss of cyclohexanol, when drying it from a water-saturated state and disposing of the water saturated with solvent, is only 0.5%, so that recycling the water phase is probably not worthwhile.

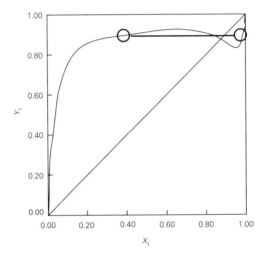

Fig. A.13 VLE diagram for water (1)–cyclohexanol (2)

ETHYLENE GLYCOL

1 Names

Ethylene glycol
Glycol
Monoethylene glycol
Ethane-1,2-diol
MEG

2 Physical properties

Molecular weight	62
Empirical formula	$C_2H_6O_2$
Boiling point (°C)	198
Freezing point (°C)	-13
Specific gravity (20/4 °C)	1.1135
Liquid expansion coefficient (per °C)	0.0007
Surface tension (at 20 °C in dyn/cm)	46.49
Absolute viscosity (at 25 °C in cP)	20

3 Fire and health hazards

Flash point (closed cup) (°C)	116
Autoignition temperature (°C)	400
Lower explosive limit (ppm)	32 000
Upper explosive limit (ppm)	–
IDLH (ppm)	–
Odour threshold (ppm)	–
TWA–TLV (ppm)	50
Saturated vapour concentration (70 °F in ppm)	53
Vapour density (relative to air)	2.15

4 Solvent properties

Hildebrand solubility parameter	17.1
Dipole moment (D)	2.2
Dielectric constant (20 °C)	37.7
Evaporation time (diethyl ether = 1.0)	1550

5 Aqueous effluent characteristics

Solubility in water (at 25 °C in % w/w)	Total
Solubility of water in (at 25 °C in % w/w)	Total
Biological oxygen demand (w/w)	0.16
Chemical oxygen demand (w/w)	1.2
Theoretical oxygen demand (w/w)	1.29

6 Handling details

Hazchem number	–
Hazard class	–
Vapour pressure (at 21 °C in mmHg)	0.12
Loss per transfer (% of liquid transferred)	0.36×10^{-4}

7 Vapour pressure constants (mmHg, log base 10)

Antoine equation	$A = 8.09083$
	$B = 2088.936$
	$C = 203.454$
Cox chart	$A = $ N/A*
	$B = $ N/A

8 Thermal information

Latent heat (cal/mol)	12 524
Specific heat (cal/mol/°C)	34.8
Heat of combustion (cal/mol)	283 092

*Not available

Azeotropes and activity coefficients of component X: ethylene glycol

Component Y	Azeotrope °C	Azeotrope % w/w X	α	$A_{XY} = \ln \gamma_X^\infty$	$A_{YX} = \ln \gamma_Y^\infty$	Source[11] Vol.	Source[11] Page
Hydrocarbons							
n-Pentane		None			5.39		
n-Hexane		None					
n-Heptane	98	3			3.46		
Cyclohexane		None					
Benzene		None					
Toluene	110	2					
Ethylbenzene	133	14					
Xylenes	135	7					
Alcohols							
Methanol		None	>20	0.61	0.07	2(a)	62
Ethanol		None	>20	1.88	0.72	2(c)	297
n-Propanol		None					
Isopropanol							
n-Butanol		None	>20	0.94	1.89	2(d)	6
Isobutanol							
sec-Butanol							
Cyclohexanol			10	0.95	1.39	2(d)	14
Methyl Cellosolve							
Ethyl Cellosolve							
Butyl Cellosolve							
Chlorinated hydrocarbons							
Methylene dichloride							
Chloroform							
Carbon tetrachloride							
1,2-Dichloroethane							
1,1,1-Trichloroethane							
Trichloroethylene							
Perchloroethylene	119	6					
Monochlorobenzene	130	6					
Ketones							
Acetone							
Methyl ethyl ketone							
MIBK							
Cyclohexanone							
N-Methylpyrrolidone							
Ethers							
Diethyl ether							
Diisopropyl ether							
1,4-Dioxane		None					
Tetrahydrofuran		None	>20	1.94	1.70	2(d)	2
Esters							
Methyl acetate							
Ethyl acetate							
Butyl acetate		None	>20	1.93	2.41	2(d)	15
Miscellaneous							
Dimethylformamide				0.06	0.60		
Dimethyl sulphoxide							
Pyridine		None					
Acetonitrile							
Furfuraldehyde		None					
Water		None	>20	−1.31	−0.43	2(c)	297

Recovery notes: ethylene glycol

The two largest uses for monoethylene glycol (MEG) are in the production of polyester fibres and in antifreeze. The former yields considerable quantities of MEG for recovery, but needs a higher quality (fibre grade) for reuse. There is a considerable attraction for the recoverer to recycle MEG to the lower quality outlet, despite the fact that it is a seasonal market. Since antifreeze concentrate is blended into water for vehicle cooling systems, the water specification does not need to be very tight for this market. MEG is hygroscopic and storage tanks need driers on their vents for very long-term storage applications.

MEG is not stable at its atmospheric pressure boiling point and develops a 'cracked' smell on prolonged heating, but if a small concentration of water is acceptable in a residue of recovered MEG it is possible to remove water with a single evaporation stage, as Table A.8 indicates. The ease with which water can be removed on heating presents a

Table A.8 Water–MEG vapour–liquid equilibrium at 760 mmHg

Boiling temperature (°C)	Mole fraction of water	
	Liquid	Vapour
197	0	0
130	0.15	0.95
120	0.28	0.97
110	0.46	0.99
100	1.00	1.00

fire and explosion hazard when carrying out maintenance involving welding on MEG–water cooling systems that have not been fully drained. The water can be evaporated from the mixture leaving MEG at a temperature within its flammable range. It can be easily separated from water at 100–150 mmHg without breakdown taking place.

METHYL CELLOSOLVE

1 **Names***

Methyl Cellosolve
Ethylene glycol monomethyl ether
Methyl glycol
2-Methoxyethanol

2 **Physical properties**

Molecular weight	76
Empirical formula	$C_3H_8O_2$
Boiling point (°C)	124
Freezing point (°C)	−85
Specific gravity (20/4 °C)	0.975
Liquid expansion coefficient (per °C)	0.00095
Surface tension (at 20 °C in dyn/cm)	29
Absolute viscosity (at 25 °C in cP)	1.72

3 **Fire and health hazards**

Flash point (closed cup) (°C)	41
Autoignition temperature (°C)	285
Lower explosive limit (ppm)	25000
Upper explosive limit (ppm)	198000
IDLH (ppm)	4500
Odour threshold (ppm)	1
TWA–TLV (ppm)	25
Saturated vapour concentration (70 °F in ppm)	8900
Vapour density (relative to air)	2.62

4 **Solvent properties**

Hildebrand solubility parameter	10.8
Dipole moment (D)	1.34
Dielectric constant (20 °C)	16.9
Evaporation time (diethyl ether = 1.0)	25.1

5 **Aqueous effluent characteristics**

Solubility in water (at 25 °C in % w/w)	Total
Solubility of water in (at 25 °C in % w/w)	Total
Biological oxygen demand (w/w)	0.27
Chemical oxygen demand (w/w)	1.70
Theoretical oxygen demand (w/w)	1.83

6 **Handling details**

Hazchem number	1188
Hazard class	2 (S)
Vapour pressure (at 21 °C in mmHg)	6.7
Loss per transfer (% of liquid transferred)	0.028×10^{-2}

7 **Vapour pressure constants (mmHg, log base 10)**

Antoine equation	$A = 7.8498$
	$B = 1793.982$
	$C = 236.877$
Cox chart	$A = $ N/A
	$B = $ N/A

8 **Thermal information**

Latent heat (cal/mol)	10260
Specific heat (cal/mol/°C)	40.6
Heat of combustion (cal/mol)	399000

*Trade names contain EM as part of the name.

Azeotropes and activity coefficients of component X: methyl Cellosolve

Component Y	Azeotrope °C	Azeotrope % w/w X	α	A_{XY} $=\ln \gamma_X^\infty$	A_{YX} $=\ln \gamma_Y^\infty$	Source[11] Vol.	Page
Hydrocarbons							
n-Pentane							
n-Hexane	Probable						
n-Heptane	93	23					
Cyclohexane	80	82	10[a]	2.90	1.68	2(b)	128
Benzene	None		7	1.80	0.83	2(b)	127
Toluene	106	25					
Ethylbenzene	117	54	3[a]	1.77	1.41	2(b)	132
Xylenes	120	55	3[a]	1.34	1.12	2(b)	133
Alcohols							
Methanol	None		>15				98
Ethanol	None						
n-Propanol	None		2.5	0.23	0.50	2(c)	490
Isopropanol							
n-Butanol	None						
Isobutanol	None						
sec-Butanol	None						
Cyclohexanol							
Ethylene glycol							
Ethyl Cellosolve							
Butyl Cellosolve							
Chlorinated hydrocarbons							
Methylene dichloride							
Chloroform							
Carbon tetrachloride							
1,2-Dichloroethane							
1,1,1-Trichloroethane							
Trichloroethylene							
Perchloroethylene	109	24					
Monochlorobenzene	119	43	2[a]	1.23	0.85	2(d)	120
Ketones							
Acetone	None		18	0.90	0.72	2(d)	113
Methyl ethyl ketone	None		5	0.57	0.68	2(b)	122
MIBK	114	25					
Cyclohexanone							
N-Methylpyrrolidone							
Ethers							
Diethyl ether							
Diisopropyl ether							
1,4-Dioxane							
Tetrahydrofuran							
Esters							
Methyl acetate							
Ethyl acetate	None		5	0.71	0.56	2(b)	126
Butyl acetate	118	48	1.7[a]	0.82	0.88	2(d)	122
Miscellaneous							
Dimethylformamide							
Dimethyl sulphoxide							
Pyridine							
Acetonitrile	None		5	0.53	0.50	2(d)	109
Furfuraldehyde	None						
Water	99	85					

[a] The relative volatility between the less volatile component of the binary mixture and the azeotrope.

Recovery notes: methyl Cellosolve

Precautions appropriate for avoiding the formation of peroxides when processing glycol ethers are covered under the notes for butyl Cellosolve. Methyl Cellosolve has similar toxic properties to the other glycol ethers. Surprisingly, propylene glycol methyl ether has a higher volatility and very similar other properties to methyl Cellosolve and can usually be used in reformulation.

Methyl Cellosolve is a highly polar solvent and is not miscible with alkanes, but is miscible with all other solvents.

Methyl Cellosolve does not form a ternary azeotrope with water and the hydrocarbons benzene, cyclohexane or toluene, although it does with xylene and ethylbenzene. As for ethyl Cellosolve, methyl Cellosolve can be dehydrated with toluene, provided a continuing use for the resulting methyl Cellosolve–toluene mixture can be found. If not, a C_7 normal/iso alkane mixture can be used with a phase separation to remove the hydrocarbon when the feed has been dried. This will best be carried out in a hybrid or batch still and, if the same unit is used as for drying, facilities for rejecting both the denser (water) phase and the less dense (heptane) phase will be needed.

ETHYL CELLOSOLVE

1 Names*
Ethyl Cellosolve
Ethylene glycol monoethyl ether
2-Ethoxyethanol

2 Physical properties

Molecular weight	90
Empirical formula	$C_4H_{10}O_2$
Boiling point (°C)	136
Freezing point (°C)	−69
Specific gravity (20/4 °C)	0.929
Liquid expansion coefficient (per °C)	0.00097
Surface tension (at 20 °C in dyn/cm)	32
Absolute viscosity (at 25 °C in cP)	2.05

3 Fire and health hazards

Flash point (closed cup) (°C)	40
Autoignition temperature (°C)	238
Lower explosive limit (ppm)	18 000
Upper explosive limit (ppm)	140 000
IDLH (ppm)	6 000
Odour threshold (ppm)	200
TWA–TLV (ppm)	50
Saturated vapour concentration (70 °F in ppm)	5 400
Vapour density (relative to air)	3.12

4 Solvent properties

Hildebrand solubility parameter	9.9
Dipole moment (D)	1.7
Dielectric constant (20 °C)	–
Evaporation time (diethyl ether = 1.0)	40

5 Aqueous effluent characteristics

Solubility in water (at 25 °C in % w/w)	Total
Solubility of water in (at 25 °C in % w/w)	Total
Biological oxygen demand (w/w)	1.58
Chemical oxygen demand (w/w)	1.9
Theoretical oxygen demand (w/w)	1.96

6 Handling details

Hazchem number	1171
Hazard class	2 (S)
Vapour pressure (at 21 °C in mmHg)	4.0
Loss per transfer (% of liquid transferred)	0.0021

7 Vapour pressure constants (mmHg, log base 10)

Antoine equation	$A = 7.81910$
	$B = 1801.900$
	$C = 230.00$
Cox chart	$A = 7.81910$
	$B = 1801.900$

8 Thermal information

Latent heat (cal/mol)	9630
Specific heat (cal/mol/°C)	50.0
Heat of combustion (cal/mol)	666 000

*Trade names for ethylene glycol monoethyl ether contain EE as part of the name

Azeotropes and activity coefficients of component X: ethyl Cellosolve

Component Y	Azeotrope °C	Azeotrope % w/w X	α	A_{XY} =ln γ_X^∞	A_{YX} =ln γ_Y^∞	Source[11] Vol.	Page
Hydrocarbons							
n-Pentane							
n-Hexane	66	5	10[a]	1.70	1.54	2(b)	295
n-Heptane	97	14					
Cyclohexane		None					
Benzene		None					
Toluene	109	11					
Ethylbenzene	126	43	2[a]	1.09	0.88	2(b)	299
Xylenes	128	50	2[a]	1.06	0.75	2(b)	301
Alcohols							
Methanol		None	>15				
Ethanol		None	>10				
n-Propanol							
Isopropanol							
n-Butanol		None					
Isobutanol		None					
sec-Butanol							
Cyclohexanol							
Ethylene glycol							
Methyl Cellosolve							
Butyl Cellosolve							
Chlorinated hydrocarbons							
Methylene dichloride							
Chloroform							
Carbon tetrachloride							
1,2-Dichloroethane							
1,1,1-Trichloroethane							
Trichloroethylene							
Perchloroethylene	116	16	2.5[a]	1.45	0.97	2(d)	396
Monochlorobenzene	127	32					
Ketones							
Acetone							
Methyl ethyl ketone							
MIBK							
Cyclohexanone							
N-Methylpyrrolidone							
Ethers							
Diethyl ether							
Diisopropyl ether							
1,4-Dioxane							
Tetrahydrofuran							
Esters							
Methyl acetate							
Ethyl acetate							
Butyl acetate	126	12	1.5[a]	0.56	0.48	2(b)	294
Miscellaneous							
Dimethylformamide							
Dimethyl sulphoxide							
Pyridine							
Acetonitrile							
Furfuraldehyde							
Water	98	87	5[a]	1.90	0.71	1	450

[a] The relative volatility between the less volatile component of the binary mixture and the azeotrope.

Recovery notes: ethyl Cellosolve

Like all glycol ethers, ethyl Cellosolve can form peroxides which are spontaneously flammable. The precautions necessary for avoiding the problems that peroxides present are covered under the notes for butyl Cellosolve.

Ethyl Cellosolve forms many azeotropes with hydrocarbons, and cyclohexane is one with the highest water-carrying capacity (in Table 7.9) that is free from this interference. However, if azeotropic drying is a long-term job, toluene can be used as an entrainer.

The possible azeotropes in the toluene–ethyl Cellosolve–water system are given in Table A.9. There is no ternary azeotrope of toluene–ethyl Cellosolve–water. If there is no continuity, the technique for using heptane isomers in a batch mode described for methyl Cellosolve can be used.

Table A.9 Azeotropes in the toluene–ethyl Cellosolve–water system

Components	Boiling point (°C)	Composition (% w/w)
Toluene–water	85	80:20
Ethyl Cellosolve–water	98	87:13
Ethyl Cellosolve–toluene	109	14:86
Ethyl Cellosolve	136	100

Because ethyl Cellosolve has aprotic properties, it cannot be dried by pervaporation using currently available membranes.

Despite the fact that ethyl Cellosolve is a product that has been in widespread use for many years, it is only recently that its property of causing aplastic anaemia has been discovered. The propylene glycol ethers, which do not have a similar health risk, provide a range of solvents with generally similar properties.

Protective clothing of neoprene and natural rubber is suitable for glycol ethers, but PVC is not recommended.

BUTYL CELLOSOLVE

1 **Names***

    Butyl Cellosolve
    Ethylene glycol monobutyl ether
    2-Butoxyethanol

2 **Physical properties**

Molecular weight	118
Empirical formula	$C_6H_{14}O_2$
Boiling point (°C)	171
Freezing point (°C)	−75
Specific gravity (20/4 °C)	0.901
Liquid expansion coefficient (per °C)	0.00092
Surface tension (at 20 °C in dyn/cm)	31.5
Absolute viscosity (at 25 °C in cP)	6.4

3 **Fire and health hazards**

Flash point (closed cup) (°C)	60
Autoignition temperature (°C)	244
Lower explosive limit (ppm)	11 000
Upper explosive limit (ppm)	106 000
IDLH (ppm)	700
Odour threshold (ppm)	12.5
TWA–TLV (ppm)	25
Saturated vapour concentration (70 °F in ppm)	860
Vapour density (relative to air)	4.07

4 **Solvent properties**

Hildebrand solubility parameter	9.5
Dipole moment (D)	–
Dielectric constant (20 °C)	–
Evaporation time (diethyl ether = 1.0)	118

5 **Aqueous effluent characteristics**

Solubility in water (at 25 °C in % w/w)	} see Figure A.14
Solubility of water in (at 25 °C in % w/w)	
Biological oxygen demand (w/w)	0.60
Chemical oxygen demand (w/w)	2.2
Theoretical oxygen demand (w/w)	2.31

6 **Handling details**

Hazchem number	2369
Hazard class	2 R
Vapour pressure (at 21 °C in mmHg)	0.7
Loss per transfer (% of liquid transferred)	4.8×10^{-4}

7 **Vapour pressure constants (mmHg, log base 10)**

Antoine equation	$A = 7.8448$
	$B = 1988.90$
	$C = 230.00$
Cox chart	$A = 7.8448$
	$B = 1988.90$

8 **Thermal information**

Latent heat (cal/mol)	10 384
Specific heat (cal/mol/°C)	55.9
Heat of combustion (cal/mol)	910 960

*Various trade names contain EB as part of the name

Azeotropes and activity coefficients of component X: butyl Cellosolve

Component Y	Azeotrope °C	Azeotrope % w/w X	α	A_{XY} $= \ln \gamma_X^\infty$	A_{YX} $= \ln \gamma_Y^\infty$	Source[11] Vol.	Page
Hydrocarbons							
n-Pentane							
n-Hexane							
n-Heptane							
Cyclohexane							
Benzene							
Toluene	None		8	0.58	0.31	2(d)	560
Ethylbenzene	Probable						
Xylenes	144	96					
Alcohols							
Methanol	None		>15				
Ethanol							
n-Propanol							
Isopropanol							
n-Butanol	None		4				
Isobutanol							
sec-Butanol							
Cyclohexanol	None						
Ethylene glycol	None						
Methyl Cellosolve							
Ethyl Cellosolve	None						
Chlorinated hydrocarbons							
Methylene dichloride							
Chloroform							
Carbon tetrachloride							
1,2-Dichloroethane							
1,1,1-Trichloroethane							
Trichloroethylene							
Perchloroethylene							
Monochlorobenzene							
Ketones							
Acetone							
Methyl ethyl ketone	None		>10	0.32	0.86	2(b)	430
MIBK							
Cyclohexanone							
N-Methylpyrrolidone							
Ethers							
Diethyl ether							
Diisopropyl ether							
1,4-Dioxane							
Tetrahydrofuran							
Esters							
Methyl acetate							
Ethyl acetate	None						
Butyl acetate	None						
Miscellaneous							
Dimethylformamide							
Dimethyl sulphoxide							
Pyridine							
Acetonitrile							
Furfuraldehyde	161	12					
Water	99	79	20[a]	5.30	0.47	1	526

[a] The relative volatility between the less volatile component of the binary mixture and the azeotrope.

Recovery notes: butyl Cellosolve

The property of butyl Cellosolve most often used in its recovery is its water solubility and its behaviour in water mixtures is therefore of great importance. As Figure A.14 shows,

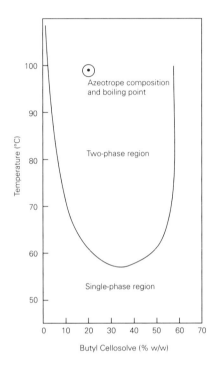

Fig. A.14 Water solubility of butyl Cellosolve vs. temperature

butyl Cellosolve is completely miscible with water at low temperatures but forms two liquid phases at certain concentrations above 57 °C. Its upper critical solution temperature is 128 °C. At the boiling point of the butyl Cellosolve–water azeotrope the condensate splits into an aqueous phase containing about 2% and an organic phase of about 57% w/w (0.17 mole fraction) of butyl Cellosolve. As reference to Figure A.15(a) will show, the organic phase can very easily be separated by distillation into the azeotrope and a dry butyl Cellosolve fraction.

The butyl Cellosolve content of the water-rich phase is low and is neither very toxic in effluent nor very expensive, so that further treatment may not be economically justifiable. However, the azeotrope can be stripped from the water easily if necessary as a blown-up version of the water-rich end of the VLE diagram shows [Figure A.15(b)].

Like other glycol ethers, butyl Cellosolve is very hygroscopic and should be protected against water pick-up from the atmosphere.

A number of accidents have occurred when butyl Cellosolve has been distilled without proper care. The ingress of air to hot glycol ethers can lead to spontaneous ignition, so vacuum must be broken using inert gas and plant gaskets must not be broken while the contents of the plant are still hot.

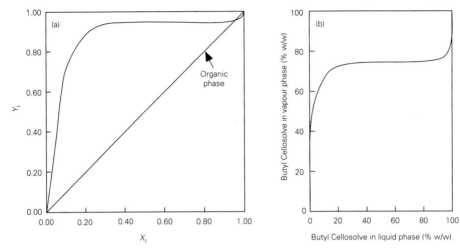

Fig. A.15 (a) VLE diagram for water (1)–butyl Cellosolve (2) at 760 mmHg; (b) VLE diagram for butyl Cellosolve–water on a weight basis

Peroxides may be present in used glycol ethers and, before heating, a test for their presence should always be carried out using a Merquant stick. If they are found, peroxides should be removed using sodium tetrahydroborate.

In storage, and immediately after distillation which will remove the usual inhibitor, inhibiting is necessary. Butylated hydroxytoluene (BHT; 2,6-di-*tert*-butyl-*p*-cresol) is a suitable inhibitor for butyl Cellosolve.

If distillation is inappropriate for removing butyl Cellosolve from water, perhaps because of the presence of corrosive salts, octan-2-ol and similar high-boiling alcohols are effective for solvent extraction.

All ethylene glycol ethers are suspected of causing aplastic anaemia, similar to the effect of benzene poisoning, and should be handled with care. Protective equipment of neoprene is suitable for gloves and aprons.

METHYLENE DICHLORIDE

1. **Names**

   Methylene chloride
   Methylene dichloride
   Dichloromethane
   MDC
   Not methyl chloride

2. **Physical properties**

Molecular weight	85
Empirical formula	CH_2Cl_2
Boiling point (°C)	40
Freezing point (°C)	−97
Specific gravity (20/4 °C)	1.326
Liquid expansion coefficient (per °C)	0.00137
Surface tension (at 20 °C in dyn/cm)	28.1
Absolute viscosity (at 25 °C in cP)	0.44

3. **Fire and health hazards**

Flash point (closed cup) (°C)	None
Autoignition temperature (°C)	605
Lower explosive limit (ppm)	130000
Upper explosive limit (ppm)	220000
IDLH (ppm)	
Odour threshold (ppm)	250
TWA–TLV (ppm)	100
Saturated vapour concentration (70 °F in ppm)	500000
Vapour density (relative to air)	2.95

4. **Solvent properties**

Hildebrand solubility parameter	9.7
Dipole moment (D)	1.8
Dielectric constant (20 °C)	9.1
Evaporation time (diethyl ether = 1.0)	1.4

5. **Aqueous effluent characteristics**

Solubility in water (at 25 °C in % w/w)	1.85
Solubility of water in (at 25 °C in % w/w)	0.15
Biological oxygen demand (w/w)	0
Chemical oxygen demand (w/w)	
Theoretical oxygen demand (w/w)	0.56

6. **Handling details**

Hazchem number	1593
Hazard class	2 Z
Vapour pressure (at 21 °C in mmHg)	382
Loss per transfer (% of liquid transferred)	0.13

7. **Vapour pressure constants (mmHg, log base 10)**

Antoine equation	$A = 7.0803$
	$B = 1138.91$
	$C = 231.45$
Cox chart	$A = 6.91821$
	$B = 1090.1$

8. **Thermal information**

Latent heat (cal/mol)	6715
Specific heat (cal/mol/°C)	23.8
Heat of combustion (cal/mol)	122352

Azeotropes and activity coefficients of component X: methylene dichloride

Component Y	Azeotrope °C	Azeotrope % w/w X	α	A_{XY} $=\ln \gamma_X^\infty$	A_{YX} $=\ln \gamma_Y^\infty$	Source[11] Vol.	Page
Hydrocarbons							
n-Pentane	31	49	2[a]	0.85	1.06	6(a)	100
n-Hexane							
n-Heptane							
Cyclohexane							
Benzene							
Toluene							
Ethylbenzene							
Xylenes					−0.23		
Alcohols							
Methanol	38	93	9[a]	0.93	2.37	2(a)	24
Ethanol	40	98	9[a]	0.46	3.34	2(a)	293
n-Propanol							
Isopropanol							
n-Butanol							
Isobutanol							
sec-Butanol							
Cyclohexanol							
Ethylene glycol							
Methyl Cellosolve							
Ethyl Cellosolve							
Butyl Cellosolve							
Chlorinated hydrocarbons							
Chloroform			1.8	−0.22	−0.42	8	202
Carbon tetrachloride			3	−0.48	−0.09	8	62
1,2-Dichloroethane			5	0.00	−0.04	8	263
1,1,1-Trichloroethane					0.17		
Trichloroethylene							
Perchloroethylene		None	11	0.11	−0.06	8	256
Monochlorobenzene							
Ketones							
Acetone		None	1.8	−0.34	−0.37		
Methyl ethyl ketone		None	2	−0.56	−1.07	3+4	261
MIBK							
Cyclohexanone							
N-Methylpyrrolidone							
Ethers							
Diethyl ether	40	70	<1.5[a]	−0.46	−0.36	3+4	492
Diisopropyl ether							
1,4-Dioxane							
Tetrahydrofuran		Possible					
Esters							
Methyl acetate		None	1.5/2	−0.50	−0.69	5	347
Ethyl acetate		None	2/5	−0.57	−0.94	5	449
Butyl acetate							
Miscellaneous							
Dimethylformamide		None	14	3.05	−0.80	8	265
Dimethyl sulphoxide		None	>20	−0.57	−0.82	8	264
Pyridine		None	11	−0.46	−0.60	8	267
Acetonitrile		None	4.5	0.19	0.15	8	258
Furfuraldehyde				−0.16	0.48		
Water	38	99	20[a]	5.75	3.89	1	1

[a]The relative volatility between the less volatile component of the binary mixture and the azeotrope.
[b]Reacts.

Recovery notes: methylene dichloride

Methylene chloride is not flammable although it will burn at high temperatures and in oxygen-enriched atmospheres. It is very stable when dry but can react with metals when wet. High temperatures increase the rate of hydrolysis and violent reactions can take place between wet MDC and aluminium, particularly in the presence of toluene and its homologues.

There is a very wide range of inhibitors available to stabilize MDC, but it is best to avoid aluminium as a material of construction when MDC may be processed or stored. Because of MDC's volatility there is always a danger that it will distil away from a less volatile inhibitor and lose its protection. Low-boiling inhibitors such as *tert*-butylamine, propylene oxide and amylene (2-methylbut-2-ene) tend to stay with MDC when it is vaporized whereas dioxane, ethanol, tetrahydrofuran, *N*-methylmorpholine and cyclohexene tend to remain in the liquid phase. The concentrations required for effective inhibition are low (50 ppm to 0.2%) and, if MDC is used as a reaction medium, it is almost always possible to find one that does not become involved in the reaction itself.

In recovering MDC by distillation, it is possible to achieve a very dry product by taking a side stream from the column 4–6 trays from the top and running the column tops through a phase separator from which the water is decanted. Although the boiling point of the MDC–water azeotrope is only 2 °C below the boiling point of dry MDC, the relative volatility between them is very large, as the VLE diagram shows (Figure 5.4). If even drier material is needed, MDC can be dried using molecular sieves, activated alumina and the Na^+ form of Amberlite IR-120. The last named can be regenerated at 120 °C and therefore does not need sophisticated air heating equipment.

MDC is used as a way of removing methanol selectively from mixtures (e.g. with acetone) and frequently has to be recovered from the methanol–MDC azeotropic mixture. This can be achieved by contacting the MDC–methanol azeotrope with an equal weight of water. The methanol partitions about 10:1 in favour of the wash water, although the wash water will contain about 2% of MDC and will need to be processed before discharge (Figure A.16).

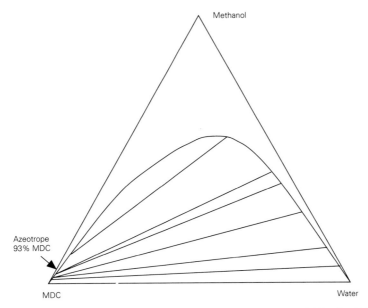

Fig. A.16 Ternary solubility diagram for methylene chloride–water–methanol (% w/w, 20 °C)

CHLOROFORM

1 Names　　　　　　　　　　　　　　　　　　　　　Chloroform
　　　　　　　　　　　　　　　　　　　　　　　　　　　Trichloromethane

2 Physical properties

Molecular weight	119
Empirical formula	$CHCl_3$
Boiling point (°C)	61
Freezing point (°C)	−63
Specific gravity (20/4 °C)	1.48
Liquid expansion coefficient (per °C)	0.00126
Surface tension (at 20 °C in dyn/cm)	27.2
Absolute viscosity (at 25 °C in cP)	0.57

3 Fire and health hazards

Flash point (closed cup) (°C)	None
Autoignition temperature (°C)	None
Lower explosive limit (ppm)	None
Upper explosive limit (ppm)	None
IDLH (ppm)	1000
Odour threshold (ppm)	300
TWA–TLV (ppm)	10
Saturated vapour concentration (70 °F in ppm)	286 000
Vapour density (relative to air)	4.13

4 Solvent properties

Hildebrand solubility parameter	9.3
Dipole moment (D)	1.1
Dielectric constant (25 °C)	4.806
Evaporation time (diethyl ether = 1.0)	1.9

5 Aqueous effluent characteristics

Solubility in water (at 25 °C in % w/w)	0.82
Solubility of water in (at 25 °C in % w/w)	0.2
Biological oxygen demand (w/w)	0.02
Chemical oxygen demand (w/w)	–
Theoretical oxygen demand (w/w)	1.35

6 Handling details

Hazchem number	1888
Hazard class	2 Z
Vapour pressure (at 21 °C in mmHg)	169
Loss per transfer (% of liquid transferred)	0.074

7 Vapour pressure constants (mmHg, log base 10)

Antoine equation	$A = 6.95465$
	$B = 1170.966$
	$C = 226.232$
Cox chart	$A = 6.97909$
	$B = 1192.6$

8 Thermal information

Latent heat (cal/mol)	7021
Specific heat (cal/mol/°C)	27.4
Heat of combustion (cal/mol)	90 758

Azeotropes and activity coefficients of component X: chloroform

Component Y	Azeotrope °C	Azeotrope % w/w X	α	$A_{XY} = \ln \gamma_X^\infty$	$A_{YX} = \ln \gamma_Y^\infty$	Source[11] Vol.	Page
Hydrocarbons							
n-Pentane		None					
n-Hexane	60	84	1.7[a]	0.32	0.57	6(a)	430
n-Heptane		None	3.5	0.36	0.29	6(b)	77
Cyclohexane		None					
Benzene		None	1.5/2	−0.48	−0.43	7	67
Toluene		None	4	−0.10	−0.24	7	352
Ethylbenzene		None					
Xylenes		None					
Alcohols							
Methanol	53	87	3[a]	0.99	2.07	2(a)	18
Ethanol	59	93	3[a]	0.69	1.48	2(a)	285
n-Propanol		None					
Isopropanol	61	96	3[a]				
n-Butanol		None	10	0.38	0.98	2(b)	136
Isobutanol		None					
sec-Butanol		None					
Cyclohexanol		None					
Ethylene glycol		None					
Methyl Cellosolve		None					
Ethyl Cellosolve		None					
Butyl Cellosolve							
Chlorinated hydrocarbons							
Methylene dichloride		None	1.8	−0.42	−0.22	8	202
Carbon tetrachloride		None	1.5	0.02	0.47	8	55
1,2-Dichloroethane				0.03	0.05		
1,1,1-Trichloroethane		None		−0.03	0.00		
Trichloroethylene		None		0.08			
Perchloroethylene		None	7	0.15	0.17	8	215
Monochlorobenzene		None	9	0.18	−0.19	8	244
Ketones							
Acetone	64	78	<1.5[a]	−0.52	−0.61	3+4	90
Methyl ethyl ketone	80	17	1.5[a]	−0.70	−0.84	3+4	260
MIBK		None	3/5	−0.66	−1.48	3+4	343
Cyclohexanone		None					
N-Methylpyrrolidone		None					
Ethers							
Diethyl ether	37	97	2[a]	−0.94	−0.82	3+4	486
Diisopropyl ether	70	36	<1.5[a]	−0.73	−0.76	3+4	537
1,4-Dioxane		None	3/5	−0.82	−1.26	3+4	441
Tetrahydrofuran	72	66					
Esters							
Methyl acetate	65	64		−0.50	−0.82	5	344
Ethyl acetate	78	28		−0.55	−1.50	5	443
Butyl acetate		None	5	−0.42	−1.08	5	574
Miscellaneous							
Dimethylformamide		None					
Dimethyl sulphoxide		None	>10	−0.91	−1.88	8	234
Pyridine		None	3/5	−1.20	−0.49	8	237
Acetonitrile		None	2	0.30	0.21	8	217
Furfuraldehyde		None	>10	−0.24	−0.27	3+4	36
Water	56	97		6.81	4.48		

[a] The relative volatility between the less volatile component of the binary mixture and the azeotrope.

Recovery notes: chloroform

In addition to the binary azeotropes listed, chloroform enters into a number of ternary azeotropes that can interfere with its recovery (Table A.10).

Table A.10 Ternary azeotropes of chloroform

Components	Composition (% w/w)	°C	Type
Chloroform–methanol–water	90.5:8.2:1.3	52.3	2 phase
Chloroform–ethanol–water	91.2:4.9:2.3	78.0[a]	2 phase
Chloroform–acetone–water	57.6:38.4:4.0		
Chloroform–methanol–acetone	47:23:30	57.5	1 phase
Chloroform–methanol–methyl acetate	52.5:21.6:25.9	56.4	1 phase
Chloroform–ethanol–acetone	70:7:23	55	1 phase
Chloroform–ethanol–hexane	56.1:9.5:34.5	57.3	1 phase
Chloroform–acetone–hexane	68.8:3.6:27.6	60.8	1 phase

[a]High-boiling azeotrope.

Chloroform is unstable in daylight (dark bottles are needed for storing samples) unless it is stabilized. The most commonly used stabilizer is ethanol added at 2% and, since this tends to be extracted with water, considerable care has to be taken to ensure that recovered material is kept protected at all times.

Although chloroform was widely used at one time for dehydrating ethanol, for which it is operationally well suited, its toxicity has caused it to be withdrawn from this use and from virtually any other application for which there is an acceptable alternative.

Chloroform should not be brought into contact with strong alkalis (e.g. NaOH) and it reacts with some organic bases.

It has one of the highest ratios of saturated vapour concentrations to IDLH (286:1) and an odour threshold so far above its TLV that its smell is no protection at all against an unhealthy working environment.

Other stabilizers which would not be leached out with water to the same extent as ethanol, but would be removed by distillation, are *tert*-butylphenol, *n*-octylphenol and thymol.

CARBON TETRACHLORIDE

1 Names
Carbon tetrachloride
CTET
Carbon tet

2 Physical properties

Molecular weight	154
Empirical formula	CCl_4
Boiling point (°C)	76
Freezing point (°C)	−23
Specific gravity (20/4 °C)	1.58
Liquid expansion coefficient (per °C)	0.00127
Surface tension (at 20 °C in dyn/cm)	27
Absolute viscosity (at 25 °C in cP)	0.97

3 Fire and health hazards

Flash point (closed cup) (°C)	None
Autoignition temperature (°C)	None
Lower explosive limit (ppm)	None
Upper explosive limit (ppm)	None
IDLH (ppm)	300
Odour threshold (ppm)	20
TWA–TLV (ppm)	10
Saturated vapour concentration (70 °F in ppm)	148 100
Vapour density (relative to air)	1.8

4 Solvent properties

Hildebrand solubility parameter	8.6
Dipole moment (D)	0
Dielectric constant (20 °C)	2.24
Evaporation time (diethyl ether = 1.0)	1.8

5 Aqueous effluent characteristics

Solubility in water (at 25 °C in % w/w)	0.077
Solubility of water in (at 25 °C in % w/w)	0.008
Biological oxygen demand (w/w)	0
Chemical oxygen demand (w/w)	
Theoretical oxygen demand (w/w)	0.21

6 Handling details

Hazchem number	1846
Hazard class	2 Z
Vapour pressure (at 21 °C in mmHg)	61.5
Loss per transfer (% of liquid transferred)	0.033

7 Vapour pressure constants (mmHg, log base 10)

Antoine equation	$A = 6.84083$
	$B = 1210.595$
	$C = 229.664$
Cox chart	$A = 7.02433$
	$B = 1267.9$

8 Thermal information

Latent heat (cal/mol)	7238
Specific heat (cal/mol/°C)	32.3
Heat of combustion (cal/mol)	61 652

Azeotropes and activity coefficients of component X: carbon tetrachloride

Component Y	Azeotrope °C	Azeotrope % w/w X	α	$A_{XY} = \ln \gamma_X^\infty$	$A_{YX} = \ln \gamma_Y^\infty$	Source[11] Vol.	Page
Hydrocarbons							
n-Pentane	None						
n-Hexane	None		1.5	0.17	0.24	6(a)	403
n-Heptane	None		2.2	0.05	0.12	6(b)	72
Cyclohexane	None		<1.5	2.69	0.03	6(c)	189
Benzene	None		<1.5	0.07	0.10	7	11
Toluene	None		2.5	0.01	1.64	7	332
Ethylbenzene	None		5	−0.29	−0.03		
Xylenes	None		8	−0.27	0.14	7	480
Alcohols							
Methanol	56	79	5[a]	1.91	2.33	2(a)	1
Ethanol	65	84	3[a]	1.62	2.05	2(a)	280
n-Propanol	73	92	5[a]	1.14	2.00	2(a)	514
Isopropanol	69	82	3[a]	1.66	1.89	2(b)	40
n-Butanol	77	3	10[a]	0.89	2.12	2(b)	135
Isobutanol	76	95	5[a]	0.90	1.83	2(d)	342
sec-Butanol	74	92			2.72		
Cyclohexanol	None						
Ethylene glycol	None						
Methyl Cellosolve							
Ethyl Cellosolve							
Butyl Cellosolve							
Chlorinated hydrocarbons							
Methylene dichloride	None		3	0.40	−0.48	8	62
Chloroform	None		1.5	0.47	0.02	8	55
1,2-Dichloroethane	75	80		0.58	0.50		
1,1,1-Trichloroethane	None		<1.5	0.10	0.22	8	81
Trichloroethylene	None		1.5	−0.05	−0.34	8	80
Perchloroethylene	None		5	0.03	0.04	8	78
Monochlorobenzene	None		5	0.18	0.38	8	166
Ketones							
Acetone	56	11	2[a]	0.82	0.83	3+4	80
Methyl ethyl ketone[b]	74	71	1.5[a]	0.39	0.58	3+4	259
MIBK							
Cyclohexanone							
N-Methylpyrrolidone							
Ethers							
Diethyl ether							
Diisopropyl ether	None		1.5	0.06	0.07	3+4	530
1,4-Dioxane	None		3	0.25	0.46	3+4	440
Tetrahydrofuran			1.5	−0.31	−0.15	3+4	429
Esters							
Methyl acetate	None		2	0.51	0.48	5	339
Ethyl acetate	74	57	1.1[a]	0.26	0.30	5	437
Butyl acetate	None		4	0.09	−0.17	5	573
Miscellaneous							
Dimethylformamide	None		>20	0.62	1.71	8	117
Dimethyl sulphoxide	None		15	1.26	2.33	8	107
Pyridine	None		5	0.37	0.07	8	141
Acetonitrile	65	83	3[a]	1.51	2.37	8	86
Furfuraldehyde	None		>10	1.29	1.33	3+4	35
Water	66	96					

[a] The relative volatility between the less volatile component of the binary mixture and the azeotrope.
[b] Reacts.

Recovery notes: carbon tetrachloride

Carbon tetrachloride (CTET) is very toxic (TLV currently 10 ppm but about to be further reduced) and is harmful as a water contaminant. It has an ozone depletion potential of 1.04, considerably higher than some of the CFCs currently being reduced in use under the terms of the Montreal Protocol. However, very small amounts of CTET are now used as a solvent, most being consumed by the production of CFCs.

CTET attacks aluminium but can be stored safely in other usual engineering metals. It reacts violently with sodium.

When heated to a high temperature or involved in a fire CTET gives off fumes of phosgene. When fire fighting, self-contained breathing apparatus should be worn.

CTET may be used in contact with air and light up to 130 °C. It hydrolyses readily in the presence of free water at the boiling point of its water azeotrope (66 °C) and will corrode mild and stainless steel under these conditions. Monel, nickel and other alloys resistant to HCl must be used. Aluminium should be rigorously excluded. Properly stabilized but containing only dissolved water, 90 °C can be reached without hydrolysis provided that aluminium, sodium, magnesium and similar substances are absent.

If silica gel is used to dry CTET it should not be regenerated as there is evidence that CTET and water react to produce phosgene at regeneration temperatures.

Nitromethane and 1,4-dioxane are used as inhibitors.

1,2-DICHLOROETHANE

1 **Names**　　　　　　　　　　　　　　　　　1,2-Dichloroethane
　　　　　　　　　　　　　　　　　　　　　　Ethylene dichloride
　　　　　　　　　　　　　　　　　　　　　　EDC

2 **Physical properties**

Molecular weight	99
Empirical formula	$C_2H_4Cl_2$
Boiling point (°C)	83.5
Freezing point (°C)	−36
Specific gravity (20/4 °C)	1.253
Liquid expansion coefficient (per °C)	0.00116
Surface tension (at 20 °C in dyn/cm)	32.23
Absolute viscosity (at 25 °C in cP)	0.9

3 **Fire and health hazards**

Flash point (closed cup) (°C)	13
Autoignition temperature (°C)	413
Lower explosive limit (ppm)	62000
Upper explosive limit (ppm)	169000
IDLH (ppm)	1000
Odour threshold (ppm)	400
TWA–TLV (ppm)	10
Saturated vapour concentration (70 °F in ppm)	94000
Vapour density (relative to air)	3.4

4 **Solvent properties**

Hildebrand solubility parameter	9.8
Dipole moment (D)	1.8
Dielectric constant (25 °C)	10.45
Evaporation time (diethyl ether = 1.0)	2.7

5 **Aqueous effluent characteristics**

Solubility in water (at 25 °C in % w/w)	0.81
Solubility of water in (at 25 °C in % w/w)	0.15
Biological oxygen demand (w/w)	0.002
Chemical oxygen demand (w/w)	1.0
Theoretical oxygen demand (w/w)	0.97

6 **Handling details**

Hazchem number	1184
Hazard class	2 YE
Vapour pressure (at 21 °C in mmHg)	71
Loss per transfer (% of liquid transferred)	0.03

7 **Vapour pressure constants (mmHg, log base 10)**

Antoine equation	$A = 7.02530$
	$B = 1271.254$
	$C = 222.927$
Cox chart	$A = 7.04532$
	$B = 1303.5$

8 **Thermal information**

Latent heat (cal/mol)	7623
Specific heat (cal/mol/°C)	30.89
Heat of combustion (cal/mol)	268983

Azeotropes and activity coefficients of component X: 1,2-dichloroethane

Component Y	Azeotrope °C	Azeotrope % w/w X	α	$A_{XY} = \ln \gamma_X^\infty$	$A_{YX} = \ln \gamma_Y^\infty$	Source[11] Vol.	Page
Hydrocarbons							
n-Pentane							
n-Hexane				0.88	1.32		
n-Heptane	81	76	2.5[a]	0.84	1.23	6(c)	444
Cyclohexane	74	50	2[a]	0.97	1.07	6(a)	158
Benzene	80	18	<1.5	0.01	0.02	7	139
Toluene		None	2	1.05	0.05	7	382
Ethylbenzene		None	5	−0.68	−0.12	7	466
Xylenes		None	7	0.07	0.23	7	490
Alcohols							
Methanol	60	67	8[a]	1.53	2.36	2(a)	47
Ethanol	70	63	2.5[a]	1.28	1.60	2(a)	299
n-Propanol	81	81	2.8[a]	1.06	1.30	2(a)	520
Isopropanol	75	57					
n-Butanol		None	5/10	0.84	1.20	2(b)	137
Isobutanol	84	84	5[a]	0.83	1.57	2(b)	272
sec-Butanol	83	88					
Cyclohexanol							
Ethylene glycol							
Methyl Cellosolve							
Ethyl Cellosolve							
Butyl Cellosolve							
Chlorinated hydrocarbons							
Methylene dichloride		None	5	−0.04	0.00	8	263
Chloroform				0.05	0.03		
Carbon tetrachloride	75	20		0.50	0.58		
1,1,1-Trichloroethane		None	<1.5	0.07	0.46	8	363
Trichloroethylene	82	61	<1.5[a]	0.34	0.31	8	351
Perchloroethylene		None	4/3	0.71	0.53	8	340
Monochlorobenzene							
Ketones							
Acetone		None	3/5	−0.07	−0.36	3+4	144
Methyl ethyl ketone		None		−0.29	−0.25		
MIBK							
Cyclohexanone							
N-Methylpyrrolidone							
Ethers							
Diethyl ether		None					
Diisopropyl ether		None					
1,4-Dioxane		None	1.5	0.30	−0.11	3+4	447
Tetrahydrofuran			1.5				
Esters							
Methyl acetate							
Ethyl acetate				0.1	>0.5		
Butyl acetate							
Miscellaneous							
Dimethylformamide							
Dimethyl sulphoxide							
Pyridine							
Acetonitrile	79	51	<1.5[a]	0.35	0.32	8	364
Furfuraldehyde				0.13	0.24		
Water	72	92		6.31	4.01		

[a]The relative volatility between the less volatile component of the binary mixture and the azeotrope.

Recovery notes: 1,2-dichloroethane (EDC)

EDC is produced in very large quantities as a raw material for the manufacture of PVC. As a result it is a relatively low-cost solvent. This simultaneously reduces the incentive for the original user to recover it and for the customer of a merchant recoverer to buy recovered rather than virgin material. However, EDC's high chlorine content makes it expensive to incinerate and this factor 'subsidises' recovery.

Unlike other chlorinated hydrocarbons used as solvents, EDC has a low flash point and cannot be considered as a safety solvent so it is used primarily in the pharmaceutical and fine chemical industry and must be recovered to a high standard of purity.

Even with low levels of water present EDC is not stable when it is heated. It hydrolyses slowly at 80 °C and rapidly at 100 °C. Since this reaction produces hydrochloric acid, equipment for distillation should not be fabricated from stainless steel unless air, light and water can be completely excluded. Under such rigid conditions EDC is stable to 160 °C. A commonly used stabilizer is diisopropylamine at a level of 0.05–0.10%, but it is a fairly reactive chemical and may not be acceptable for some reactions where EDC is used.

The combination of a low TLV with a fairly high vapour pressure and an odour that is difficult to detect make EDC dangerous from the toxicity point of view.

The flash point of EDC and its range of flammability mean that it is advisable to store it under inert gas or nitrogen blanketing.

1,1,1-TRICHLOROETHANE

1 Names

1,1,1-Trichloroethane
Methylchloroform
Chlorothene
Not 1,1,2-trichloroethane

2 Physical properties

Molecular weight	133
Empirical formula	$C_2H_3Cl_3$
Boiling point (°C)	74
Freezing point (°C)	−38
Specific gravity (20/4 °C)	1.329
Liquid expansion coefficient (per °C)	0.0013
Surface tension (at 20 °C in dyn/cm)	30
Absolute viscosity (at 25 °C in cP)	0.65

3 Fire and health hazards

Flash point (closed cup) (°C)	None
Autoignition temperature (°C)	537
Lower explosive limit (ppm)	65 000
Upper explosive limit (ppm)	155 000
IDLH (ppm)	1 000
Odour threshold (ppm)	100
TWA–TLV (ppm)	350
Saturated vapour concentration (70 °F in ppm)	170 000
Vapour density (relative to air)	2.4

4 Solvent properties

Hildebrand solubility parameter	7.7
Dipole moment (D)	1.7
Dielectric constant (25 °C)	7.0
Evaporation time (diethyl ether = 1.0)	2.4

5 Aqueous effluent characteristics

Solubility in water (at 25 °C in % w/w)	0.13
Solubility of water in (at 25 °C in % w/w)	0.05
Biological oxygen demand (w/w)	–
Chemical oxygen demand (w/w)	–
Theoretical oxygen demand (w/w)	0.48

6 Handling details

Hazchem number	2831
Hazard class	2 Z
Vapour pressure (at 21 °C in mmHg)	109
Loss per transfer (% of liquid transferred)	0.059

7 Vapour pressure constants (mmHg, log base 10)

Antoine equation	$A = 6.90633$
	$B = 1211.31$
	$C = 226.816$
Cox chart	$A = 7.01846$
	$B = 1257.7$

8 Thermal information

Latent heat (cal/mol)	7780
Specific heat (cal/mol/°C)	31.9
Heat of combustion (cal/mol)	232915

Azeotropes and activity coefficients of component X: 1,1,1-trichloroethane

Component Y	Azeotrope °C	Azeotrope % w/w X	α	$A_{XY} = \ln \gamma_X^\infty$	$A_{YX} = \ln \gamma_Y^\infty$	Source[11] Vol.	Source[11] Page
Hydrocarbons							
n-Pentane							
n-Hexane	67	29	<1.5[a]	0.29	0.34	6(a)	473
n-Heptane							
Cyclohexane							
Benzene		None	<1.5	0.03	0.03	7	121
Toluene							
Ethylbenzene							
Xylenes							
Alcohols							
Methanol	54	78		1.70	1.93		
Ethanol							
n-Propanol							
Isopropanol							
n-Butanol							
Isobutanol							
sec-Butanol							
Cyclohexanol							
Ethylene glycol							
Methyl Cellosolve							
Ethyl Cellosolve							
Butyl Cellosolve							
Chlorinated hydrocarbons							
Methylene dichloride				0.17			
Chloroform							
Carbon tetrachloride		None	⩽1.5	0.00	−0.03	8	82
1,2-Dichloroethane		None	<1.5	0.46	0.07	8	363
Trichloroethylene							
Perchloroethylene							
Monochlorobenzene							
Ketones							
Acetone							
Methyl ethyl ketone							
MIBK							
Cyclohexanone							
N-Methylpyrrolidone							
Ethers							
Diethyl ether							
Diisopropyl ether							
1,4-Dioxane							
Tetrahydrofuran							
Esters							
Methyl acetate							
Ethyl acetate							
Butyl acetate							
Miscellaneous							
Dimethylformamide							
Dimethyl sulphoxide							
Pyridine							
Acetonitrile							
Furfuraldehyde							
Water	65	96		8.67	5.43		

[a] The relative volatility between the less volatile component of the binary mixture and the azeotrope.

Recovery notes: 1,1,1-trichloroethane

The properties of 1,1,1-trichloroethane are generally similar to those of trichloroethylene and perchloroethylene, and it is widely used in mechanical engineering and other industries where the standards taken for granted in the chemical industry do not apply.

Its stabilizer content tends to be high (3–7%) and additional stabilizer may be needed if the solvent is recovered from large amounts of water or from active carbon filters. The need for extra stabilizer can be detected by scratching the surface of an aluminium test coupon immersed in a beaker of solvent where air cannot reach it, and observing if the scratches give rise to a dark red coloration.

Each manufacturer of 1,1,1-trichloroethane has a different stabilizer formulation, but if material must be re-stabilized without their assistance, the following mixtures have been used:

1.
   - 0.1–1.0% butene oxide
   - 0.5–5.0% nitromethane
   - 0.1–1.0% dimethoxyethane
   - 2.0%   *tert*-butanol

2.
   - 0.5% tetrahydrofuran
   - 1.5% *tert*-butanol
   - 1.5% 3-hydroxy-3-methylbutyne

Other materials that may be found in inhibitor mixtures are 1,4-dioxane, isobutanol, *sec*-butanol, toluene and *N*-methylpyrrole.

The high percentage of additives and their generally hydrophilic nature make the solubility of water in commercially available 1,1,1-trichloroethane much higher than in the pure solvent. Saturated water contents about twice the figure for pure solvent are typical. Loss of water-soluble additives also occurs if large quantities of water are in contact with the solvent.

1,1,1-Trichloroethane used in degreasing is usually recovered when 25% of the mixture is oil or grease. It should not normally be steam distilled but, provided it has a proper inhibitor system, it can be heated to 120 °C in stripping from the residue without the solvent decomposing. A further 10% of the residue at this point is likely to be solvent and this can be mostly recovered by vacuum distillation.

If the solvent mixture for recovery is acidic it should be distilled after adding 1% w/w soda ash. Strong alkalis such as caustic soda or potash should never be used. To squeeze the last available solvent from a residue, a similar addition of soda ash should be made before steam stripping. Such residues often foam on boiling and an anti-foam addition may be necessary.

1,1,1-Trichloroethane recovered by steam stripping will need re-inhibiting or blending off with at least four times its volume of virgin material. Even then the mixture should be checked for acid acceptance and aluminium attack.

The complex inhibitor used in 1,1,1-trichloroethane will not necessarily distil over in a batch distillation as a single entity or evenly throughout the charge. It is therefore inadvisable to distil into drums from a batch distillation, but rather to distil into a tank which will hold the complete batch and in which the inhibitor can be mixed evenly.

Methods of test and test kits can be obtained from 1,1,1-trichloroethane manufacturers.

As a degreaser 1,1,1-trichloroethane is in competition with trichloroethylene, but it gives rise to lower losses in use, has an appreciably higher TLV and is more stable at welding temperatures. However, unlike most other chlorinated hydrocarbons (except carbon tetrachloride), it has a significantly high ozone depletion potential. Although at 0.15 this is lower than that of the chlorofluorocarbons (CFCs), the scale of usage and discharge to the atmosphere is large enough to make the damage 1,1,1-trichloroethane does to the ozone layer of the same order of magnitude as the CFCs. In the present climate of legislation and public opinion it would seem unwise to choose 1,1,1-trichloroethane as the solvent basis for any long-term new product or operation.

The most suitable protective clothing for use in handling 1,1,1-trichloroethane is neoprene.

For storing and handling stabilized 1,1,1-trichloroethane the usual metals for plant construction are satisfactory, although acidic material will attack aluminium and the welds and stressed areas of stainless steel. 1,1,1-Trichloroethane will react with magnesium even when stabilized.

For continuous operations involving the steam distillation of 1,1,1-trichloroethane, monel or nickel may be needed but for occasional use heavy-gauge mild steel will be adequate, although the recovered distillate may be brown.

If recovered 1,1,1-trichloroethane with a very low water content is needed, the best way to remove water down to the 20 ppm level is by contacting the solvent with ion-exchange resin. The solubility of water in 1,1,1-trichloroethane is increased by the stabilizer content.

TRICHLOROETHYLENE

1. **Names**

   Trichloroethylene
   Triclene
   Trike
   Not trichloroethane

2. **Physical properties**

Molecular weight	131
Empirical formula	C_2HCl_3
Boiling point (°C)	87
Freezing point (°C)	−86
Specific gravity (20/4 °C)	1.464
Liquid expansion coefficient (per °C)	0.00117
Surface tension (at 20 °C in dyn/cm)	29.5
Absolute viscosity (at 25 °C in cP)	0.566

3. **Fire and health hazards**

Flash point (closed cup) (°C)	None
Autoignition temperature (°C)	420
Lower explosive limit (ppm)	80 000
Upper explosive limit (ppm)	105 000
IDLH (ppm)	500
Odour threshold (ppm)	5
TWA–TLV (ppm)	50
Saturated vapour concentration (70 °F in ppm)	80 000
Vapour density (relative to air)	4.55

4. **Solvent properties**

Hildebrand solubility parameter	8.0
Dipole moment (D)	0.9
Dielectric constant (20 °C)	3.42
Evaporation time (diethyl ether = 1.0)	3.6

5. **Aqueous effluent characteristics**

Solubility in water (at 25 °C in % w/w)	0.11
Solubility of water in (at 25 °C in % w/w)	0.033
Biological oxygen demand (w/w)	–
Chemical oxygen demand (w/w)	–
Theoretical oxygen demand (w/w)	0.61

6. **Handling details**

Hazchem number	1710
Hazard class	2 Z
Vapour pressure (at 21 °C in mmHg)	60.3
Loss per transfer (% of liquid transferred)	0.029

7. **Vapour pressure constants (mmHg, log base 10)**

Antoine equation	$A = 6.51827$
	$B = 1018.603$
	$C = 192.731$
Cox chart	$A = N/A$
	$B = N/A$

8. **Thermal information**

Latent heat (cal/mol)	7467
Specific heat (cal/mol/°C)	30.1
Heat of combustion (cal/mol)	206 436

Azeotropes and activity coefficients of component X: trichloroethylene

Component Y	Azeotrope °C	Azeotrope % w/w X	α	$A_{XY} = \ln \gamma_X^\infty$	$A_{YX} = \ln \gamma_Y^\infty$	Source[11] Vol.	Page
Hydrocarbons							
n-Pentane							
n-Hexane		None	4	0.38	0.43	6(a)	462
n-Heptane		None					
Cyclohexane	80	17	1.5[a]	0.42	0.25	6(a)	155
Benzene		None	<1.5	0.01	0.04	7	116
Toluene		None	2	−0.08	−0.21	7	370
Ethylbenzene							
Xylenes							
Alcohols							
Methanol	59	62	7[a]	2.08	1.97	2(a)	40
Ethanol	71	72	3[a]	1.57	2.02	2(a)	295
n-Propanol	82	83	2.5[a]	0.97	1.70	2(a)	518
Isopropanol	75	70		1.39	1.15		
n-Butanol	87	97					
Isobutanol	85	91	3.5[a]	0.91	1.35	2(d)	345
sec-Butanol	84	85					
Cyclohexanol		None					
Ethylene glycol							
Methyl Cellosolve							
Ethyl Cellosolve							
Butyl Cellosolve							
Chlorinated hydrocarbons							
Methylene dichloride				0.38			
Chloroform							
Carbon tetrachloride		None	<1.5	−0.35	−0.05	8	81
1,2-Dichloroethane	82	39	<1.5[a]	0.31	0.34	8	81
1,1,1-Trichloroethane				0.00	0.02		
Perchloroethylene		None	3	0.09	1.48	8	326
Monochlorobenzene							
Ketones							
Acetone							
Methyl ethyl ketone		None	<1.5	0.06	0.22	3+4	26
MIBK							
Cyclohexanone							
N-Methylpyrrolidone							
Ethers							
Diethyl ether							
Diisopropyl ether							
1,4-Dioxane		None					
Tetrahydrofuran							
Esters							
Methyl acetate							
Ethyl acetate		None	<1.5	−0.16	0.09	5	450
Butyl acetate		None	2/3	−0.30	−0.51	5	575
Miscellaneous							
Dimethylformamide							
Dimethyl sulphoxide							
Pyridine							
Acetonitrile	75	71	2[a]	1.20	1.25	8	350
Furfuraldehyde		None	>10	0.37	2.90	3+4	37
Water	73	94					

[a] The relative volatility between the less volatile component of the binary mixture and the azeotrope.

Recovery notes: trichloroethylene

Because of its wide usage as a degreasing solvent, the most commonly met contaminant of trichloroethylene is a mixture of high-boiling hydrocarbons. Provided this 'waste' has not been contaminated with other solvents, the recovery of trichloroethylene for further degreasing use is easy. It can either be by steam distillation or by vacuum distillation or a combination of both techniques. The distillate will form two easily separating phases with a very low concentration of water in the recovered solvent.

Provided that the solvent–oil mixture has not been contaminated with high-boiling chlorine compounds it should be possible to use the residue from recovery as a fuel. However, trichloroethylene is very effective for cleaning chlorinated paraffin waxes from equipment and the chances of ending up with a residue containing too much chlorine to burn in anything but a chemical incinerator are appreciable.

In trying to strip out the last traces of trichloroethylene temperatures should not exceed 120 °C, as above this temperature it may begin to decompose, generating HCl.

Each manufacturer has an inhibitor package which usually includes an acid acceptor to mop up any acid formed in use or redistillation, an antioxidant and a metal deactivator to protect aluminium against attack. The various additives that give chemical protection are usually compounds that will not be fractionated out of the solvent when it is being recovered and may be present to a total of 2% in virgin material. Since some trichloroethylene is lost during use, recycling and recovery of the make-up of virgin material usually bring with it enough inhibitor to maintain coverage.

If for any reason trichloroethylene needs to be inhibited without the assistance of a manufacturer, the technical literature quotes:

- against oxidation: 0.01–0.02% of 1-ethoxy-2-iminoethane;
- against heat and oxidation: pyrrole plus a 1,2-epoxide;
- general: tetrahydrothiophene and an amine or a phenol or a 1,2-epoxide;
- general: 0.05–1.0% of methacrylonitrile.

Caustic soda or potash should never be used to remove acidity from trichloroethylene since they react together to produce dichloroacetylene, which is spontaneously flammable. Milder alkalis such as soda ash or sodium hydrogencarbonate can be used safely.

The alkali or alkaline earth metals can react with chlorinated solvents in general and such solvents are not suitable reaction media for them.

Although aluminium can be degreased with inhibited trichloroethylene, the use of aluminium tanks and plant for handling it is not safe.

In a general-purpose solvent recovery operation, the high density (1.46) of trichloroethylene should always be borne in mind and tanks only designed and tested to store water should not be overloaded by filling with trichloroethylene.

The inhalation of trichloroethylene vapour has led to addiction and to serious harm to the addict. Supervisors of operatives handling trichloroethylene should be on the alert for such bad practice.

PERCHLOROETHYLENE

1 **Names**

Perchloroethylene
Tetrachloroethylene
Perk
Tetrachloroethene

2 **Physical properties**

Molecular weight	166
Empirical formula	C_2Cl_4
Boiling point (°C)	121
Freezing point (°C)	-23
Specific gravity (20/4 °C)	1.622
Liquid expansion coefficient (per °C)	0.00102
Surface tension (at 20 °C in dyn/cm)	32
Absolute viscosity (at 25 °C in cP)	0.88

3 **Fire and health hazards**

Flash point (closed cup) (°C)	None
Autoignition temperature (°C)	None
Lower explosive limit (ppm)	None
Upper explosive limit (ppm)	None
IDLH (ppm)	400
Odour threshold (ppm)	5
TWA–TLV (ppm)	50
Saturated vapour concentration (70 °F in ppm)	23 000
Vapour density (relative to air)	5.8

4 **Solvent properties**

Hildebrand solubility parameter	4.5
Dipole moment (D)	0
Dielectric constant (20 °C)	2.3
Evaporation time (diethyl ether = 1.0)	11

5 **Aqueous effluent characteristics**

Solubility in water (at 25 °C in % w/w)	0.015
Solubility of water in (at 25 °C in % w/w)	0.008
Biological oxygen demand (w/w)	0.06
Chemical oxygen demand (w/w)	
Theoretical oxygen demand (w/w)	0.39

6 **Handling details**

Hazchem number	1897
Hazard class	22
Vapour pressure (at 21 °C in mmHg)	16
Loss per transfer (% of liquid transferred)	0.013

7 **Vapour pressure constants (mmHg, log base 10)**

Antoine equation	$A = 7.62930$
	$B = 1803.960$
	$C = 258.976$
Cox chart	$A = N/A$
	$B = N/A$

8 **Thermal information**

Latent heat (cal/mol)	8316
Specific heat (cal/mol/°C)	34.9
Heat of combustion (cal/mol)	N/A

Azeotropes and activity coefficients of component X: perchloroethylene

Component Y	Azeotrope °C	Azeotrope % w/w X	α	$A_{XY} = \ln \gamma_X^\infty$	$A_{YX} = \ln \gamma_Y^\infty$	Source[11] Vol.	Page
Hydrocarbons							
n-Pentane	None						
n-Hexane	None		10	0.34	0.41	6(a)	453
n-Heptane							
Cyclohexane							
Benzene	None		4	0.23	0.25	7	112
Toluene	None						
Ethylbenzene	None						
Xylenes	None						
Alcohols							
Methanol	64	41	7	2.47	2.82	2(a)	37
Ethanol	77	37	15^a	1.80	1.79	2(c)	285
n-Propanol	94	52	5^a	1.57	1.73	2(a)	517
Isopropanol	82	30		1.68	1.75	2(b)	42
n-Butanol	109	70	2^a	1.33	1.14	2(d)	154
Isobutanol	103	60	2.5^a	1.22	1.30	2(d)	344
sec-Butanol	97	43	3.5^a	1.38	1.20	2(d)	240
Cyclohexanol							
Ethylene glycol	119	94					
Methyl Cellosolve	109	76					
Ethyl Cellosolve	116	84	2.5^a	0.97	1.45	2(d)	396
Butyl Cellosolve	None						
Chlorinated hydrocarbons							
Methylene dichloride	None		11	−0.06	0.11	8	256
Chloroform	None		7	0.17	0.15	8	215
Carbon tetrachloride	None		5	0.04	0.03	8	78
1,2-Dichloroethane	None		4/3	0.53	0.71	8	340
1,1,1-Trichloroethane							
Trichloroethylene	None		3	1.48	0.09	8	326
Monochlorobenzene							
Ketones							
Acetone	None						
Methyl ethyl ketone	None						
MIBK	114	48					
Cyclohexanone							
N-Methylpyrrolidone							
Ethers							
Diethyl ether							
Diisopropyl ether	None						
1,4-Dioxane							
Tetrahydrofuran							
Esters							
Methyl acetate							
Ethyl acetate							
Butyl acetate	121	79					
Miscellaneous							
Dimethylformamide							
Dimethyl sulphoxide							
Pyridine	113	52	1.5^a	0.62	0.70	8	346
Acetonitrile	None		10	1.59	1.37	8	342
Furfuraldehyde							
Water	88	84					

^{a}The relative volatility between the less volatile component of the binary mixture and the azeotrope.

Recovery notes: perchloroethylene

Of all the normally used solvents, perchloroethylene has the highest density. It is therefore the most likely to cause problems in the overloading of pump motors and structures carrying storage vessels. If recovery is proposed on a general-purpose plant, a careful survey of the effect that excess density may have must be made. A pallet of 4×200 l drums will weigh almost 1.5 Te.

As with other chlorinated hydrocarbons containing stabilizers, it is inadvisable to have water in long-term contact with perchloroethylene and when it is being handled in a recovery operation, facilities should exist for detecting water lying as a separate phase on the solvent's surface and removing it if it is found.

Conventional water-finding paste will not work on a 'top' layer of water because as the dipstick or gauging tape passes through the water layer, the colour change is triggered over the whole wetted length. Because hydrocarbon-based greases, such as vaseline, are very soluble in chlorinated hydrocarbons they will be stripped from any part of the dipstick immersed in the solvent whereas they will not be dissolved off that part that is only wetted by water.

Drain cocks on the side of the tank to allow top phase water to be drained off the solvent surface should be fitted at intervals on storage vessels where water might accumulate. Water that has been lying on the surface of solvent for a prolonged period is likely to be mildly acidic. It is, of course, better to keep water out of storage tanks and process units and a phase separator (*see* 1,1,1-trichloroethane) at the head of any distillation column is desirable.

Among the chlorinated C_2 hydrocarbons, perchloroethylene is the one that gives off most phosgene when heated by flame cutting or contact with a very hot surface. Since it is very difficult for an operative to detect this gas before he may have received a harmful, or even fatal, dose, plant cleaning before hot work is important.

Despite perchloroethylene's low fire hazard, vapour that may arise in recovery operations should not be able to reach very hot surfaces (e.g. boiler shell). If electrical heating is used for small recovery units the heating element should never be exposed while the current is on.

The high molecular weight of perchloroethylene means that its vapour can travel a long way in drains, basements and other places where air is relatively undisturbed. Since it is very sparingly miscible in water, an undetected layer of solvent can lie underwater in a drainage system, evolving vapour at concentrations many times the TLV.

Perchloroethylene, provided that it is properly stabilized, can be used in the presence of air, light and water up to 140 °C. If it should ever be necessary to inhibit without the assistance of the solvent's original manufacturer, the literature records the following stabilizers: pyrrole or its derivatives plus an epoxy compound; 0.005–1.0% of diallylamine; 0.005–1.0% of tripropylamine; 0.01–1.0% of 3-chloropropyne; or 0.01–1.0% of 1,4-dichlorobut-1-yne.

Perchloroethylene will decompose if adsorbed on activated carbon and then steam stripped in the regeneration, unless it is adequately stabilized.

Perchloroethylene has a zero ozone depletion potential and a very low photochemical activity.

Aluminium should not be used in plants that handle perchloroethylene and protective clothing should be made of neoprene or poly(vinyl alcohol).

MONOCHLOROBENZENE

1 Names

Monochlorobenzene
Chlorobenzene
MCB
Dowtherm E

2 Physical properties

Molecular weight	113
Empirical formula	C_6H_5Cl
Boiling point (°C)	132
Freezing point (°C)	-46
Specific gravity (20/4 °C)	1.106
Liquid expansion coefficient (per °C)	0.00098
Surface tension (at 20 °C in dyn/cm)	33
Absolute viscosity (at 25 °C in cP)	0.8

3 Fire and health hazards

Flash point (closed cup) (°C)	29
Autoignition temperature (°C)	640
Lower explosive limit (ppm)	13 000
Upper explosive limit (ppm)	71 000
IDLH (ppm)	2400
Odour threshold (ppm)	0.2
TWA–TLV (ppm)	75
Saturated vapour concentration (70 °F in ppm)	13 500
Vapour density (relative to air)	3.9

4 Solvent properties

Hildebrand solubility parameter	9.5
Dipole moment (D)	1.6
Dielectric constant (20 °C)	5.62
Evaporation time (diethyl ether = 1.0)	10.0

5 Aqueous effluent characteristics

Solubility in water (at 25 °C in % w/w)	0.049
Solubility of water in (at 25 °C in % w/w)	0.033
Biological oxygen demand (w/w)	0.03
Chemical oxygen demand (w/w)	0.4
Theoretical oxygen demand (w/w)	2.05

6 Handling details

Hazchem number	1134
Hazard class	2 Y
Vapour pressure (at 21 °C in mmHg)	9.5
Loss per transfer (% of liquid transferred)	0.53×10^{-3}

7 Vapour pressure constants (mmHg, log base 10)

Antoine equation	$A = 7.17294$
	$B = 1549.200$
	$C = 229.260$
Cox chart	$A = 7.18576$
	$B = 1558.4$

8 Thermal information

Latent heat (cal/mol)	8814
Specific heat (cal/mol/°C)	35.0
Heat of combustion (cal/mol)	757 100

Azeotropes and activity coefficients of component X: monochlorobenzene

Component Y	Azeotrope °C	Azeotrope % w/w X	α	$A_{XY} = \ln \gamma_X^\infty$	$A_{YX} = \ln \gamma_Y^\infty$	Source[11] Vol.	Page
Hydrocarbons							
n-Pentane		None					
n-Hexane		None	11	0.60	0.46	6(a)	529
n-Heptane		None	3	0.51	0.61	6(b)	119
Cyclohexane		None	5	0.76	0.29	6(a)	202
Benzene		None	5	0.37	0.03	7	243
Toluene		None	2	−0.03	0.00	7	416
Ethylbenzene		None	<1.5	−0.01	0.01	7	469
Xylenes		None	<1.5	−0.13	−0.09	7	508
Alcohols							
Methanol		None	>20	2.16	1.69	2(a)	204
Ethanol		None	20	1.75	1.53	2(a)	397
n-Propanol	97	17	5[a]	1.26	1.12	2(a)	552
Isopropanol		None	10/3	1.56	1.23	2(d)	64
n-Butanol	115	44	3[a]	0.89	0.93	2(b)	174
Isobutanol	107	37	3[a]	0.93	1.02	2(d)	357
sec-Butanol		None	5	0.98	1.04	2(d)	258
Cyclohexanol		None	3	0.46	0.95	2(b)	393
Ethylene glycol	130	94					
Methyl Cellosolve	119	57	2[a]	0.85	1.23	2(d)	120
Ethyl Cellosolve	127	68					
Butyl Cellosolve		None					
Chlorinated hydrocarbons							
Methylene dichloride		None					
Chloroform		None	9	−0.19	0.18	8	244
Carbon tetrachloride		None	6	0.38	0.18	8	166
1,2-Dichloroethane		None					
1,1,1-Trichloroethane		None					
Trichloroethylene		None					
Perchloroethylene		None					
Ketones							
Acetone		None	11	0.45	0.30	3+4	192
Methyl ethyl ketone		None	6	0.44	0.25	3+4	283
MIBK							
Cyclohexanone		None					
N-Methylpyrrolidone		None					
Ethers							
Diethyl ether		None					
Diisopropyl ether		None					
1,4-Dioxane		None					
Tetrahydrofuran		None					
Esters							
Methyl acetate		None	10	0.29	0.16	5	374
Ethyl acetate		None	5	0.84	0.17	5	492
Butyl acetate							
Miscellaneous							
Dimethylformamide							
Dimethyl sulphoxide							
Pyridine		None					
Acetonitrile		None	10/3	1.01	1.21	8	381
Furfuraldehyde		None					
Water	90	72					

[a] The relative volatility between the less volatile component of the binary mixture and the azeotrope.

Recovery notes: monochlorobenzene

Monochlorobenzene (MCB) is so stable in the absence of water that it has been used as a heat-transfer liquid (Dowtherm E). At its boiling point in the presence of water it hydrolyses slightly so long-term boiling in a batch distillation can result in acid distillates. Since water is very sparingly soluble in MCB, it can usually be decanted before processing.

MCB has little to recommend it as an azeotropic entrainer for water since it is of similar flammability to the more stable aromatic hydrocarbons of similar volatility (xylene, ethylbenzene) and is also marginally less acceptable environmentally.

MCB forms ideal binary systems with both toluene and ethylbenzene and these are therefore suitable as a test mixture for columns. The toluene mixture is appropriate for columns of 5–20 actual trays and the ethylbenzene for 15–40 actual trays. The latter is also suitable for testing at pressures of 100 mmHg and more.

Although MCB is very stable, it does react with sodium, which cannot be used for drying it.

The very low mutual solubility of MCB with water allows it to be used for extracting ketones from water particularly when they boil at a lower temperature than MCB, since then comparatively large quantities of MCB can be cycled through liquid–liquid extraction and stripping without having to be evaporated.

ACETONE

1 Names

Acetone
Dimethyl ketone
Propan-2-one

2 Physical properties

Molecular weight	58
Empirical formula	C_3H_6O
Boiling point (°C)	56
Freezing point (°C)	−95
Specific gravity (20/4 °C)	0.790
Liquid expansion coefficient (per °C)	0.001
Surface tension (at 20 °C in dyn/cm)	23.3
Absolute viscosity (at 25 °C in cP)	0.33

3 Fire and health hazards

Flash point (closed cup) (°C)	−18
Autoignition temperature (°C)	465
Lower explosive limit (ppm)	26 000
Upper explosive limit (ppm)	128 000
IDLH (ppm)	20 000
Odour threshold (ppm)	100
TWA–TLV (ppm)	750
Saturated vapour concentration (70 °F in ppm)	342 760
Vapour density (relative to air)	2.0

4 Solvent properties

Hildebrand solubility parameter	10.0
Dipole moment (D)	2.9
Dielectric constant (20 °C)	20.7
Evaporation time (diethyl ether = 1.0)	1.8

5 Aqueous effluent characteristics

Solubility in water (at 25 °C in % w/w)	Infinite
Solubility of water in (at 25 °C in % w/w)	Infinite
Biological oxygen demand (w/w)	7.22
Chemical oxygen demand (w/w)	2.0
Theoretical oxygen demand (w/w)	2.21

6 Handling details

Hazchem number	1090
Hazard class	2 YE
Vapour pressure (at 21 °C in mmHg)	194
Loss per transfer (% of liquid transferred)	0.077

7 Vapour pressure constants (mmHg, log base 10)

Antoine equation	$A = 7.11714$
	$B = 1210.596$
	$C = 229.664$
Cox chart	$A = 7.18990$
	$B = 1232.4$

8 Thermal information

Latent heat (cal/mol)	7076
Specific heat (cal/mol/°C)	23.78
Heat of combustion (cal/mol)	394 864

Azeotropes and activity coefficients of component X: acetone

Component Y	Azeotrope °C	Azeotrope % w/w X	α	A_{XY} $=\ln \gamma_X^\infty$	A_{YX} $=\ln \gamma_Y^\infty$	Source[11] Vol.	Page
Hydrocarbons							
n-Pentane	32	20	5[a]	1.74	1.39	3+4	187
n-Hexane	50	59	4[a]	1.47	1.64	3+4	225
n-Heptane	56	90	8[a]	1.17	1.70	3+4	242
Cyclohexane	53	67	5[a]	1.60	1.46	3+4	213
Benzene		None	2/3	0.47	0.33	3+4	197
Toluene		None		0.60	0.44	3+4	236
Ethylbenzene		None					
Xylenes		None					
Alcohols							
Methanol	56	88	1.5/2[a]	0.61	0.74	2(a)	76
Ethanol		None	2.8	0.62	0.58	2(a)	320
n-Propanol							
Isopropanol		None	3	0.89	0.84	2(b)	44
n-Butanol		None	10	0.21	0.49	2(b)	140
Isobutanol		None	10	0.43	0.34	2(d)	348
sec-Butanol							
Cyclohexanol		None	>20	1.34	1.54	2(d)	510
Ethylene glycol							
Methyl Cellosolve		None	18	0.72	0.90	2(d)	113
Ethyl Cellosolve							
Butyl Cellosolve							
Chlorinated hydrocarbons							
Methylene dichloride		None	1.8	−0.37	−0.34		
Chloroform	64	22	<1.5[a]	−0.61	−0.52	3+4	90
Carbon tetrachloride	56	89	2[a]	0.84	0.82	3+4	80
1,2-Dichloroethane		None	3/5	−0.36	−0.07	3+4	144
1,1,1-Trichloroethane							
Trichloroethylene							
Perchloroethylene							
Monochlorobenzene		None	11	0.30	0.45	3+4	192
Ketones							
Methyl ethyl ketone[b]		None	2	0.06	0.21	3+4	174
MIBK		None					
Cyclohexanone		None					
N-Methylpyrrolidone		None		0.28			
Ethers							
Diethyl ether		None	3	0.84	0.40	3+4	177
Diisopropyl ether	54	61					
1,4-Dioxane							
Tetrahydrofuran	64	92		0.33	0.38		
Esters							
Methyl acetate	55	50		0.27	0.23		
Ethyl acetate		None	2	0.17	0.12	3+4	176
Butyl acetate		None	10	0.24	0.47	3+4	219
Miscellaneous							
Dimethylformamide		None	>10	0.26	1.89	3+4	164
Dimethyl sulphoxide		None	>10	0.61	0.74	3+4	154
Pyridine		None	8	0.20	0.74	3+4	181
Acetonitrile		None	2/3	0.03	0.05	3+4	143
Furfuraldehyde							
Water		None	10/2	2.29	1.35	1	193

[a] The relative volatility between the less volatile component of the binary mixture and the azeotrope.
[b] Can react.

Recovery notes: acetone

Acetone is sufficiently reactive to pose serious problems if it needs to be recovered to a high purity by distillation. Some commonly used inorganic dehydrating agents such as activated alumina and barium hydroxide, and also mildly acidic conditions, can accelerate acetone's condensation when warm to diacetone alcohol, which can in turn dehydrate to yield mesityl oxide. Not only can this cause a loss of acetone from a fractionating system (since both of these derivatives of acetone are high boiling) but it increases the amount of water that may need to be removed in the recovery.

Calcium chloride reacts with acetone so it is not a suitable dehydrating agent, but potassium carbonate, anhydrous sodium sulphate (Drierite) and 4A molecular sieves can be safely used to dry acetone to 0.1% water or less.

All normal metallic materials of construction are suitable for handling acetone but Viton rubbers swell and disintegrate on contact with it.

Acetone can be stripped easily from water but cannot be scrubbed economically from air using water (Figure A.17). Acetone can be removed from air on activated carbon but

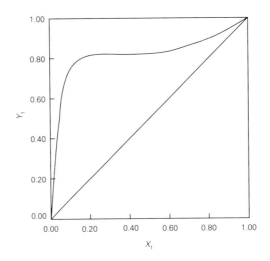

Fig. A.17 VLE diagram for acetone (1)–water (2) at 760 mmHg

there are problems with acetone and other ketones due to the formation of hot spots in the carbon bed.

Distillation of acetone from water presents no difficulty if very dry acetone is not required, but for a recovered acetone of less than 2% water a considerable reflux ratio is needed and the separation of water becomes progressively more difficult at pressures above atmospheric.

The assumption of equal molal overflow (acetone 7076 vs. water 9270 cal/mol) is not valid enough to use a standard vapour–liquid diagram for a McCabe Thiele solution of acetone–water fractionation. Using 'corrected' molecular weights as in Table A.11 allows a graphical solution.

Because of the ease of stripping acetone from water and the comparative difficulty of producing a dry distillate, acetone is particularly well suited to recovery in a batch still rather than a continuous fractionating column. Typically the numbers of theoretical trays required are as given in Table A.12.

Table A.11 'Corrected' molecular weights for acetone

Mole fraction (adjusted) of acetone in liquid	Mole fraction (adjusted) of acetone in vapour
0	0
0.0074	0.201
0.0149	0.354
0.0376	0.552
0.0761	0.696
0.116	0.745
0.156	0.766
0.241	0.784
0.331	0.794
0.426	0.807
0.527	0.819
0.634	0.837
0.748	0.867
0.870	0.914
0.934	0.951
1.000	1.000

Table A.12 Theoretical trays and reflux ratios required for batch drying of acetone

Water content of distillate (% w/w)	Reflux ratio	Theoretical trays
0.14	5	40
0.16	5	25
0.24	5	14
0.27	3	25
0.35	3	14

METHYL ETHYL KETONE

1. **Names**

    Methyl ethyl ketone
    Butan-2-one
    MEK

2. **Physical properties**

Molecular weight	72
Empirical formula	C_4H_8O
Boiling point (°C)	80
Freezing point (°C)	−87
Specific gravity (20/4 °C)	0.805
Liquid expansion coefficient (per °C)	0.001
Surface tension (at 20 °C in dyn/cm)	24.6
Absolute viscosity (at 25 °C in cP)	0.41

3. **Fire and health hazards**

Flash point (closed cup) (°C)	−6
Autoignition temperature (°C)	516
Lower explosive limit (ppm)	26 000
Upper explosive limit (ppm)	128 000
IDLH (ppm)	3 000
Odour threshold (ppm)	10
TWA–TLV (ppm)	200
Saturated vapour concentration (70 °F in ppm)	112 000
Vapour density (relative to air)	2.50

4. **Solvent properties**

Hildebrand solubility parameter	9.3
Dipole moment (D)	3.18
Dielectric constant (20 °C)	15.4
Evaporation time (diethyl ether = 1.0)	2.5

5. **Aqueous effluent characteristics**

Solubility in water (at 25 °C in % w/w)	26
Solubility of water in (at 25 °C in % w/w)	12.0
Biological oxygen demand (w/w)	2.14
Chemical oxygen demand (w/w)	2.4
Theoretical oxygen demand (w/w)	2.44

6. **Handling details**

Hazchem number	1193
Hazard class	2 YE
Vapour pressure (at 21 °C in mmHg)	75.3
Loss per transfer (% of liquid transferred)	0.037

7. **Vapour pressure constants (mmHg, log base 10)**

Antoine equation	$A = 7.06356$
	$B = 1261.340$
	$C = 221.969$
Cox chart	$A = 7.22242$
	$B = 1345.9$

8. **Thermal information**

Latent heat (cal/mol)	7848
Specific heat (cal/mol/°C)	35.9
Heat of combustion (cal/mol)	582 552

Azeotropes and activity coefficients of component X: methyl ethyl ketone

Component Y	Azeotrope °C	Azeotrope % w/w X	α	A_{XY} $=\ln \gamma_X^\infty$	A_{YX} $=\ln \gamma_Y^\infty$	Source[11] Vol.	Page
Hydrocarbons							
n-Pentane		None					
n-Hexane	64	29	3[a]	1.18	1.03	3+4	301
n-Heptane	77	70	5[a]	1.32	1.16	3+4	311
Cyclohexane	72	40	1.5/2[a]	1.26	0.89	3+4	296
Benzene	78	45	>1.5[a]	0.24	0.16	3+4	284
Toluene		None	3	0.30	0.45	3+4	308
Ethylbenzene		None	8	0.32	0.78	3+4	316
Xylenes		None	8		0.53		
Alcohols							
Methanol	64	30	2.5[a]	0.72	0.72	2(a)	133
Ethanol	74	61	<1.5	0.68	0.57	2(a)	342
n-Propanol		None	2.2	0.48	0.49	2(c)	496
Isopropanol	78	68	1.5[a]	0.38	0.45	2(b)	54
n-Butanol		None	5	0.28	0.60	2(b)	142
Isobutanol							
sec-Butanol		None	2.2	0.31	0.33	2(d)	249
Cyclohexanol							
Ethylene glycol							
Methyl Cellosolve		None	5	0.68	0.57	2(b)	122
Ethyl Cellosolve							
Butyl Cellosolve		None	>10	0.86	0.32	2(b)	430
Chlorinated hydrocarbons							
Methylene dichloride		None	2	−1.07	−0.56	3+4	261
Chloroform	80	17	1.5/2[a]	−0.84	−0.70	3+4	260
Carbon tetrachloride	74	29	<1.5[a]	0.58	0.39	3+4	259
1,2-Dichloroethane				−0.29	−0.25		
1,1,1-Trichloroethane							
Trichloroethylene		None	<1.5	0.22	0.06	3+4	26
Perchloroethylene							
Monochlorobenzene		None	6	0.25	0.44	3+4	283
Ketones							
Acetone[b]		None	2	0.21	0.06	3+4	174
MIBK		None	3	0.07	0.01	3+4	300
Cyclohexanone							
N-Methylpyrrolidone				0.37			
Ethers							
Diethyl ether							
Diisopropyl ether							
1,4-Dioxane							
Tetrahydrofuran							
Esters							
Methyl acetate		None	2	0.02	0.02	3+4	271
Ethyl acetate	77	12	<1.5[a]	0.68	0.08	3+4	278
Butyl acetate							
Miscellaneous							
Dimethylformamide							
Dimethyl sulphoxide							
Pyridine							
Acetonitrile							
Furfuraldehyde							
Water	73	89	2[a]	3.37	1.83	1	363

[a] The relative volatility between the less volatile component of the binary mixture and the azeotrope.
[b] Can react.

Recovery notes: methyl ethyl ketone

MEK does not decompose at moderate temperatures but in the presence of acids, which catalyse the reaction, it will condense to form a dimer. It can react with ethylene glycol, DMF and DMAc, which disqualifies these as solvents in extractive distillation. MEK also reacts with chloroform. Although it is possible to scrub MEK from air using water, its high activity in aqueous solutions means that such scrubbing is not very effective and carbon bed adsorption is usually required to reach acceptable air discharge quality. At ambient temperature (Figure A.18), it is necessary to operate at a low loading of solvent

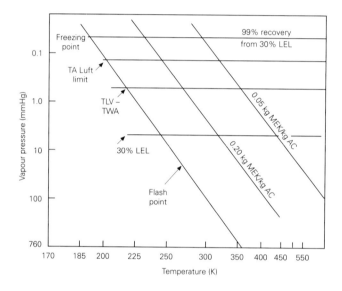

Fig. A.18 Limits for MEK recovery by activated carbon (AC) and low temperature

on the carbon bed, and this in turn means that steam regeneration of the bed is necessary.

Since MEK is not wholly stable when adsorbed on activated carbon and this may lead to dangerous hot spots in the bed, steam regeneration is the preferred method for both safety and efficiency. It does mean, however, that recovery is likely to include drying.

MEK can be stripped from water easily since it has a very high activity coefficient in dilute aqueous solution (Figure A.19) but its mutual solubility with water makes the recovery of dry MEK from an aqueous mixture difficult.

Drying by azeotropic distillation

Figure A.20 shows that, even if the phase separation takes place at the most favourable temperature, it is not an effective way of removing water from the MEK–water azeotrope at atmospheric pressure. Even by doing the distillation at 100 psia, where the

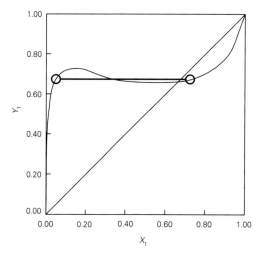

Fig. A.19 VLE diagram for MEK (1)–water (2) at 760 mmHg

water content of the azeotrope is 19–20%, drying without an entrainer is not an easy method of water removal.

Drying using an entrainer

Table A.13 shows the effectiveness of entrainers similar to those used for drying ethanol and isopropanol. In all three cases the MEK content of the water phase separating at the column top is low enough to consider sending it to effluent treatment rather than recycling it to try to improve the yield of the process.

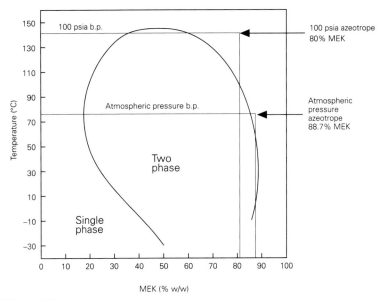

Fig. A.20 Water–MEK solubility

Table A.13 Entrainers for drying MEK

Entrainer	Benzene	Cyclohexane	Hexane
	Ternary azeotrope (% w/w)		
MEK	26.1 (17.5)[a]	60	22
Water	8.8 (9.9)[a]	5	4
Entrainer	65.1 (73.6)[a]	35	74
Boiling point (°C)	68.5 (68.9)[a]	63.6	55
	Entrainer-rich phase (% w/w)		
MEK	28.1	37.0	23.0
Water	0.6	0.6	0.4
Entrainer	71.3	62.4	76.6
Density	0.992	0.98	0.98
	Water-rich phase (% w/w)		
MEK loss in water phase (%)[b]	0.62	1.42	1.51
Closest boiling binary	Water–benzene 69.5 °C	Water–cyclohexane 69.5 °C	Water–hexane 61.6 °C

[a] Alternative values quoted in the literature are given in parentheses.
[b] Assuming feed is MEK–water azeotrope containing 11.3% w/w water.

Unfortunately, all three entrainers have azeotropes with MEK that are difficult to separate from MEK itself and only hexane, which has a low water-carrying power, is suitable to make a pure MEK product.

If the MEK drying operation is a continuing one, the MEK–hexane mixture can be kept for subsequent use, but it is almost impossible to recover the hexane free from MEK or vice versa on a small scale and it is likely that the mixture may be of no further use.

Salting-out

Addition of inorganic salts to the MEK–water azeotrope causes a phase separation that is a possible means of drying MEK provided that facilities exist for distilling the solvent containing a residue of salt. The salt is most likely to be a chloride.

Thus passing the azeotrope through a column of rock salt at 25 °C yields an organic phase containing 3.8% of water and an aqueous phase with 3% of MEK. The latter is considerably less than the MEK content of the water from using an azeotropic entrainer and will normally be disposed of with a loss of only 0.35% of the MEK being dried, but the salt-containing MEK will need refractionation. Other salts, although they are more costly, are more effective than NaCl. Calcium chloride will dry MEK to 0.7% w/w water and lithium chloride to 0.3% w/w water.

If chlorides cannot be considered because of corrosion problems, sodium acetate will give results similar to NaCl.

In general, low temperature favours the dehydrating properties of salts, but in the case of MEK, LiCl is little affected by temperature in the range 20–60 °C.

Liquid–liquid extraction

Chlorinated hydrocarbons are useful solvents for extracting MEK from water but many of them (trichloroethylene, carbon tetrachloride, chloroform) form azeotropes with MEK and it is difficult to recover pure MEK from the extract.

Monochlorobenzene (MCB) is suitable and, because it is never evaporated in passing through the extraction and stripping stages, it can be used liberally. A 50:50 mixture of MEK–water azeotrope and MCB yields (Table A.14) an MEK containing 0.46% of water after stripping from MCB. The aqueous phase, if it were not recycled for further recovery, would take to waste 1.4% of the MEK in the feed. However, for drying small quantities of MEK contaminated with a lot of water, the use of MCB extraction can avoid the need for a preliminary fractionation to produce the azeotrope. Thus a 50:50 mixture of MEK–water would yield MEK with 1.3% of water, although with a loss of 8.5% of the MEK in the feed. MCB can be easily stripped of MEK and makes the operation attractive for a one-off recovery.

Table A.14 Composition of phases for a 50:50 mixture of MCB and MEK–water

	Feed (% w/w)	Aqueous phase (% w/w)	MCB phase (% w/w)
MEK	44.5	9.0	43.7
Water	5.5	89.08	0.2
MCB	50.0	0.12	55.1

Extractive distillation

A further way of drying MEK is by extractive distillation using butyl Cellosolve as the solvent. Very high ratios of butyl Cellosolve to wet MEK feed must be used to break the MEK–water azeotrope and to achieve a relative volatility of MEK to water of greater than 1.5. Since the solvent to feed ratio is between 30:1 and 50:1 only columns specifically designed for a high liquid capacity would be suitable.

Pervaporation

Provided that a water content of 0.2% w/w is satisfactory, it seems likely that pervaporation is the simplest technique for drying MEK–water azeotrope and it has the great advantage of not introducing another component, whether it be an inorganic salt or another solvent, into the system.

All metals usually used in recovery plants are suitable for handling MEK but the Viton synthetic rubbers are unsuitable for gaskets, diaphragms and hoses.

MEK is classified under Rule 66 as not photochemically reactive.

METHYL ISOBUTYL KETONE

1 Names
    Methyl isobutyl ketone
    4-Methylpentan-2-one
    MIBK

2 Physical properties

Molecular weight	100
Empirical formula	$C_6H_{12}O$
Boiling point (°C)	116
Freezing point (°C)	−84
Specific gravity (20/4 °C)	0.801
Liquid expansion coefficient (per °C)	0.00094
Surface tension (at 20 °C in dyn/cm)	23.6
Absolute viscosity (at 25 °C in cP)	0.61

3 Fire and health hazards

Flash point (closed cup) (°C)	13
Autoignition temperature (°C)	459
Lower explosive limit (ppm)	14 000
Upper explosive limit (ppm)	75 000
IDLH (ppm)	
Odour threshold (ppm)	25
TWA–TLV (ppm)	100
Saturated vapour concentration (70 °F in ppm)	21 700
Vapour density (relative to air)	3.47

4 Solvent properties

Hildebrand solubility parameter	8.4
Dipole moment (D)	2.8
Dielectric constant (20 °C)	13.1
Evaporation time (diethyl ether = 1.0)	5.6

5 Aqueous effluent characteristics

Solubility in water (at 25 °C in % w/w)	1.7
Solubility of water in (at 25 °C in % w/w)	1.9
Biological oxygen demand (w/w)	2.06
Chemical oxygen demand (w/w)	2.2
Theoretical oxygen demand (w/w)	2.72

6 Handling details

Hazchem number	1245
Hazard class	
Vapour pressure (at 21 °C in mmHg)	16.5
Loss per transfer (% of liquid transferred)	0.011

7 Vapour pressure constants (mmHg, log base 10)

Antoine equation	$A = 6.67272$
	$B = 1168.408$
	$C = 191.944$
Cox chart	$A = 7.27155$
	$B = 1519.2$

8 Thermal information

Latent heat (cal/mol)	8500
Specific heat (cal/mol/°C)	47.0
Heat of combustion (cal/mol)	736 400

Azeotropes and activity coefficients of component X: MIBK

Component Y	Azeotrope °C	Azeotrope % w/w X	α	A_{XY} $=\ln \gamma_X^\infty$	A_{YX} $=\ln \gamma_Y^\infty$	Source[11] Vol.	Page
Hydrocarbons							
n-Pentane				0.62			
n-Hexane				0.64			
n-Heptane	98	13					
Cyclohexane		None	4	1.07	0.39	3+4	354
Benzene		None	5	0.07	0.03	3+4	351
Toluene	111	3	<1.5[a]	0.16	0.12	3+4	356
Ethylbenzene							
Xylenes				1.20			
Alcohols							
Methanol		None	6	1.20	0.78	2(a)	248
Ethanol		None	5	0.74	1.00	2(c)	423
n-Propanol	94	65					
Isopropanol		None	4	0.54	0.38	2(b)	90
n-Butanol	114	70	<1.5[a]	0.53	1.07	2(b)	196
Isobutanol	108	9					
sec-Butanol							
Cyclohexanol							
Ethylene glycol							
Methyl Cellosolve	114	75					
Ethyl Cellosolve							
Butyl Cellosolve		None					
Chlorinated hydrocarbons							
Methylene dichloride							
Chloroform		None	3/5	−1.48	−0.66	3+4	343
Carbon tetrachloride							
1,2-Dichloroethane							
1,1,1-Trichloroethane							
Trichloroethylene							
Perchloroethylene	114	52					
Monochlorobenzene							
Ketones							
Acetone							
Methyl ethyl ketone		None	3	0.01	0.07	3+4	300
Cyclohexanone							
N-Methylpyrrolidone							
Ethers							
Diethyl ether							
Diisopropyl ether							
1,4-Dioxane							
Tetrahydrofuran							
Esters							
Methyl acetate							
Ethyl acetate							
Butyl acetate							
Miscellaneous							
Dimethylformamide							
Dimethyl sulphoxide							
Pyridine	115	40					
Acetonitrile							
Furfuraldehyde							
Water	88	76		5.9	1.85		

[a] The relative volatility between the less volatile component of the binary mixture and the azeotrope.

Recovery notes: methyl isobutyl ketone

Because of its very low solubility in water (Table A.15), MIBK cannot be scrubbed from air using water but it can be adsorbed with activated carbon and regenerated with steam from the carbon bed.

If the application of the recovered solvent is not too critical for water content, it may be possible, as in the case of hydrocarbons, to reuse water-saturated MIBK particularly if the phase separation takes place at a low temperature.

Table A.15 Solubility of MIBK and water

Temperature (°C)	MIBK in water (% w/w)	Water in MIBK (% w/w)
10		1.30
20	1.90	1.55
30	1.65	1.80
40	1.50	2.08
50	1.40	2.35
60	1.35	2.70

If drier MIBK is needed the solvent can be easily dried by distilling off the water azeotrope and separating the water phase, which is not worth recovering further.

CYCLOHEXANONE

1 Names

Cyclohexanone
Cyclohexyl ketone
Sextone

2 Physical properties

Molecular weight	98
Empirical formula	$C_6H_{10}O$
Boiling point (°C)	156
Freezing point (°C)	−32
Specific gravity (20/4 °C)	0.948
Liquid expansion coefficient (per °C)	0.0009
Surface tension (at 20 °C in dyn/cm)	34.5
Absolute viscosity (at 25 °C in cP)	2.2

3 Fire and health hazards

Flash point (closed cup) (°C)	43
Autoignition temperature (°C)	420
Lower explosive limit (ppm)	11 000
Upper explosive limit (ppm)	94 000
IDLH (ppm)	5000
Odour threshold (ppm)	0.12
TWA–TLV (ppm)	25
Saturated vapour concentration (70 °F in ppm)	5500
Vapour density (relative to air)	3.40

4 Solvent properties

Hildebrand solubility parameter	9.9
Dipole moment (D)	3.1
Dielectric constant (20 °C)	18.2
Evaporation time (diethyl ether = 1.0)	40

5 Aqueous effluent characteristics

Solubility in water (at 25 °C in % w/w)	2.3
Solubility of water in (at 25 °C in % w/w)	8.0
Biological oxygen demand (w/w)	1.23
Chemical oxygen demand (w/w)	2.6
Theoretical oxygen demand (w/w)	2.61

6 Handling details

Hazchem number	1915
Hazard class	3 Y
Vapour pressure (at 21 °C in mmHg)	3.1
Loss per transfer (% of liquid transferred)	0.0017

7 Vapour pressure constants (mmHg, log base 10)

Antoine equation	$A = 7.47050$
	$B = 1832.200$
	$C = 244.200$
Cox chart	$A = 7.32768$
	$B = 1716.5$

8 Thermal information

Latent heat (cal/mol)	9016
Specific heat (cal/mol/°C)	43.8
Heat of combustion (cal/mol)	839 860

Azeotropes and activity coefficients of component X: cyclohexanone

Component Y	Azeotrope[a] °C	Azeotrope[a] % w/w X	α	$A_{XY} = \ln\gamma_X^\infty$	$A_{YX} = \ln\gamma_Y^\infty$	Source[11] Vol.	Source[11] Page
Water	96	45	10[b]	3.60	1.76	1	511
Toluene		None	4		−0.09	3+4	339
Cyclohexane		None	>10	1.00	0.81	3+4	337
Cyclohexanol		None	<1.5	0.11	0.11	2(b)	395

[a] Cyclohexanone does not form azeotropes with xylenes, furfuraldehyde, monochlorobenzene and dioxane.
[b] The relative volatility between the less volatile component of the binary mixture and the azeotrope.

Recovery notes: cyclohexanone

Cyclohexanone is slightly unstable at its atmospheric pressure boiling point and should be distilled under vacuum or steam distilled to avoid any decomposition.

Its water azeotrope readily splits into two liquid phases and, if water saturated cyclohexanone is being dried, the amount of solvent lost in the water phase (approximately 0.2% of the charge) is normally not worth recovering.

Cyclohexanone is classified as not being photochemically reactive under Rule 66.

Cyclohexanone is not stable when adsorbed on activated carbon and can oxidize to adipic acid. Since adipic acid has a boiling point of over 300 °C, it cannot be removed from the activated carbon bed by steaming and the capacity of the bed is quickly reduced.

N-METHYL-2 PYRROLIDONE

1 **Names**

N-Methyl-2 pyrrolidone
M-Pyrol
NMP

2 **Physical properties**

Molecular weight	99
Empirical formula	C_5H_9NO
Boiling point (°C)	202
Freezing point (°C)	−24
Specific gravity (20/4 °C)	1.03
Liquid expansion coefficient (per °C)	0.0008
Surface tension (at 20 °C in dyn/cm)	40.7
Absolute viscosity (at 25 °C in cP)	1.8

3 **Fire and health hazards**

Flash point (closed cup) (°C)	95
Autoignition temperature (°C)	346
Lower explosive limit (ppm)	21 800
Upper explosive limit (ppm)	122 400
IDLH (ppm)	
Odour threshold (ppm)	
TWA–TLV (ppm)	
Saturated vapour concentration (70 °F in ppm)	460
Vapour density (relative to air)	3.44

4 **Solvent properties**

Hildebrand solubility parameter	11.0
Dipole moment (D)	4.1
Dielectric constant (25 °C)	32.2
Evaporation time (diethyl ether = 1.0)	N/A

5 **Aqueous effluent characteristics**

Solubility in water (at 25 °C in % w/w)	Total
Solubility of water in (at 25 °C in % w/w)	Total
Biological oxygen demand (w/w)	
Chemical oxygen demand (w/w)	
Theoretical oxygen demand (w/w)	2.18

6 **Handling details**

Hazchem number	
Hazard class	
Vapour pressure (at 21 °C in mmHg)	0.35
Loss per transfer (% of liquid transferred)	0.18×10^{-3}

7 **Vapour pressure constants (mmHg, log base 10)**

Antoine equation	$A = 8.27890$
	$B = 2570.30$
	$C = 273.150$
Cox chart	$A =$
	$B =$

8 **Thermal information**

Latent heat (cal/mol)	12 600
Specific heat (cal/mol/°C)	39.6
Heat of combustion (cal/mol)	719 000

Azeotropes and activity coefficients of component X: *N*-methyl-2-pyrrolidone

Component Y	Azeotrope °C	Azeotrope % w/w X	α	A_{XY} $=\ln \gamma_X^\infty$	A_{YX} $=\ln \gamma_Y^\infty$	Source[11] Vol.	Page
Hydrocarbons							
n-Pentane					2.44		
n-Hexane					2.60		
n-Heptane				2.89	2.79		
Cyclohexane					2.10		
Benzene					0.30		
Toluene					0.51		
Ethylbenzene					0.53		
Xylenes			>10				
Alcohols							
Methanol					−0.64		
Ethanol					−0.41		
n-Propanol							
Isopropanol					−0.28		
n-Butanol							
Isobutanol							
sec-Butanol							
Cyclohexanol							
Ethylene glycol							
Methyl Cellosolve							
Ethyl Cellosolve							
Butyl Cellosolve							
Chlorinated hydrocarbons							
Methylene dichloride							
Chloroform							
Carbon tetrachloride							
1,2-Dichloroethane							
1,1,1-Trichloroethane							
Trichloroethylene							
Perchloroethylene							
Monochlorobenzene							
Ketones							
Acetone					0.28		
Methyl ethyl ketone					0.37		
MIBK							
Cyclohexanone							
Ethers							
Diethyl ether							
Diisopropyl ether							
1,4-Dioxane							
Tetrahydrofuran							
Esters							
Methyl acetate					0.48		
Ethyl acetate							
Butyl acetate							
Miscellaneous							
Dimethylformamide							
Dimethyl sulphoxide							
Pyridine							
Acetonitrile							
Furfuraldehyde							
Water	None		10	0.46	0.17	1	379

Recovery notes: *N*-methyl-2-pyrrolidone

Neutral NMP is chemically extremely stable up to 315 °C and this allows it to undergo prolonged batch distillation and to be stripped out to a low mole fraction from resins and polymers, but because it is fully miscible with water it cannot be steam distilled. It can be recovered from mixtures with resins (e.g. PVC) by throwing the resin out of solution by adding water and by recovering NMP from its aqueous solution.

NMP is hygroscopic and if long storage is planned precautions to avoid contact with air are required.

NMP is an aprotic solvent and cannot be dried with the common pervaporation membranes and it is a good solvent for most polymers so that Teflon gaskets are required for handling it. All common metals are satisfactory in contact with NMP but it goes brown on prolonged storage with mild steel.

While at room temperature NMP is stable in contact with acids or alkalis, unlike many other aprotic solvents. At 80 °C hydrolysis in either acidic or alkaline conditions is appreciable and a limit of pH 11.5 is the maximum for alkaline stability.

Because of its thermal stability and complete miscibility with all solvents and its lack of reported azeotropes, NMP is worth consideration as an extractive distillation solvent in recovery operations.

316 Appendix 1

DIETHYL ETHER

1	**Names**	Diethyl ether
		Ethyl ether
		Ether
		Ethyl oxide
		Sulphuric ether

2 **Physical properties**

Molecular weight	74
Empirical formula	$C_4H_{10}O$
Boiling point (°C)	34.5
Freezing point (°C)	−116
Specific gravity (20/4 °C)	0.715
Liquid expansion coefficient (per °C)	0.0011
Surface tension (at 20 °C in dyn/cm)	17
Absolute viscosity (at 25 °C in cP)	0.24

3 **Fire and health hazards**

Flash point (closed cup) (°C)	−45
Autoignition temperature (°C)	160
Lower explosive limit (ppm)	18 500
Upper explosive limit (ppm)	360 000
IDLH (ppm)	19 000
Odour threshold (ppm)	1
TWA–TLV (ppm)	400
Saturated vapour concentration (70 °F in ppm)	610 000
Vapour density (relative to air)	2.57

4 **Solvent properties**

Hildebrand solubility parameter	7.4
Dipole moment (D)	1.3
Dielectric constant (20 °C)	4.34
Evaporation time (diethyl ether = 1.0)	1.0

5 **Aqueous effluent characteristics**

Solubility in water (at 25 °C in % w/w)	6.9
Solubility of water in (at 25 °C in % w/w)	1.3
Biological oxygen demand (w/w)	0.03
Chemical oxygen demand (w/w)	
Theoretical oxygen demand (w/w)	2.59

6 **Handling details**

Hazchem number	1155
Hazard class	3 YE
Vapour pressure (at 21 °C in mmHg)	462
Loss per transfer (% of liquid transferred)	0.25

7 **Vapour pressure constants (mmHg, log base 10)**

Antoine equation	$A = 6.98472$
	$B = 1090.64$
	$C = 231.20$
Cox chart	$A = 7.00353$
	$B = 1088.4$

8 **Thermal information**

Latent heat (cal/mol)	6216
Specific heat (cal/mol/°C)	41.3
Heat of combustion (cal/mol)	598 068

Azeotropes and activity coefficients of component X: diethyl ether

Component Y	Azeotrope °C	Azeotrope % w/w X	α	A_{XY} $=\ln \gamma_X^\infty$	A_{YX} $=\ln \gamma_Y^\infty$	Source[11] Vol.	Page
Hydrocarbons							
n-Pentane	33	60					
n-Hexane			3				
n-Heptane							
Cyclohexane							
Benzene		None	5	−0.32	−0.06	3+4	516
Toluene			13		0.21		
Ethylbenzene							
Xylenes				0.09			
Alcohols							
Methanol		None	7	1.12	1.52	2(a)	170
Ethanol		None		0.91	1.25	2(a)	374
n-Propanol							
Isopropanol			6/7				
n-Butanol							
Isobutanol							
sec-Butanol							
Cyclohexanol							
Ethylene glycol							
Methyl Cellosolve							
Ethyl Cellosolve							
Butyl Cellosolve							
Chlorinated hydrocarbons							
Methylene dichloride	40	30	<1.5[a]	−0.36	−0.46	3+4	492
Chloroform	37	3	2[a]	−0.82	−0.94	3+4	486
Carbon tetrachloride							
1,2-Dichloroethane							
1,1,1-Trichloroethane							
Trichloroethylene							
Perchloroethylene							
Monochlorobenzene							
Ketones							
Acetone		None	3	0.40	0.84	3+4	177
Methyl ethyl ketone							
MIBK							
Cyclohexanone							
N-Methylpyrrolidone							
Ethers							
Diisopropyl ether							
1,4-Dioxane							
Tetrahydrofuran							
Esters							
Methyl acetate							
Ethyl acetate		None	5	−0.03	0.08	3+4	513
Butyl acetate							
Miscellaneous							
Dimethylformamide							
Dimethyl sulphoxide							
Pyridine							
Acetonitrile		None	10	1.14	0.88	3+4	499
Furfuraldehyde							
Water	34	99	>20[a]	4.49	3.53	1	344

[a] The relative volatility between the less volatile component of the binary mixture and the azeotrope.

Recovery notes: diethyl ether

Diethyl ether presents serious hazards of fire and explosion and must be treated with great care.

Autoignition

Diethyl ether has a low autoignition temperature (160 °C) and this means that its vapour can be caused to explode by contact with a pipe carrying steam over 75 psig. Since the gap between its LEL and UEL is large, both dilute and concentrated vapours will explode and its high vapour density means that vapour will not disperse readily and is liable to spread along pipetracks and through unsealed drains for long distances.

Much electrical equipment, although flameproof, will attain surface temperatures over 160 °C when in use and careful inspection of the certification of such equipment must be made before diethyl ether is handled in a plant not specifically built for the purpose.

Diethyl ether may be formed from ethanol during a process and, particularly in a batch distillation when very volatile diethyl ether may be concentrated in the first cut, even low concentrations of diethyl ether in a solvent recovery feedstock may represent a hazard.

Peroxides

Ether peroxide can form quickly when diethyl ether is exposed to light in a clear glass bottle, but even in the absence of light when diethyl ether is stored in contact with air. When the concentration of peroxide has reached a sufficient level the liquid may explode violently, particularly when heated to about 90 °C. The effect of heating is to turn a moderately stable hydroperoxide into a highly unstable alkylidene peroxide.

It follows that it is very dangerous to heat diethyl ether, and particularly to distil it to dryness, if there is a possibility of peroxides being present. Even if it is believed that appropriate inhibitors have always been present in the ether, it should be checked for peroxide content before distillation and the peroxides, if present, must be destroyed before heating takes place. The use of Merquant sticks is an easy and reliable means of checking for the presence of peroxides.

At the end of distillation, the plant must not fill with air until any ether-containing residues have been cooled to room temperature. If air reaches hot diethyl ether of a high concentration, an explosion may take place and inert gas with no oxygen in it should be used to vent the plant.

Diethyl ether should be inhibited against peroxide formation in storage and use with an inhibitor of which pyrogallol (0.2% w/w), hydroquinone or other phenols and diphenylamine are possible choices. Storage under nitrogen is very desirable. Samples should be stored in brown bottles with a minimum of ullage. Since all the inhibitors are very much less volatile than diethyl ether, newly distilled material will be uninhibited and should be treated without delay.

If an ether mixture must be distilled and is found to contain peroxides, these must be decomposed before distillation starts. Possible routes for doing this are agitation with potassium iodide solution, agitation and distillation in the presence of potassium hydroxide and permanganate or contacting in a column with silica gel or alumina. The last method has the advantage of not introducing water into the feedstock. If it should happen that a distillation residue does, despite all the precautions, contain peroxides, it can be disposed of safely by adding the residue very slowly to a stirred vessel containing 5% sodium hydroxide solution.

Drying

The azeotrope formed by water and diethyl ether is single phase. Unlike ethanol and several other solvents which have single-phase water azeotropes, it is impracticable to form a low-boiling ternary azeotrope to remove water because the boiling point of diethyl ether is so low that the condensing of any such system would be very difficult.

Molecular sieves, activated alumina and calcium chloride dry diethyl ether very satisfactorily but the first two are difficult to regenerate given the low temperatures demanded to avoid autoignition and the need to avoid the use of air. A number of other chemicals are suitable (*see* Chapter 7) provided that they are not harmful to the subsequent use of the diethyl ether. Pervaporation is also a suitable way of drying diethyl ether down to 0.1% w/w water.

There is a very large use of diethyl ether for cold starting formulations for diesel engines and in many cases this is a better outlet for recovered diethyl ether than attempting to return it to reagent or pharmaceutical quality. The shelf life of diethyl ether in an aerosol can needs to be very long and impurities that could corrode the can internally must be carefully avoided.

Refining

Alcohols can be washed out of diethyl ether using water. Diethyl ether behaves in this respect very like a hydrocarbon.

Toxicity

The smell of diethyl ether is, to most people, a good warning of its presence in air in harmful concentrations. The odour threshold is well below its TLV. It was for many years used as an anaesthetic and it is liable to cause drowsiness.

Diethyl ether tends to degrease the skin and PVC gloves should be worn when handling it.

DIISOPROPYL ETHER

1 **Names**

Diisopropyl ether
Isopropyl ether
DIPE

2 **Physical properties**

Molecular weight	102
Empirical formula	$C_6H_{14}O$
Boiling point (°C)	68
Freezing point (°C)	−86
Specific gravity (20/4 °C)	0.724
Liquid expansion coefficient (per °C)	0.00108
Surface tension (at 20 °C in dyn/cm)	18
Absolute viscosity (at 25 °C in cP)	0.33

3 **Fire and health hazards**

Flash point (closed cup) (°C)	−21
Autoignition temperature (°C)	430
Lower explosive limit (ppm)	14 000
Upper explosive limit (ppm)	79 000
IDLH (ppm)	10 000
Odour threshold (ppm)	0.2
TWA–TLV (ppm)	250
Saturated vapour concentration (70 °F in ppm)	200 000
Vapour density (relative to air)	3.58

4 **Solvent properties**

Hildebrand solubility parameter	6.9
Dipole moment (D)	1.2
Dielectric constant (20 °C)	
Evaporation time (diethyl ether = 1.0)	1.6

5 **Aqueous effluent characteristics**

Solubility in water (at 25 °C in % w/w)	1.2
Solubility of water in (at 25 °C in % w/w)	0.62
Biological oxygen demand (w/w)	0.19
Chemical oxygen demand (w/w)	
Theoretical oxygen demand (w/w)	2.83

6 **Handling details**

Hazchem number	1159
Hazard class	3 YE
Vapour pressure (at 21 °C in mmHg)	116
Loss per transfer (% of liquid transferred)	0.09

7 **Vapour pressure constants (mmHg, log base 10)**

Antoine equation	$A = 6.84953$
	$B = 1139.34$
	$C = 231.742$
Cox chart	$A = 7.09624$
	$B = 1256.2$

8 **Thermal information**

Latent heat (cal/mol)	6936
Specific heat (cal/mol/°C)	51.7
Heat of combustion (cal/mol)	957 780

Azeotropes and activity coefficients of component X: diisopropyl ether

Component Y	Azeotrope °C	Azeotrope % w/w X	α	A_{XY} $=\ln \gamma_X^\infty$	A_{YX} $=\ln \gamma_Y^\infty$	Source[11] Vol.	Page
Hydrocarbons							
n-Pentane							
n-Hexane	67	53					
n-Heptane		None	2/3	1.43	0.05	3+4	559
Cyclohexane		None	1.5	0.01	0.04	3+4	555
Benzene		None	<1.5	0.20	0.16	3+4	553
Toluene		None	4	0.14	0.14	3+4	558
Ethylbenzene		None	10	0.15	0.29	3+4	563
Xylenes							
Alcohols							
Methanol	59	24	2[a]	1.42	1.18	2(a)	261
Ethanol	64	83	3[a]	1.39	1.35	2(a)	458
n-Propanol		None	5	1.27	1.25	2(a)	586
Isopropanol	66	85	3[a]	1.06	1.43	2(b)	101
n-Butanol		None	10	0.85	1.22	2(b)	203
Isobutanol							
sec-Butanol							
Cyclohexanol							
Ethylene glycol							
Methyl Cellosolve							
Ethyl Cellosolve							
Butyl Cellosolve							
Chlorinated hydrocarbons							
Methylene dichloride							
Chloroform	70	64	1.5[a]	−0.76	−0.73	3+4	537
Carbon tetrachloride		None	1.5	0.07	0.06	3+4	530
1,2-Dichloroethane							
1,1,1-Trichloroethane							
Trichloroethylene							
Perchloroethylene							
Monochlorobenzene							
Ketones							
Acetone	54	39					
Methyl ethyl ketone							
MIBK							
Cyclohexanone							
N-Methylpyrrolidone							
Ethers							
Diethyl ether							
1,4-Dioxane							
Tetrahydrofuran							
Esters							
Methyl acetate							
Ethyl acetate							
Butyl acetate							
Miscellaneous							
Dimethylformamide							
Dimethyl sulphoxide							
Pyridine							
Acetonitrile							
Furfuraldehyde							
Water	62	95	10[a]	1.54	3.00	1	525

[a] The relative volatility between the less volatile component of the binary mixture and the azeotrope.

Recovery notes: diisopropyl ether

Unlike its homologue, diethyl ether, diisopropyl ether (DIPE) has a comparatively high autoignition temperature. Indeed, it can be blended into motor gasoline as an octane improver. Hence precautions are not required to avoid a vapour explosion due to contact with hot surfaces.

DIPE does, however, form peroxides even more rapidly than diethyl ether and the steps to be taken to avoid exposure to oxygen in air are similar to those described for diethyl ether. There are a number of effective inhibitors of peroxide formation, including morpholine, ethylenediamine and N-benzyl-p-aminophenol. Inhibitors used commercially include N-benzyl-4-aminobiphenyl (20 ppm) and diethylenetriamine (50 ppm). Hydroquinone, resorcinol and pyrocatechol at a level of 10 ppm are all effective for storage of DIPE for up to 6 months. The peroxides that form in DIPE during prolonged storage eventually undergo chemical change to become cyclic peroxides of acetone. Not only do these decompose violently when heated but also they are sensitive to impact. If a solid phase of these peroxides comes out of solution in a drum after long storage then there is a risk of an explosion on moving or even unscrewing the cap of the drum and expert assistance should be called upon for safe disposal. Long-term storage of a partly filled drum of uninhibited DIPE is especially dangerous and should be avoided.

DIPE is comparatively non-toxic and, provided that it is kept inhibited, it is not difficult to handle so that its use as an entraining agent for removing water by azeotropic distillation is straightforward and it can be very effectively dehydrated itself because its water azeotrope (4.6% w/w water) splits into two phases with little DIPE in the aqueous phase. Apart from its tendency to form peroxides, DIPE is stable and can be used as an entraining agent over a prolonged period. As the VLE diagram (Figure A.21) shows, approach to the azeotrope from both directions is easy.

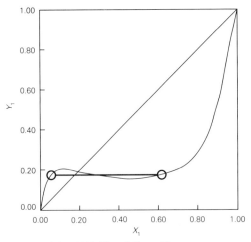

Fig. A.21 VLE diagram for water (1)–DIPE (2) at 760 mmHg

Peroxides in DIPE can be destroyed by treatment with tin(II) chloride or triethylenetetramine. They can be removed by passing the solvent through a Dowex 1 ion-exchange column or an alumina column. The rate of formation of peroxides is accelerated if DIPE and other ethers are stored wet.

1,4-DIOXANE

1 Names

1,4-Dioxane
p-Dioxane
Dioxane
Diethylene oxide

2 Physical properties

Molecular weight	88
Empirical formula	$C_4H_8O_2$
Boiling point (°C)	101
Freezing point (°C)	12
Specific gravity (20/4 °C)	1.0336
Liquid expansion coefficient (per °C)	0.0012
Surface tension (at 20 °C in dyn/cm)	40
Absolute viscosity (at 25 °C in cP)	1.3

3 Fire and health hazards

Flash point (closed cup) (°C)	12
Autoignition temperature (°C)	180
Lower explosive limit (ppm)	20 000
Upper explosive limit (ppm)	222 000
IDLH (ppm)	200
Odour threshold (ppm)	170
TWA–TLV (ppm)	25
Saturated vapour concentration (70 °F in ppm)	38 000
Vapour density (relative to air)	3.06

4 Solvent properties

Hildebrand solubility parameter	10.0
Dipole moment (D)	0.4
Dielectric constant (20 °C)	2.21
Evaporation time (diethyl ether = 1.0)	7.3

5 Aqueous effluent characteristics

Solubility in water (at 25 °C in % w/w)	Total
Solubility of water in (at 25 °C in % w/w)	Total
Biological oxygen demand (w/w)	0
Chemical oxygen demand (w/w)	–
Theoretical oxygen demand (w/w)	1.82

6 Handling details

Hazchem number	1185
Hazard class	2 SE
Vapour pressure (at 21 °C in mmHg)	28.6
Loss per transfer (% of liquid transferred)	0.013

7 Vapour pressure constants (mmHg, log base 10)

Antoine equation	$A = 7.43155$
	$B = 1554.679$
	$C = 240.337$
Cox chart	$A = 7.19047$
	$B = 1426.5$

8 Thermal information

Latent heat (cal/mol)	8510
Specific heat (cal/mol/°C)	35.6
Heat of combustion (cal/mol)	566 720
Heat of fusion (cal/g)	35
Cryoscopic constant (kg °C/mol)	4.63

Azeotropes and activity coefficients of component X: 1,4-dioxane

Component Y	Azeotrope °C	Azeotrope % w/w X	α	$A_{XY} = \ln \gamma_X^\infty$	$A_{YX} = \ln \gamma_Y^\infty$	Source[11] Vol.	Source[11] Page
Hydrocarbons							
n-Pentane							
n-Hexane	60	2	5[a]	1.13	1.23	3+4	471
n-Heptane	92	44	2[a]	0.97	1.30	3+4	478
Cyclohexane	80	25	3[a]	1.10	0.95	3+4	468
Benzene	82	12	1.8[a]	0.10	0.14	3+4	465
Toluene	102	80	<1.5[a]	0.22	0.18	3+4	477
Ethylbenzene							
Xylenes							
Alcohols							
Methanol	78	9	5[a]	0.78	0.51	2(a)	148
Ethanol	78	9	3[a]	0.80	0.65	2(a)	348
n-Propanol	95	45	1.5[a]	0.59	0.55	2(a)	533
Isopropanol		None	2/3	0.58	0.25	2(b)	56
n-Butanol		None	2	0.19	0.41	2(b)	147
Isobutanol	101	96	<1.5[a]	0.18	0.35	2(b)	279
sec-Butanol	99	60					
Cyclohexanol		None					
Ethylene glycol							
Methyl Cellosolve							
Ethyl Cellosolve							
Butyl Cellosolve							
Chlorinated hydrocarbons							
Methylene dichloride							
Chloroform		None	3/5	−1.26	−0.82	3+4	441
Carbon tetrachloride		None	3	0.46	0.25	3+4	440
1,2-Dichloroethane		None	1.5	−0.11	0.30	3+4	447
1,1,1-Trichloroethane							
Trichloroethylene		None					
Perchloroethylene							
Monochlorobenzene							
Ketones							
Acetone				0.30			
Methyl ethyl ketone							
MIBK							
Cyclohexanone		None					
N-Methylpyrrolidone							
Ethers							
Diethyl ether							
Diisopropyl ether		None					
Tetrahydrofuran							
Esters							
Methyl acetate							
Ethyl acetate		None	2/3	0.07	0.68	3+4	455
Butyl acetate							
Miscellaneous							
Dimethylformamide		None	10	0.27	0.53	3+4	454
Dimethyl sulphoxide		None	>10	0.50	0.89	3+4	450
Pyridine		None					
Acetonitrile		None					
Furfuraldehyde		None					
Water	88	82		2.03	1.35	1	383

[a]The relative volatility between the less volatile component of the binary mixture and the azeotrope.

Recovery notes: 1,4-dioxane

Dioxane presents acute problems with peroxide formation. These form in contact with air and the reaction is accelerated by light and heat. Even in unopened containers a shelf life of more than 6 months should not be assumed.

It should be stored under nitrogen (*not* air depleted of oxygen but still containing, say, 3% O_2) and should be inhibited at all times. The most commonly used inhibitor is di-*tert*-butyl-*p*-cresol at 25 ppm, but this is left in the residue when dioxane is evaporated and should be replaced as soon as possible. Commercial dioxane contains traces of peroxide even when dispatched by the manufacturer and evaporation to dryness should be avoided if possible.

Sodium hydroxide, tin(II) chloride and iron(II) sulphate can all destroy dioxane peroxides, but to avoid boiling to dryness in recovery, a heel of hydrocarbon should be considered to keep the peroxides in solution at the end of a distillation. C_8 or C_9 aromatic hydrocarbons which do not azeotrope with dioxane and have a fairly good solvent power might be suitable for this duty. Dioxane is very hygroscopic.

As the VLE diagram (Figure A.22) shows, the water–dioxane azeotrope is separated

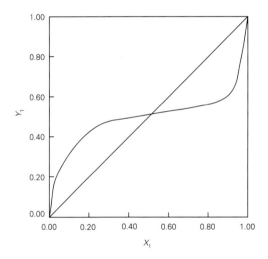

Fig. A.22 VLE diagram for water (1)–dioxane (2) at 760 mmHg

easily from both water and from the solvent. However, the laboratory techniques used for drying (molecular sieves, barium oxide, magnesium sulphate and potassium hydroxide) are all rather expensive without a recovery system. Chloroform is an effective azeotropic entrainer and its toxicity is not an automatic disqualification because dioxane itself needs to be handled with very great care.

Dioxane's odour is mild and not unpleasant, which adds to the danger of handling. Its TLV is much below its odour threshold and its IDLH is not reliably above its odour threshold. It is a suspected carcinogen in high doses and it can be absorbed through the skin in toxic amounts.

The high melting point of dioxane, coupled with its toxicity problems, means that storage and handling equipment should be lagged and traced meticulously since the clearance of blockages in a safe manner presents major difficulties. Tank vents should be traced and storage tanks should be kept at a very steady temperature to avoid solvent vapour being ejected into the atmosphere by tank breathing. Hoses should be carefully drained after use.

TETRAHYDROFURAN

1. **Names**

   Tetrahydrofuran
   THF
   Tetramethylene oxide
   1,4-Epoxybutane

2. **Physical properties**

Molecular weight	68
Empirical formula	C_4H_8O
Boiling point (°C)	66
Freezing point (°C)	−109
Specific gravity (20/4 °C)	0.888
Liquid expansion coefficient (per °C)	0.0011
Surface tension (at 20 °C in dyn/cm)	28
Absolute viscosity (at 25 °C in cP)	0.55

3. **Fire and health hazards**

Flash point (closed cup) (°C)	−15
Autoignition temperature (°C)	321
Lower explosive limit (ppm)	23 000
Upper explosive limit (ppm)	118 000
IDLH (ppm)	–
Odour threshold (ppm)	30
TWA–TLV (ppm)	200
Saturated vapour concentration (70 °F in ppm)	230 000
Vapour density (relative to air)	2.36

4. **Solvent properties**

Hildebrand solubility parameter	9.9
Dipole moment (D)	1.75
Dielectric constant (20 °C)	7.6
Evaporation time (diethyl ether = 1.0)	2.3

5. **Aqueous effluent characteristics**

Solubility in water (at 25 °C in % w/w)	} See Figure A.24
Solubility of water in (at 25 °C in % w/w)	
Biological oxygen demand (w/w)	–
Chemical oxygen demand (w/w)	–
Theoretical oxygen demand (w/w)	2.59

6. **Handling details**

Hazchem number	2056
Hazard class	2 SE
Vapour pressure (at 21 °C in mmHg)	133
Loss per transfer (% of liquid transferred)	0.055

7. **Vapour pressure constants (mmHg, log base 10)**

Antoine equation	$A = 6.99515$
	$B = 1202.29$
	$C = 226.254$
Cox chart	$A = 7.09092$
	$B = 1246.2$

8. **Thermal information**

Latent heat (cal/mol)	6664
Specific heat (cal/mol/°C)	31.3
Heat of combustion (cal/mol)	566 440

Azeotropes and activity coefficients of component X: tetrahydrofuran

Component Y	Azeotrope °C	Azeotrope % w/w X	α	$A_{XY} = \ln \gamma_X^\infty$	$A_{YX} = \ln \gamma_Y^\infty$	Source[11] Vol.	Page
Hydrocarbons							
n-Pentane		None	2.5				
n-Hexane	63	50					
n-Heptane		None					
Cyclohexane	60	97		0.49	0.50		
Benzene		None					
Toluene		None					
Ethylbenzene		None					
Xylenes		None					
Alcohols							
Methanol	61	69	2[a]	0.76	0.90	2(a)	140
Ethanol	66	90	2[a]	0.43	0.66	2(c)	328
n-Propanol		None	3	0.19	0.32	2(c)	497
Isopropanol		None	2	0.33	0.33	2(b)	55
n-Butanol		None	6/7	0.21	0.15	2(b)	146
Isobutanol		None					
sec-Butanol		None					
Cyclohexanol		None					
Ethylene glycol		None	>20	1.70	1.94	2(d)	3
Methyl Cellosolve							
Ethyl Cellosolve							
Butyl Cellosolve							
Chlorinated hydrocarbons							
Methylene dichloride		Possible					
Chloroform	72	34					
Carbon tetrachloride		None	1.5	−0.15	0.31	3+4	429
1,2-Dichloroethane		None		−0.34	−0.53		
1,1,1-Trichloroethane							
Trichloroethylene							
Perchloroethylene		None					
Monochlorobenzene		None					
Ketones							
Acetone	64	8		0.38	0.33		
Methyl ethyl ketone							
MIBK							
Cyclohexanone							
N-Methylpyrrolidone							
Ethers							
Diethyl ether							
Diisopropyl ether							
1,4-Dioxane							
Esters							
Methyl acetate							
Ethyl acetate							
Butyl acetate							
Miscellaneous							
Dimethylformamide							
Dimethyl sulphoxide		None	10	1.05	1.46	3+4	434
Pyridine							
Acetonitrile		None		0.81	0.60		
Furfuraldehyde							
Water	64	96	20[a]	3.44	1.60	1	370

[a] The relative volatility between the less volatile component of the binary mixture and the azeotrope.

Recovery notes: tetrahydrofuran

THF is very difficult to scrub from air using water but can be very readily stripped from water. Monoethylene glycol, as Table 2.4 indicates, scrubs THF effectively.

Great care should be taken when handling THF because of its propensity to form peroxides which explode very violently when heated to dryness. It is good practice never to evaporate to dryness even when all the precautions detailed below have been taken. THF should only be handled without inhibitor in solution in very exceptional circumstances. Peroxides can be detected by the use of Merquant sticks and if it is proposed to distil a parcel of THF it should be routinely tested in this way. If peroxides are present they can be decomposed with NaOH.

Because of its common use as a solvent for Grignard reagents, techniques for drying THF are very important.

1. Azeotropic distillation using n-pentane as an entrainer is a time-consuming process because of the low 'water-carrying power' of the pentane–water azeotrope, which is only 1.44% w/w (0.87% v/v). In addition, the azeotrope's low boiling point leads to restrictions to the rate of boil-up on equipment that is general purpose and not specifically designed for the duty. However, as Figure 7.2 illustrates, it is possible to use the pentane as an extraction solvent for THF from water as well as an entrainer. Only about half the pentane in the overheads is required for reflux. The other half can be used to contact the feed in a single stage of liquid–liquid extraction. Assuming the feed is close to the THF–water azeotrope (say 5% water), over half the water in the feed will form a lower phase containing negligible THF. This can therefore be discharged to effluent and will not need to be removed in the slow distillation operation.

2. High-pressure distillation. The THF–water azeotrope at 100 psig contains 12% water compared with only 4.6% at atmospheric pressure (Figure A.23), so that by alternate

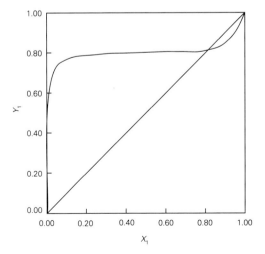

Fig. A.23 VLE diagram for THF (1)–water (2) at 760 mmHg

distillations at high and low pressure it is possible to have a residue of dry THF followed by a residue of water to be discharged. This route clearly requires unusual equipment although the VLE diagram indicates that very few plates are needed for either separation (*see* Figure A.23).

3 Caustic soda contacting. Contacting the THF azeotrope with solid caustic soda gives a two-phase mixture with very little THF in the saturated caustic layer, although this may be worth recovering. The upper THF layer is sufficiently far from its azeotropic composition to yield more than 50% of dry THF on distilling. Care must be taken when boiling residues of very concentrated NaOH that caustic cracking does not damage the reboiler or kettle.

4 Molecular sieves. Type 5A sieves dry THF very satisfactorily to levels of about 200 ppm, which is suitable for Grignard reagent work. If sieve regeneration is warranted, good-quality inert gas low in oxygen should be used.

After distillation it will be necessary to re-inhibit since the appropriate inhibitors are all much less volatile than THF itself and react with NaOH and so will remain with the still residue. *tert*-Butylcatechol and hydroquinone are commonly used inhibitors.

Whereas THF is miscible with water in all proportions at ambient temperature, it forms a two-phase mixture at 70 °C and displays other unusual behaviour in mixtures with water (Figure A.24). One of the practical effects of this is that over a mixture containing only 5% THF the vapour will be explosive. A very large excess of water is therefore called for in ensuring safety when washing away a spillage.

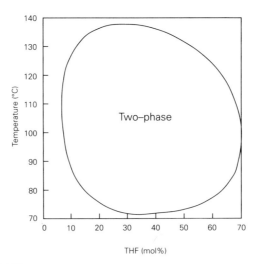

Fig. A.24 THF–water solubility vs. temperature

METHYL ACETATE

1. **Names** — Methyl acetate

2. **Physical properties**

Molecular weight	74
Empirical formula	$C_3H_6O_2$
Boiling point (°C)	57
Freezing point (°C)	−98
Specific gravity (20/4 °C)	0.927
Liquid expansion coefficient (per °C)	0.0014
Surface tension (at 20 °C in dyn/cm)	24
Absolute viscosity (at 25 °C in cP)	0.37

3. **Fire and health hazards**

Flash point (closed cup) (°C)	−10
Autoignition temperature (°C)	500
Lower explosive limit (ppm)	31 000
Upper explosive limit (ppm)	160 000
IDLH (ppm)	10 000
Odour threshold (ppm)	50
TWA–TLV (ppm)	200
Saturated vapour concentration (70 °F in ppm)	290 000
Vapour density (relative to air)	2.57

4. **Solvent properties**

Hildebrand solubility parameter	9.6
Dipole moment (D)	1.7
Dielectric constant (25°C)	6.7
Evaporation time (diethyl ether = 1.0)	2.0

5. **Aqueous effluent characteristics**

Solubility in water (at 25 °C in % w/w)	24.5
Solubility of water in (at 25 °C in % w/w)	8.2
Biological oxygen demand (w/w)	–
Chemical oxygen demand (w/w)	–
Theoretical oxygen demand (w/w)	1.51

6. **Handling details**

Hazchem number	1231
Hazard class	2 SE
Vapour pressure (at 21 °C in mmHg)	171
Loss per transfer (% of liquid transferred)	0.074

7. **Vapour pressure constants (mmHg, log base 10)**

Antoine equation	$A = 7.06524$
	$B = 1157.63$
	$C = 219.726$
Cox chart	$A = 7.25014$
	$B = 1254.0$

8. **Thermal information**

Latent heat (cal/mol)	7178
Specific heat (cal/mol/°C)	36.9
Heat of combustion (cal/mol)	381 100

Azeotropes and activity coefficients of component X: methyl acetate

Component Y	Azeotrope °C	Azeotrope % w/w X	α	A_{XY} =ln γ_X^∞	A_{YX} =ln γ_Y^∞	Source[11] Vol.	Page
Hydrocarbons							
n-Pentane	34	22					
n-Hexane	52	61					
n-Heptane	56	96					
Cyclohexane	56	80	5[a]	1.25	1.28	5	393
Benzene		None	2	0.22	0.31	5	375
Toluene							
Ethylbenzene							
Xylenes							
Alcohols							
Methanol	54	81	3.5[a]	1.06	1.00	2(a)	92
Ethanol	57	97	2.8[a]	0.62	0.62	2(a)	330
n-Propanol		None	6	1.22	1.00	2(a)	530
Isopropanol		None	5	0.77	0.93	2(b)	50
n-Butanol							
Isobutanol							
sec-Butanol							
Cyclohexanol							
Ethylene glycol							
Methyl Cellosolve							
Ethyl Cellosolve							
Butyl Cellosolve							
Chlorinated hydrocarbons							
Methylene dichloride		None	1.5/2	−0.69	−0.50	5	347
Chloroform	65	31		−0.82	−0.50	5	344
Carbon tetrachloride		None	2	0.48	0.51	5	339
1,2-Dichloroethane							
1,1,1-Trichloroethane							
Trichloroethylene							
Perchloroethylene							
Monochlorobenzene		None	10	0.16	0.29	5	374
Ketones							
Acetone	55	50	>1.5[a]	0.23	0.27	3+4	159
Methyl ethyl ketone		None	2	0.02	0.02	3+4	271
MIBK							
Cyclohexanone							
N-Methylpyrrolidone				0.48			
Ethers							
Diethyl ether		None					
Diisopropyl ether							
1,4-Dioxane							
Tetrahydrofuran							
Esters							
Ethyl acetate		None	2	0.07	−0.19	5	357
Butyl acetate		None	10	0.22	0.20	5	397
Miscellaneous							
Dimethylformamide							
Dimethyl sulphoxide							
Pyridine		None					
Acetonitrile		None	2/3	0.11	0.39	5	354
Furfuraldehyde							
Water	56	95	>10	3.16	2.14	1	264

[a]The relative volatility between the less volatile component of the binary mixture and the azeotrope.

Recovery notes: methyl acetate

A considerable proportion of 'technical' methyl acetate is an 80:20 mixture of methyl acetate and methanol, which is mostly derived as a by-product from the production of poly(vinyl alcohol). The properties of this mixture, both chemical and toxicological, are different from those of pure methyl acetate and the two grades should be treated as different products.

Pure methyl acetate is often produced from the technical product since the market for the latter, once a cheap substitute for acetone in gun wash, has been reduced since the concentration of methanol in such products was restricted.

Removing the highly polar methanol from its azeotrope with methyl acetate is an application for extractive distillation. Monoethylene glycol (MEG) has been shown to increase the value of γ^∞ from 1.0 to 7.2.

The resistance to hydrolysis under acidic, alkaline and neutral conditions of methyl acetate is of the same order of magnitude as that of ethyl acetate, with the former being slightly less stable at all values of pH.

As can be seen from the binary solubility diagram (Figure A.25), the water azeotrope

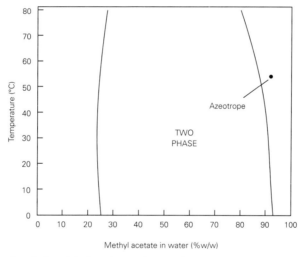

Fig. A.25 Solubility of methyl acetate in water vs. temperature

of methyl acetate is single phase at all temperatures between 0 °C and its boiling point. It is not possible, therefore, to dry methyl acetate by the easiest means of distillation and phase separation.

The VLE diagram for methyl acetate–water (Figure A.26) shows that ordinary fractionation also cannot be used for separating very dry product from the azeotrope if a feed of, say, 2.5% w/w water content were achieved, since the value for α in this composition area is virtually unity.

Examination of Table 3.3 shows that methyl acetate is considerably more hydrophilic than ethyl acetate, and it therefore does not lend itself to a single stage extraction with a high-boiling alkane although, since the water phase is small, it can be recycled to the column used for stripping methyl acetate azeotrope.

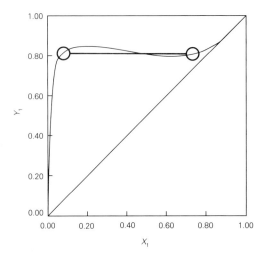

Fig. A.26 VLE diagram for methyl acetate (1)–water (2) at 760 mmHg

Pervaporation is well suited to removing 8% of water from a solvent stream, but the currently available membranes are not resistant to acetic acid, so any hydrolysis is likely to damage them.

Molecular sieves will dry the methyl acetate azeotrope satisfactorily, but regeneration of the sieves would be economically essential from such a high water content.

For handling large quantities of methyl acetate, the most effective method of dehydration is extractive distillation using MEG as the entrainer. This can be done simultaneously with the removal of methanol if necessary.

Provided that a powerful enough column is available to separate methylene chloride from methyl acetate, the former can be used as an azeotropic entrainer for water, provided that traces of methylene chloride are acceptable in the finished dried product.

ETHYL ACETATE

1 Names
Ethyl acetate
Acetic ester

2 Physical properties

Molecular weight	88
Empirical formula	$C_4H_8O_2$
Boiling point (°C)	77
Freezing point (°C)	−84
Specific gravity (20/4 °C)	0.902
Liquid expansion coefficient (per °C)	0.0014
Surface tension (at 20 °C in dyn/cm)	24
Absolute viscosity (at 25 °C in cP)	0.455

3 Fire and health hazards

Flash point (closed cup) (°C)	−4
Autoignition temperature (°C)	484
Lower explosive limit (ppm)	22 000
Upper explosive limit (ppm)	115 000
IDLH (ppm)	10 000
Odour threshold (ppm)	3.6
TWA–TLV (ppm)	400
Saturated vapour concentration (70 °F in ppm)	100 000
Vapour density (relative to air)	3.04

4 Solvent properties

Hildebrand solubility parameter	9.1
Dipole moment (D)	1.7
Dielectric constant (20 °C)	6.02
Evaporation time (diethyl ether = 1.0)	2.9

5 Aqueous effluent characteristics

Solubility in water (at 25 °C in % w/w)	7.7
Solubility of water in (at 25 °C in % w/w)	3.3
Biological oxygen demand (w/w)	1.20
Chemical oxygen demand (w/w)	1.5
Theoretical oxygen demand (w/w)	1.82

6 Handling details

Hazchem number	1.173
Hazard class	3(Y)E
Vapour pressure (at 21 °C in mmHg)	86
Loss per transfer (% of liquid transferred)	0.05

7 Vapour pressure constants (mmHg, log base 10)

Antoine equation	$A = 7.10179$
	$B = 1244.950$
	$C = 217.881$
Cox chart	$A = 7.30648$
	$B = 1358.7$

8 Thermal information

Latent heat (cal/mol)	7744
Specific heat (cal/mol/°C)	40.4
Heat of combustion (cal/mol)	536 888

Azeotropes and activity coefficients of component X: ethyl acetate

Component Y	Azeotrope °C	Azeotrope % w/w X	α	A_{XY} $=\ln \gamma_X^\infty$	A_{YX} $=\ln \gamma_Y^\infty$	Source[11] Vol.	Page
Hydrocarbons							
n-Pentane							
n-Hexane	65	38	2[a]	0.98	0.89	5	514
n-Heptane				1.30			
Cyclohexane	73	54	1.5[a]	0.98	0.83	3+4	506
Benzene	77	94	<1.5[a]	0.12	0.08	5	496
Toluene		None	3	0.19	0.21	5	520
Ethylbenzene		None	8	0.06	0.25	5	538
Xylenes		None	9	0.47	0.46	5	541
Alcohols							
Methanol	62	54	2.5[a]	1.02	1.62	2(a)	154
Ethanol	72	69	1.8[a]	0.78	0.85	2(a)	352
n-Propanol		None	2.8[a]	0.52	0.65	2(a)	536
Isopropanol	75	75	1.5[a]	0.87	0.89	2(b)	59
n-Butanol		None	7	0.59	0.82	2(b)	148
Isobutanol							
sec-Butanol							
Cyclohexanol		None	>20	0.47	1.42	2(d)	511
Ethylene glycol							
Methyl Cellosolve		None	5	0.56	0.71	2(b)	126
Ethyl Cellosolve							
Butyl Cellosolve							
Chlorinated hydrocarbons							
Methylene dichloride		None	2/5	−0.94	−0.57	5	449
Chloroform	78	22	1.5	−1.50	−0.55	5	443
Carbon tetrachloride	74	43	1.7	0.30	0.26	5	437
1,2-Dichloroethane	74	43		>0.5	0.1		
1,1,1-Trichloroethane							
Trichloroethylene		None	<1.5	0.09	0.16	5	450
Perchloroethylene							
Monochlorobenzene		None	5	0.17	0.84	5	492
Ketones							
Acetone		None	2	0.12	0.17	3+4	176
Methyl ethyl ketone	77	88	<1.5[a]	0.68	0.68	3+4	278
MIBK							
Cyclohexanone							
N-Methylpyrrolidone							
Ethers							
Diethyl ether		None	5	0.09	−0.03	3+4	513
Diisopropyl ether							
1,4-Dioxane		None	2/3	0.68	0.07	3+4	455
Tetrahydrofuran							
Esters							
Methyl acetate		None	2	−0.19	0.07	5	357
Butyl acetate							
Miscellaneous							
Dimethylformamide		None		0.6	0.55		
Dimethyl sulphoxide		None	>10	1.11	1.30	5	461
Pyridine							
Acetonitrile	75	77	1.5[a]	0.43	0.40	5	455
Furfuraldehyde							
Water	70	92	10[a]	3.82	1.97	1	393

[a] The relative volatility between the less volatile component of the binary mixture and the azeotrope.

Recovery notes: ethyl acetate

Of all the widely used solvents, ethyl acetate is probably the least stable. It hydrolyses at ambient temperature in storage in the presence of water and in process it does so at significant rates whether in low or high pH conditions.

If recapture from air takes place on an activated carbon bed, it is likely that the ethyl acetate arising from regeneration of the bed will be mostly in aqueous solution with a small upper layer of water-saturated solvent. As the VLE diagram (Figure A.27) shows,

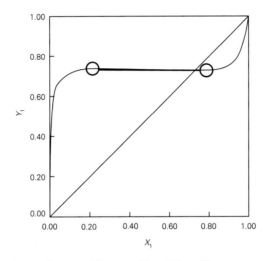

Fig. A.27 VLE diagram for ethyl acetate (1)–water (2) at 760 mmHg

ethyl acetate is very readily stripped from water in a single-stage evaporation to yield the water azeotrope. Since, whatever the method of stripping, some hydrolysis will have occurred, the recovered mixture will consist of a mixture presenting complex separation problems (Table A.16).

Table A.16 Components of wet ethyl acetate

Components	B.p. (°C)	Ethanol (% w/w)	Ethyl acetate (% w/w)	Water (% w/w)
Ethanol–ethyl acetate–water[a]	70.2	8.4	82.6	9.0
Ethyl acetate–water[b]	70.4	—	91.9	8.1
Ethanol–ethyl acetate	71.8	31.0	69.0	—
Ethyl acetate	77.1	—	100.0	—
Water	100.0	—	—	100.0
Acetic acid	118.0	—	—	—

[a] Single phase at 70 °C, two phase at 0 °C.
[b] Two phase.

The ternary azeotrope is only just two phase at low temperature (Figure A.28), so there is no practical way of using a phase separation to remove water even if it were acceptable to recover ethyl acetate with ethanol present.

To remove both water and ethanol simultaneously, with the added advantage of

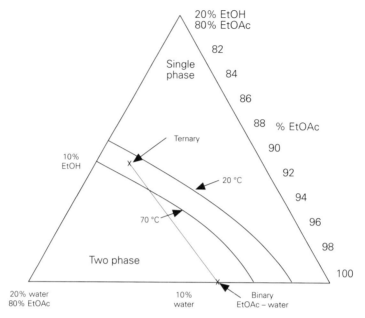

Fig. A.28 Ternary solubility diagram showing the temperature dependence of the phase behaviour of ethyl acetate–water–ethanol, indicating that the ternary azeotrope is two phase at 20 °C and single phase at 70 °C

removing any acetic acid from the system, both extraction and extractive distillation can be employed while azeotropic distillation can be used to remove just the ethanol and water.

1. Extractive distillation. As can be seen from Table 3.3, there is a large difference in polarity between ethyl acetate and ethanol with water being even more polar. Monoethylene glycol, propylene glycol, DMSO and butanediol all give relative volatilities substantially over 2.

2. Ethyl acetate is miscible in all proportions with hydrocarbons whereas water and ethanol are not, and acetic acid partitions very strongly to an aqueous rather than a hydrocarbon phase.

   Using a C_{10} n-alkane/isoalkane hydrocarbon as the extraction solvent for ethyl acetate from water one obtains at 25 °C in a single-stage contact the phases shown in Table A.17.

Table A.17 Phases produced when n-decane is used to extract ethyl acetate from water-saturated ethyl acetate using a 9:1 volume ratio of hydrocarbon to solvent

Compound	Hydrocarbon phase (% w/w)	Water phase (% w/w)
Ethyl acetate	10.0	2.30
Ethanol	0.001	0.24
Water[a]	0.008	96.96 (by diff.)
Hydrocarbon	90.0	ND
Acetic acid	ND	0.5

[a]The water-saturated hydrocarbon containing no ethyl acetate or ethanol contained 0.002% water.

The stripping of ethyl acetate from C_{10} hydrocarbon (b.p. 168–170 °C) is very easy.

3   Azeotropic distillation. The choice of an entrainer for this mixture is very limited because of the number of azeotropes formed by ethyl acetate with low-boiling hydrocarbons.

n-Pentane, which azeotropes with both water and ethanol, can be used to break the ternary azeotrope, but if more than 9% water is present in the mixture to be treated this is a very slow operation.

2,2-Dimethylbutane carries about three times more water than *n*-pentane and its water azeotrope condenses at about 50 °C (cf. *n*-pentane–water, 34.6 °C), thus making a higher boil-up possible with a given condenser. Unfortunately, 2,2-dimethylbutane is difficult to obtain.

Dichloromethane is also a possible entrainer and is slightly better than pentane.

In all circumstances in which water is present, ethyl acetate can hydrolyse. Under distillation conditions the acetic acid resulting from the reaction is rapidly removed down the fractionating column and equilibrium will never be reached.

In storage at 25 °C an equilibrium mixture would only contain about 35% ethyl acetate, although reaction is slow in a neutral solution.

If ethyl acetate is in solution in another solvent in alkaline conditions, the rate of hydrolysis is very dependent on the solvent. Water or DMSO speeds up hydrolysis whereas acetone, dioxane and ethanol induce very much lower rates.

The rate of hydrolysis is very dependent on temperature, approximately doubling for each 20 °C in the range encountered in solvent recovery. To avoid losses and the contamination of the overhead product, vacuum as low as the condenser will allow should be used. A low hold-up reboiler (e.g. a wiped-film evaporator) should be considered and batch distillation will seldom be the best choice for recovery.

Storage of contaminated ethyl acetate in mild steel containers is likely to result in corrosion of the tank and some acceleration in the rate of hydrolysis.

n-BUTYL ACETATE

1. **Names**

   n-Butyl acetate
   n-Butyl ethanoate

2. **Physical properties**

Molecular weight	116
Empirical formula	$C_6H_{12}O_2$
Boiling point (°C)	126
Freezing point (°C)	−74
Specific gravity (20/4 °C)	0.883
Liquid expansion coefficient (per °C)	0.00121
Surface tension (at 20 °C in dyn/cm)	25.1
Absolute viscosity (at 25 °C in cP)	0.73

3. **Fire and health hazards**

Flash point (closed cup) (°C)	22
Autoignition temperature (°C)	421
Lower explosive limit (ppm)	17 000
Upper explosive limit (ppm)	150 000
IDLH (ppm)	10 000
Odour threshold (ppm)	50
TWA–TLV (ppm)	150
Saturated vapour concentration (70 °F in ppm)	14 200
Vapour density (relative to air)	11.8

4. **Solvent properties**

Hildebrand solubility parameter	8.6
Dipole moment (D)	1.8
Dielectric constant (25 °C)	–
Evaporation time (diethyl ether = 1.0)	11.8

5. **Aqueous effluent characteristics**

Solubility in water (at 25 °C in % w/w)	0.7
Solubility of water in (at 25 °C in % w/w)	1.3
Biological oxygen demand (w/w)	1.15
Chemical oxygen demand (w/w)	1.72
Theoretical oxygen demand (w/w)	2.21

6. **Handling details**

Hazchem number	1123
Hazard class	3 YE
Vapour pressure (at 21 °C in mmHg)	10.6
Loss per transfer (% of liquid transferred)	0.0076

7. **Vapour pressure constants (mmHg, log base 10)**

Antoine equation	$A = 7.02845$
	$B = 1368.50$
	$C = 204.00$
Cox chart	$A = 7.44951$
	$B = 1626.5$

8. **Thermal information**

Latent heat (cal/mol)	8584
Specific heat (cal/mol/°C)	53.4
Heat of combustion (cal/mol)	846 104

Azeotropes and activity coefficients of component X: *n*-butyl acetate

Component Y	Azeotrope °C	Azeotrope % w/w X	α	A_{XY} $=\ln \gamma_X^\infty$	A_{YX} $=\ln \gamma_Y^\infty$	Source[11] Vol.	Page
Hydrocarbons							
n-Pentane							
n-Hexane							
n-Heptane	None		3/1.5	0.99	0.50	5	591
Cyclohexane	None		5	0.76	0.38	5	585
Benzene	None		4.5	−0.08	−0.69	5	583
Toluene	None		1.5	−0.87	−0.04	5	587
Ethylbenzene							
Xylenes							
Alcohols							
Methanol	None		18	1.21	1.05	2(b)	216
Ethanol	None		10	1.10	0.71	2(c)	426
n-Propanol	94	60					
Isopropanol	80	48					
n-Butanol	116	37	1.5[a]	0.81	0.43	2(b)	197
Isobutanol							
sec-Butanol							
Cyclohexanol							
Ethylene glycol	None		>20	2.41	1.93	2(d)	15
Methyl Cellosolve	118	52	1.7[a]	0.88	0.82	2(d)	122
Ethyl Cellosolve	126	88	1.5[a]	0.48	0.56	2(b)	294
Butyl Cellosolve							
Chlorinated hydrocarbons							
Methylene dichloride							
Chloroform	None		5	−1.08	−0.42	5	574
Carbon tetrachloride	None		4	−0.17	0.09	5	573
1,2-Dichloroethane							
1,1,1-Trichloroethane							
Trichloroethylene	None		2/3	−0.51	−0.30	5	575
Perchloroethylene	120	21					
Monochlorobenzene							
Ketones							
Acetone	None		10	0.47	0.24	3+4	219
Methyl ethyl ketone							
MIBK							
Cyclohexanone							
N-Methylpyrrolidone							
Ethers							
Diethyl ether							
Diisopropyl ether							
1,4-Dioxane							
Tetrahydrofuran							
Esters							
Methyl acetate	None		>10	0.20	0.22	5	397
Ethyl acetate							
Miscellaneous							
Dimethylformamide							
Dimethyl sulphoxide							
Pyridine							
Acetonitrile	None		4.5	0.62	0.83	5	577
Furfuraldehyde	None		3/5	0.45	0.59	3+4	46
Water	90	73		6.92	2.27	1	515

[a] The relative volatility between the less volatile component of the binary mixture and the azeotrope.

Recovery notes: *n*-butyl acetate

n-Butyl acetate is much more resistant to hydrolysis than ethyl acetate although, in the presence of water, it is sensible to distil at a low temperature (and pressure).

n-Butyl acetate is also much easier to separate from water because water is only soluble in it to 1.3% at 25 °C whereas its water azeotrope contains 27% water. *n*-Butyl acetate is so insoluble in water that it is usually not worth trying to recover the solvent from a saturated solution in water unless its smell, which is strong but not generally thought of as unpleasant, causes a neighbourhood nuisance.

The ternary azeotrope of *n*-butyl acetate, *n*-butanol and water (Table A.18), which boils at 91 °C, also separates so that little solvent is lost in the water phase.

Table A.18 Ternary azeotrope of *n*-butyl acetate, *n*-butanol and water

Component	Azeotrope (% w/w)	Solvent phase (% w/w)	Aqueous phase (% w/w)
n-Butyl acetate	63	86	1
n-Butanol	8	11	2
Water	29	3	97

Protective gloves used when handling *n*-butyl acetate should not be made of PVC, but neoprene is suitable.

DIMETHYLFORMAMIDE

1 Names   Dimethylformamide
                                                   DMF

2 Physical properties

Molecular weight	73
Empirical formula	C_3H_7NO
Boiling point (°C)	153
Freezing point (°C)	−61
Specific gravity (20/4 °C)	0.945
Liquid expansion coefficient (per °C)	0.00096
Surface tension (at 20 °C in dyn/cm)	35
Absolute viscosity (at 25 °C in cP)	0.82

3 Fire and health hazards

Flash point (closed cup) (°C)	62
Autoignition temperature (°C)	445
Lower explosive limit (ppm)	22 000
Upper explosive limit (ppm)	160 000
IDLH (ppm)	3500
Odour threshold (ppm)	100
TWA–TLV (ppm)	10
Saturated vapour concentration (70 °F in ppm)	3700
Vapour density (relative to air)	2.53

4 Solvent properties

Hildebrand solubility parameter	12.1
Dipole moment (D)	3.8
Dielectric constant (25°C)	36.7
Evaporation time (diethyl ether = 1.0)	>44

5 Aqueous effluent characteristics

Solubility in water (at 25 °C in % w/w)	Total
Solubility of water in (at 25 °C in % w/w)	Total
Biological oxygen demand (w/w)	0.90
Chemical oxygen demand (w/w)	–
Theoretical oxygen demand (w/w)	1.86

6 Handling details

Hazchem number	2265
Hazard class	2 P
Vapour pressure (at 21 °C in mmHg)	2. 8
Loss per transfer (% of liquid transferred)	0.021

7 Vapour pressure constants (mmHg, log base 10)

Antoine equation	$A = 7.10850$
	$B = 1537.780$
	$C = 210.390$
Cox chart	$A =$
	$B =$

8 Thermal information

Latent heat (cal/mol)	10074
Specific heat (cal/mol/°C)	35.5
Heat of combustion (cal/mol)	464 500

Azeotropes and activity coefficients of component X: dimethylformamide

Component Y	Azeotrope °C	Azeotrope % w/w X	α	A_{XY} =ln γ_X^∞	A_{YX} =ln γ_Y^∞	Source[11] Vol.	Page
Hydrocarbons							
n-Pentane		None					
n-Hexane		None	20	2.55	2.60	6(c)	332
n-Heptane	97	5	20[a]	2.19	1.82	6(b)	98
Cyclohexane		None	20	2.67	2.04	6(c)	200
Benzene		None	12	0.25	0.36	7	183
Toluene		None	4	0.96	0.31	7	390
Ethylbenzene	134	15	2[a]				
Xylenes	136	17	2[a]	0.93	0.99	7	481
Alcohols							
Methanol		None	20	−0.62	−0.21	2(a)	114
Ethanol		None		−0.32	−0.51	2(c)	372
n-Propanol		None					
Isopropanol		None		0.16	0.12		
n-Butanol		None		−0.32	−0.24		
Isobutanol							
sec-Butanol							
Cyclohexanol							
Ethylene glycol		None		0.60	0.06	2(b)	8
Methyl Cellosolve		None					
Ethyl Cellosolve		None					
Butyl Cellosolve		None					
Chlorinated hydrocarbons							
Methylene dichloride		None	14	3.05	−0.80	8	265
Chloroform		None					
Carbon tetrachloride		None	20	1.71	0.62	8	117
1,2-Dichloroethane		None					
1,1,1-Trichloroethane		None					
Trichloroethylene		None					
Perchloroethylene		None					
Monochlorobenzene		None					
Ketones							
Acetone		None	>10	1.89	0.26	3+4	164
Methyl ethyl ketone		None					
MIBK		None					
Cyclohexanone		None					
N-Methylpyrrolidone		None					
Ethers							
Diethyl ether		None					
Diisopropyl ether		None					
1,4-Dioxane		None	10	0.53	0.27	3+4	454
Tetrahydrofuran		None					
Esters							
Methyl acetate		None					
Ethyl acetate		None		0.55	0.6		
Butyl acetate		None					
Miscellaneous							
Dimethyl sulphoxide		None	8	0.10	0.20	8	407
Pyridine		None					
Acetonitrile		None					
Furfuraldehyde		None					
Water		None	6	0.82	0.16	1	276

[a]The relative volatility between the less volatile component of the binary mixture and the azeotrope.

Recovery notes: dimethylformamide

1 Removal from air

Since DMF is miscible with water in all proportions and is a comparatively high-boiling solvent, it can be removed from an air stream by water scrubbing down to levels at which there is no economic incentive for recovery, and no health or environmental problem is presented by its presence in air discharged to atmosphere (Figure A.29).

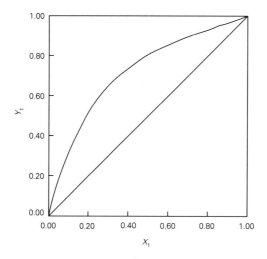

Fig. A.29 VLE diagram for water (1)–DMF (2) at 760 mmHg

2 Recovery from water solution

It is technically possible to make a separation of water and DMF by fractionation and the commonly achieved tops and bottoms specifications are 500 ppm DMF and 0.1% water, respectively, from a feed containing about 20% DMF.

Fractionation under vacuum (about 150 mmHg) improves both the yield and purity of the recovered DMF and columns with about 50 actual (say 33 theoretical) trays and reflux ratios of 0.5–1 are typical of normal industrial practice.

Because, even with a low reflux, the amount of water to be evaporated is large and the fuel costs are substantial, alternatives to fractionation have been investigated.

a Liquid extraction

The volatile chlorohydrocarbons (MDC, chloroform and carbon tetrachloride) all show favourable partition coefficients for extracting DMF from water when their comparatively low latent heats and large relative volatility with respect to DMF are taken into account.

MDC is much the most favourable, provided that the available condenser cooling water is sufficiently cold for the condensing of MDC–water (38 °C at atmospheric pressure).

The presence of an inorganic salt in the aqueous mixture from which the DMF must be extracted makes a considerable improvement in the partition coefficient, and therefore favours extraction over fractionation. However, if the salt needs to be recovered also, the economics are not so clear.

There may also be a problem if the salt solution is concentrated so that the density of the chlorohydrocarbon phase carrying its burden of DMF may be very close if not equal to the aqueous salt solution, making the operation of a liquid–liquid extraction column difficult. This can be overcome by using a mixture of MDC and an aromatic hydrocarbon to reduce the density of the organic phase at the cost of a reduced partition coefficient.

b Vapour compression

The steam at the top of a DMF–water fractionating column is almost pure and can therefore be considered for recompression if site conditions provide cheap motive power to drive a turbine for boosting its pressure.

c Two-column operation

Although the fractionation of water from DMF is better done at reduced pressure, a low hold-up system does not suffer very seriously from decomposition at atmospheric pressure. This allows two columns to be used in parallel (Figures A.30 and A.31).

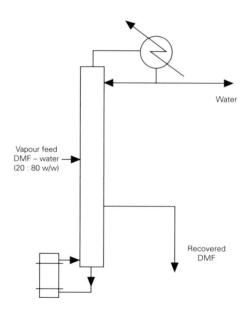

Fig. A.30 Single-column distillation

In this arrangement, the first column handles about 40% of the total feed. Therefore, with a boil-up of 1.5×0.32 of the total feed, it provides, via its condenser, the heat to evaporate the feed of the second column. The latter operates under vacuum to provide the temperature difference for the condenser/vaporizer.

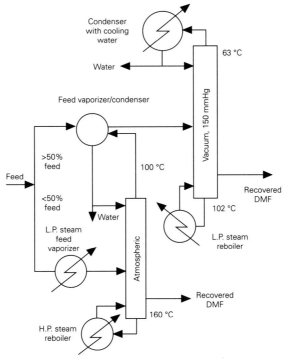

Fig. A.31 Parallel-column distillation

The amount of high-pressure steam required for recovery of dry DMF is about 40%, and the low-pressure steam is about 68% of that needed for single-column, atmospheric pressure operation.

The two-column operating mode is unsuitable for DMF–water mixtures that contain inorganic solutes and/or gross contamination with polymers or other heavy organic solutes. These involve a preliminary feed evaporation at low pressure, but not normally at low temperature, in the evaporator.

d Drying by freezing

Any non-evaporation method for removing water can be expected to make economies in the use of steam. Using direct contact refrigeration at about −20 °C, a DMF–water mixture of 50% DMF can be made which shows a 75% reduction in the water to be removed per unit of DMF from feed initially containing 80% water and 20% DMF. Whether this approach is economic or not depends on the site costs for steam and refrigeration.

Since DMF is an atropic solvent, pervaporation cannot be employed to recover it using the membranes available at present, but the advantages of a hydrophobic membrane to pass DMF and retain water would be very great if one could be developed.

All the possible recovery routes outlined have involved an eventual fractionation stage, and will therefore face a common problem because DMF is not wholly stable in the presence of water and at any pH far from neutral.

Some dissociation is likely to take place, polluting the water distillate with dimethylamine and causing a build-up of formic acid in the residue. A side-stream product of DMF either on a batch or a continuous column is desirable. High pH conditions lead to rapid decomposition of DMF and must be avoided. A high-boiling azeotrope of formic acid and DMF exists and residues will always be acidic. Stainless steel 316 is a suitable material of construction. Brass and zinc are corroded by hot DMF and colour pick-up occurs in mild steel. PVC hoses and gloves are unsuitable for handling DMF.

For certain applications, it is important to recover DMF with very low levels of formic acid and dimethylamine and such levels are best achieved by treatment with ion-exchange resins. If a vapour side stream is taken below the column feed it is likely to be acidic, but this can be corrected by adding a small stream of sodium hydrogencarbonate to the column above the vapour offtake.

DMF is a powerful liver poison and any person making skin contact with it will be affected, however quickly they wash. The noticeable symptom is severe flushing when drinking even a small amount of alcohol. Recovery will take about a week and, during that period, alcohol should be avoided completely. Long-term exposure to DMF vapour will produce the same symptoms and should be treated as an urgent indication that ventilation is unsatisfactory.

DMF causes damage to the foetus in laboratory animals, and if women of child-bearing age are employed where they might be exposed to DMF liquid or vapour, consideration should be given to special precautions.

Appendix 1   349

DIMETHYL SULPHOXIDE

1 **Names**   Dimethyl sulphoxide
                                                                         Dimethyl sulfoxide
                                                                         Dimso
                                                                         DMSO

2 **Physical properties**

    Molecular weight   78
    Empirical formula C_2H_6OS
    Boiling point (°C)   189
    Freezing point (°C)   18.5
    Specific gravity (20/4 °C)   1.101
    Liquid expansion coefficient (per °C)   0.00088
    Surface tension (at 20 °C in dyn/cm)
    Absolute viscosity (at 25 °C in cP)   2.0

3 **Fire and health hazards**

    Flash point (closed cup) (°C)   95
    Autoignition temperature (°C)   255
    Lower explosive limit (ppm)   30000
    Upper explosive limit (ppm)   420000
    IDLH (ppm)
    Odour threshold (ppm)
    TWA–TLV (ppm)   1000
    Saturated vapour concentration (70 °F in ppm)   650
    Vapour density (relative to air)   2.71

4 **Solvent properties**

    Hildebrand solubility parameter   13.0
    Dipole moment (D)   3.96
    Dielectric constant (20 °C)   46.6
    Evaporation time (diethyl ether = 1.0)   1500

5 **Aqueous effluent characteristics**

    Solubility in water (at 25 °C in % w/w)   Total
    Solubility of water in (at 25 °C in % w/w)   Total
    Biological oxygen demand (w/w)
    Chemical oxygen demand (w/w)
    Theoretical oxygen demand (w/w)   2.05

6 **Handling details**

    Hazchem number
    Hazard class
    Vapour pressure (at 21 °C in mmHg)   0.7
    Loss per transfer (% of liquid transferred)   0.0003

7 **Vapour pressure constants (mmHg, log base 10)**

    Antoine equation $A = 6.88076$
                                                     $B = 1541.52$
                                                     $C = 191.797$
    Cox chart $A =$
                                                     $B =$

8 **Thermal information**

    Latent heat (cal/mol)   12636
    Specific heat (cal/mol/°C)   36.5
    Heat of combustion (cal/mol)   471900
    Heat of fusion (cal/mol)   3221

Azeotropes and activity coefficients of component X: dimethyl sulphoxide

Component Y	Azeotrope °C	Azeotrope % w/w X	α	$A_{XY} = \ln \gamma_X^\infty$	$A_{YX} = \ln \gamma_Y^\infty$	Source[11] Vol.	Source[11] Page
Hydrocarbons							
n-Pentane							
n-Hexane					4.32		
n-Heptane							
Cyclohexane							
Benzene	None		10	1.18	0.99	7	169
Toluene	None		10	1.91	1.31	7	386
Ethylbenzene							
Xylenes							
Alcohols							
Methanol	None		>20	−1.40	−1.09	2(c)	62
Ethanol							
n-Propanol							
Isopropanol							
n-Butanol	None		6	−1.18	0.04	2(d)	163
Isobutanol	None		10	−1.41	−0.33	2(b)	275
sec-Butanol							
Cyclohexanol							
Ethylene glycol	None		10				
Methyl Cellosolve							
Ethyl Cellosolve							
Butyl Cellosolve							
Chlorinated hydrocarbons							
Methylene dichloride	None		>20	−0.82	−0.57	8	264
Chloroform	None		>20	−1.88	−0.91	8	234
Carbon tetrachloride	None		>20	2.33	1.26	8	107
1,2-Dichloroethane							
1,1,1-Trichloroethane							
Trichloroethylene							
Perchloroethylene							
Monochlorobenzene							
Ketones							
Acetone	None		>10	0.74	0.61	3+4	154
Methyl ethyl ketone							
MIBK							
Cyclohexanone							
N-Methylpyrrolidone							
Ethers							
Diethyl ether							
Diisopropyl ether							
1,4-Dioxane	None		>10	0.89	0.50	3+4	450
Tetrahydrofuran	None		>10	1.46	1.05	3+4	434
Esters							
Methyl acetate							
Ethyl acetate	None		>20	1.30	1.11	5	461
Butyl acetate							
Miscellaneous							
Dimethylformamide	None		8	0.20	0.10	8	407
Pyridine							
Acetonitrile							
Furfuraldehyde							
Water	None			−1.46	−0.83	1	119

Recovery notes: dimethyl sulphoxide

DMSO is one of the most difficult solvents to recover and there have been a number of cases in different parts of the world in which plants recovering or producing DMSO have been damaged in exothermic incidents.

From the purely mechanical viewpoint, pure DMSO's freezing point of 18.5 °C (which falls to 0 °C at 90% w/w DMSO–10% w/w water) means that pipelines to handle it and vents on vessels containing it must be traced and lagged. Tanks need coils or, as a minimum, an outflow heater. If the contents of a tank reach about 30 °C, DMSO will tend to evaporate from the liquid surface and solidify on the walls and roof. This can cause problems in stock-taking and may present a hazard if a tank has to be entered for cleaning. Internal inspection of the tank roof is vital before entry.

DMSO is not very toxic in itself but it migrates through human (or animal) skin very readily, taking with it its solute, if any. There is therefore a risk that unknown residual materials in DMSO for recovery may be absorbed if an operator's skin is wetted and all skin contact should be avoided. Pregnant women are normally advised not to handle DMSO.

If DMSO is ingested or absorbed the individual will have a foul-smelling breath as it is metabolized internally. This bad odour is also very noticeable if DMSO is digested in biological effluent treatment plants.

DMSO has a reputation for easing joint stiffness and pain if applied externally to the skin and may be pilfered in small quantities if this use were known to those working with it.

DMSO is very hygroscopic and stops to exclude damp air from storage tanks are necessary.

DMSO is not stable at its boiling point and should be distilled at low pressure (about 20 mmHg is suitable). Since it has a high value of α with most contaminants, the column is likely to be short and the reflux ratio low so the column pressure drop will not be large.

It is most important to test by differential thermal analysis any DMSO mixture before attempting to process it on a plant scale and to operate at not less than 20 °C below the temperature where an exotherm has been detected. Alkaline hydrolysis of esters is about ten times faster in the presence of DMSO than in protic solvents. Inorganic and organic halides can react explosively with DMSO although zinc oxide appears to inhibit this reaction.

Under acidic conditions DMSO is thought to form formaldehyde, which can then polymerize exothermically.

DMSO can be dried with calcium hydride or molecular sieves (4A, 5A and 13X), but reacts explosively with magnesium chlorate and other perchlorates. Alumina not only dries DMSO, but also the small amounts of impurities formed during distillation can be removed with it or with activated carbon.

Paraffinic hydrocarbons from C_5 to C_{20} are not fully miscible with DMSO and are very sparingly soluble in a DMSO–water mixture. This provides a route for removing some impurities from DMSO if they prove hard to separate by fractionation or are dangerous to distil in DMSO solution.

DMSO is a good solvent for many inorganics and for polymers except polystyrene. Teflon gaskets should be used when DMSO must be handled.

PYRIDINE

1 Names   Pyridine

2 Physical properties

Molecular weight	79
Empirical formula	C_5H_6N
Boiling point (°C)	115
Freezing point (°C)	−42
Specific gravity (20/4 °C)	0.983
Liquid expansion coefficient (per °C)	0.0007
Surface tension (at 20 °C in dyn/cm)	36.6
Absolute viscosity (at 25 °C in cP)	0.88

3 Fire and health hazards

Flash point (closed cup) (°C)	20
Autoignition temperature (°C)	522
Lower explosive limit (ppm)	18 000
Upper explosive limit (ppm)	124 000
IDLH (ppm)	3600
Odour threshold (ppm)	0.03
TWA–TLV (ppm)	15
Saturated vapour concentration (70 °F in ppm)	22 000
Vapour density (relative to air)	2.74

4 Solvent properties

Hildebrand solubility parameter	10.7
Dipole moment (D)	2.3
Dielectric constant (20 °C)	12.9
Evaporation time (diethyl ether = 1.0)	12.7

5 Aqueous effluent characteristics

Solubility in water (at 25 °C in % w/w)	Total
Solubility of water in (at 25 °C in % w/w)	Total
Biological oxygen demand (w/w)	1.47
Chemical oxygen demand (w/w)	0.05
Theoretical oxygen demand (w/w)	3.03

6 Handling details

Hazchem number	1252
Hazard class	2 WE
Vapour pressure (at 21 °C in mmHg)	16.6
Loss per transfer (% of liquid transferred)	0.0073

7 Vapour pressure constants (mmHg, log base 10)

Antoine equation	$A = 7.01328$
	$B = 1356.93$
	$C = 212.655$
Cox chart	$A =$
	$B =$

8 Thermal information

Latent heat (cal/mol)	8374
Specific heat (cal/mol/°C)	33.6
Heat of combustion (cal/mol)	631 368

Azeotropes and activity coefficients of component X: pyridine

Component Y	Azeotrope °C	Azeotrope % w/w X	α	A_{XY} $=\ln \gamma_X^\infty$	A_{YX} $=\ln \gamma_Y^\infty$	Source[11] Vol.	Page
Hydrocarbons							
n-Pentane							
n-Hexane							
n-Heptane	96	25	2.8[a]	1.54	0.64	6(b)	113
Cyclohexane		None	4	1.71	1.42	6(a)	171
Benzene		None	3	0.21	0.24	7	221
Toluene	110	22	<1.5[a]	0.51	0.29	7	406
Ethylbenzene		None					
Xylenes		None	2/3	0.33	0.29	7	482
Alcohols							
Methanol		None	5	0.07	0.04	2(a)	180
Ethanol		None	3	−0.05	0.05	2(c)	355
n-Propanol		None	1.5	−0.25	−0.13	2(c)	512
Isopropanol				−0.07	0.11		
n-Butanol	119	30	<1.5[a]	−0.52	−0.12	2(b)	166
Isobutanol		None	1.7	−0.47	−0.21	2(d)	350
sec-Butanol	89	68	<1.5[a]	−0.43	−0.12	2(d)	255
Cyclohexanol							
Ethylene glycol							
Methyl Cellosolve							
Ethyl Cellosolve							
Butyl Cellosolve							
Chlorinated hydrocarbons							
Methylene dichloride[b]		None	11	−0.60	−0.46	8	267
Chloroform		None	3/5	−0.49	−1.20	8	237
Carbon tetrachloride		None	5	0.70	−0.38	8	141
1,2-Dichloroethane							
1,1,1-Trichloroethane							
Trichloroethylene							
Perchloroethylene	113	48	1.5[a]	0.70	0.62	8	346
Monochlorobenzene		None					
Ketones							
Acetone		None	8	0.74	0.20	3+4	8
Methyl ethyl ketone							
MIBK	115	60					
Cyclohexanone							
N-Methylpyrrolidone							
Ethers							
Diethyl ether							
Diisopropyl ether							
1,4-Dioxane		None					
Tetrahydrofuran							
Esters							
Methyl acetate							
Ethyl acetate							
Butyl acetate							
Miscellaneous							
Dimethylformamide							
Dimethyl sulphoxide							
Acetonitrile							
Furfuraldehyde							
Water	94	57		3.43	0.76		

[a]The relative volatility between the less volatile component of the binary mixture and the azeotrope.
[b]Reacts.

Recovery notes: pyridine

Of the solvents reviewed in this book, pyridine is the most reactive chemically and the most costly. It is a relatively strong base (pK_a 5.25) and reacts quickly with all strong acids. The salts of these acids, particularly pyridine sulphate and chloride, are much more water soluble than pyridine itself and are stable up to 100 °C. They thus provide a means of extracting pyridine from solvents that are not water miscible, e.g. hydrocarbons and chlorinated hydrocarbons, since the salts can be transferred to an aqueous phase for subsequent springing and recovery of pyridine from water. The formation of pyridine salts can also be used to remove pyridine from an aqueous solution which includes close-boiling solvents, e.g. isobutanol. In these cases, after the salt has been formed, it will remain in aqueous solution while the other solvents are stripped off by distillation or steam stripping.

However, the recovery of pyridine from water is expensive in heat and plant time if it is done by distillation because of the very high water content (43% w/w) of the pyridine–water azeotrope (Figure A.32 and Table A.19). The traditional azeotropic

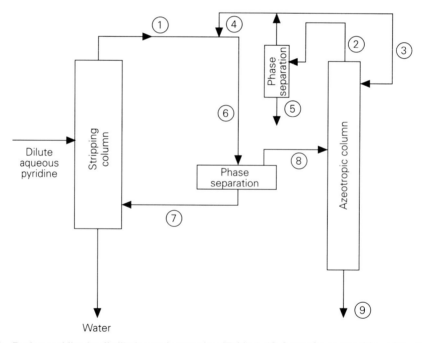

Fig. A.32 Drying pyridine by distillation and extraction (Table A.19 shows the composition of the flows)

entrainer for drying pyridine in the coal-chemical industry, where much of it is made, is benzene, since it is easy to remove from pyridine once the pyridine is dry. Toluene is more economical as an entrainer but harder to remove from dry pyridine because of a pyridine–toluene azeotrope. The presence of inorganic salts reduces the mutual solubility of water and pyridine and about half the water in the azeotrope can be salted out with sodium sulphate.

The solubility of the benzene–pyridine–water system is such that a treatment of the pyridine–water azeotrope with benzene prior to azeotropic distillation similar to that described for THF–water (Figure 7.2) shows a considerable advantage over a straightforward entrainer distillation. This can be compared with straight azeotropic distillation to dry the azeotrope, which requires an overhead of 398 kg of benzene to dry 52.8 kg of pyridine or 51% more than the combined phase separation and azeotropic distillation route.

Table A.19 Composition of streams in Figure A.32 in kg

Stream	Pyridine	Water	Benzene	Total
1	57.0	43.0	0	100.0
2	0	26.3	263.0	289.3
3	0	0	181.8	181.8[a]
4	0	0	81.2	81.2
5	0	26.3	0	26.3
6	57.0	43.0	81.2	181.2
7	4.2	16.7	0	20.9
8	52.8	26.3	81.2	160.3
9	52.8	0	0	52.8

[a] Reflux ratio 2.24:1.

The presence of inorganic salts makes a substantial difference to the partition of pyridine between aqueous and organic phases (Table A.20). This can be useful when recovering pyridine from aqueous solutions after springing it from sulphate or chloride salts.

Table A.20 Solubility of pyridine in benzene, water and saline solution at 25 °C in % w/w

Benzene layer	Water layer	8% NaCl–92% water layer
15.4	7.0	3.0
30.1	15.2	6.6
40.6	27.8	12.0
52.5	43.1	26.9

Pyridine has a strong and unpleasant smell and it can be detected by nose at concentrations below 1 ppm in water. While this smell can be removed by the addition of small amounts of acid, it will return if effluent treated in this way is neutralized at some later stage.

Many people find that smoking becomes unpleasant if they are exposed to concentrations of pyridine below its TWA–TLV of 15 ppm.

ACETONITRILE

1 Names

Acetonitrile
Methyl cyanide
Cyanomethane
Ethane nitrile

2 Physical properties

Molecular weight	41
Empirical formula	C_2H_3N
Boiling point (°C)	81.6
Freezing point (°C)	-44
Specific gravity (20/4 °C)	0.782
Liquid expansion coefficient (per °C)	0.0014
Surface tension (at 20 °C in dyn/cm)	29.1
Absolute viscosity (at 25 °C in cP)	0.375

3 Fire and health hazards

Flash point (closed cup) (°C)	6
Autoignition temperature (°C)	524
Lower explosive limit (ppm)	44 000
Upper explosive limit (ppm)	160 000
IDLH (ppm)	4000
Odour threshold (ppm)	40
TWA–TLV (ppm)	40
Saturated vapour concentration (70 °F in ppm)	96 000
Vapour density (relative to air)	1.42

4 Solvent properties

Hildebrand solubility parameter	11.9
Dipole moment (D)	3.2
Dielectric constant (20 °C)	37.5
Evaporation time (diethyl ether = 1.0)	2.04

5 Aqueous effluent characteristics

Solubility in water (at 25 °C in % w/w)	Total
Solubility of water in (at 25 °C in % w/w)	Total
Biological oxygen demand (w/w)	1.22
Chemical oxygen demand (w/w)	2.0
Theoretical oxygen demand (w/w)	2.21

6 Handling details

Hazchem number	1648
Hazard class	2 WE
Vapour pressure (at 21 °C in mmHg)	71
Loss per transfer (% of liquid transferred)	0.02

7 Vapour pressure constants (mmHg, log base 10)

Antoine equation	$A = 7.33986$
	$B = 1482.29$
	$C = 250.523$
Cox chart	$A = 7.12578$
	$B = 1322.7$

8 Thermal information

Latent heat (cal/mol)	7134
Specific heat (cal/mol/°C)	22.14
Heat of combustion (cal/mol)	304 220

Azeotropes and activity coefficients of component X: acetonitrile

Component Y	Azeotrope °C	Azeotrope % w/w X	α	A_{XY} =ln γ_X^∞	A_{YX} =ln γ_Y^∞	Source[11] Vol.	Page
Hydrocarbons							
n-Pentane	35	10	>20[a]	3.27	2.07	6(a)	101
n-Hexane	57	28					
n-Heptane	69	46	10[a]	2.25	2.95	6(b)	84
Cyclohexane	62	33					
Benzene	73	34	2[a]	1.13	0.92	7	123
Toluene	81	80	4[a]	1.21	1.24	7	373
Ethylbenzene	None		10	0.94	1.59	7	465
Xylenes	None		5/10	0.80	1.71	7	499
Alcohols							
Methanol	64	81	3[a]	0.88	1.09	2(a)	43
Ethanol	73	44	1.5[a]	0.69	1.39	2(a)	298
n-Propanol	81	72					
Isopropanol	75	52					
n-Butanol	None		6	1.29	1.35	2(d)	156
Isobutanol	None		5	0.44	0.98	2(d)	346
sec-Butanol	81	86	2.5[a]	0.81	0.99	2(d)	241
Cyclohexanol							
Ethylene glycol							
Methyl Cellosolve	None		5	0.50	0.53	2(d)	109
Ethyl Cellosolve							
Butyl Cellosolve							
Chlorinated hydrocarbons							
Methylene dichloride	None		4.5	0.15	0.19	8	258
Chloroform	None		2	0.21	0.39	8	217
Carbon tetrachloride	65	17	3[a]	2.37	1.51	8	86
1,2-Dichloroethane	79	49	<1.5[a]	0.32	0.35	8	364
1,1,1-Trichloroethane							
Trichloroethylene	75	29	2[a]	1.25	1.20	8	350
Perchloroethylene	None		10	1.37	1.59	8	342
Monochlorobenzene	None		10/3	1.21	1.01	8	381
Ketones							
Acetone	None		2/3	0.05	0.03	3+4	143
Methyl ethyl ketone							
MIBK							
Cyclohexanone							
N-Methylpyrrolidone							
Ethers							
Diethyl ether	None		10	0.88	1.14	3+4	499
Diisopropyl ether							
1,4-Dioxane	None						
Tetrahydrofuran	None			0.60	0.81		
Esters							
Methyl acetate	None		2/3	0.39	0.11	5	354
Ethyl acetate	75	23	<1.5[a]	0.40	0.43	5	455
Butyl acetate	None		4.5	0.83	0.62	5	577
Miscellaneous							
Dimethylformamide	None						
Dimethyl sulphoxide	None						
Pyridine	None						
Furfuraldehyde	None						
Water	76	84	10[a]	2.30	1.56	1	78

[a] The relative volatility between the less volatile component of the binary mixture and the azeotrope.

Recovery notes: acetonitrile

Despite the fact that one of its less commonly used names is methyl cyanide, acetonitrile (ACN) is not particularly toxic although its smell is not an adequate indication of its presence at its TLV of 40 ppm. It is particularly harmful to the eyes and great care should be taken in wearing goggles when handling ACN.

It will hydrolyse to acetic acid and ammonia in the presence of aqueous strong bases and if traces of organic bases are present in feed or recovered ACN, these should be removed by Amberlite IRC-50 or a similar ion-exchange resin.

For removing small concentrations of water, calcium chloride, silica gel or 3A molecular sieves can be used. ACN is an aprotic solvent so that the common currently available pervaporation membranes are not suitable for drying it.

A characteristic of ACN is that it forms azeotropes with most organics that are not miscible with water and boil below ACN (Table A.21). This means that it is difficult to recover the azeotropic entrainers once they have been used to remove water from ACN. This is not a problem if long-term drying of an ACN stream is involved and for this benzene, trichloroethylene and diisopropyl ether can be used.

Table A.21 Boiling points of binary and ternary azeotropes of ACN, water and entrainer

	Ternary		Binaries	
ACN (% w/w)	20.5		29	83.7
Water (% w/w)	6.4	6.3		16.3
Trichloroethylene (% w/w)	73.1	93.7	71	
Boiling point (°C)	59	73.1	74.6	76.5
ACN (% w/w)	23.3		34	83.7
Water (% w/w)	3.2	9		16.3
Benzene (% w/w)	68.5	91	66	
Boiling point (°C)	66	69.4	73	76.5
ACN (% w/w)	13	17		83.7
Water (% w/w)	5		4.6	16.3
DIPE (% w/w)	82	83	95.4	
Boiling point (°C)	59	61.7	62.2	76.5

The separation of the ternary azeotrope into two phases produces the mixtures shown in Table A.22. The aqueous phase contains enough ACN and, in the case of benzene, enough entrainer to justify recycling to recover the organic content.

For modest-sized parcels of wet ACN, the need to dispose of the entrainer–ACN azeotropes after the recovery campaign may represent an unacceptable cost. Chloroform and methylene chloride both form binary water azeotropes without forming a ternary with ACN, but the former has the disadvantage of toxicity and the latter has a very low water-carrying capacity at a low condensing temperature.

ACN has the unusual property of forming azeotropes with aliphatic hydrocarbons boiling from 36 to 180 °C and these azeotropes are two phase.

ACN is produced as a by-product of acrylonitrile manufacture and its yield from the process is small. As a result its price can be very volatile. Many of its industrial uses

Table A.22 Compositions of aqueous and organic phases of ternaries of ACN, water and entrainer (% w/w)

	Benzene	DIPE	Trichloroethylene
Organic phase:			
Entrainer	75.3	85.5	78.6
ACN	24.2	13.0	20.8
Water	0.5	1.5	0.6
Density (g/cm^3)	0.841	0.742	1.254
Aqueous phase:			
Entrainer	4.2	1.0	0.2
ACN	15.8	13.0	16.1
Water	80	86.0	83.7
Density (g/cm^3)	0.955	0.976	0.975

involve mixtures with other solvents (e.g. toluene, isopropanol) that form binary azeotropes with ACN that are very difficult to separate, involving techniques not economically viable when ACN prices are low. When ACN is in short supply and its price is high, extractive distillation is justified despite the specialized equipment needed.

ACN can be stored in all normal metals used for plant construction except copper.

FURFURALDEHYDE

1. **Names**
  Furfuraldehyde
  Furfural
  2-Furaldehyde
  Fural

2. **Physical properties**

Molecular weight	96
Empirical formula	$C_5H_4O_2$
Boiling point (°C)	162
Freezing point (°C)	−37
Specific gravity (20/4 °C)	1.1598
Liquid expansion coefficient (per °C)	0.00106
Surface tension (at 20 °C in dyn/cm)	45
Absolute viscosity (at 25 °C in cP)	1.4

3. **Fire and health hazards**

Flash point (closed cup) (°C)	62
Autoignition temperature (°C)	315
Lower explosive limit (ppm)	21 000
Upper explosive limit (ppm)	193 000
IDLH (ppm)	250
Odour threshold (ppm)	8
TWA–TLV (ppm)	2
Saturated vapour concentration (70 °F in ppm)	2400
Vapour density (relative to air)	3.33

4. **Solvent properties**

Hildebrand solubility parameter	5.46
Dipole moment (D)	3.6
Dielectric constant (20 °C)	41.9
Evaporation time (diethyl ether = 1.0)	

5. **Aqueous effluent characteristics**

Solubility in water (at 25 °C in % w/w)	8.4
Solubility of water in (at 25 °C in % w/w)	5.0
Biological oxygen demand (w/w)	0.77
Chemical oxygen demand (w/w)	
Theoretical oxygen demand (w/w)	1.67

6. **Handling details**

Hazchem number	1199
Hazard class	
Vapour pressure (at 21 °C in mmHg)	1.81
Loss per transfer (% of liquid transferred)	0.81×10^{-3}

7. **Vapour pressure constants (mmHg, log base 10)**

Antoine equation	$A = 8.40200$
	$B = 2338.49$
	$C = 261.638$
Cox chart	$A = $ N/A
	$B = $ N/A

8. **Thermal information**

Latent heat (cal/mol)	9216
Specific heat (cal/mol/°C)	36.5
Heat of combustion (cal/mol)	560 736

Azeotropes and activity coefficients of component X: furfuraldehyde

Component Y	Azeotrope °C	Azeotrope % w/w X	α	$A_{XY} = \ln \gamma_X^\infty$	$A_{YX} = \ln \gamma_Y^\infty$	Source[11] Vol.	Page
Hydrocarbons							
n-Pentane							
n-Hexane							
n-Heptane	98	5	>10[a]	2.40	2.19	3+4	50
Cyclohexane		None	>10	2.33	1.99	3+4	45
Benzene		None	>10	0.59	0.49	3+4	44
Toluene		None	5/10	0.66	1.28	3+4	48
Ethylbenzene	132	5	3[a]	1.09	0.93	3+4	51
Xylenes	139	10	3[a]	1.05	1.02	3+4	52
Alcohols							
Methanol		None	>20	0.00	0.00	2(c)	140
Ethanol		None	>20	1.66	1.37	2(a)	385
n-Propanol							
Isopropanol							
n-Butanol							
Isobutanol							
sec-Butanol							
Cyclohexanol	157	5					
Ethylene glycol							
Methyl Cellosolve		None					
Ethyl Cellosolve							
Butyl Cellosolve	161	88					
Chlorinated hydrocarbons							
Methylene dichloride				0.48	−0.16		
Chloroform		None	10	−0.27	−0.24	3+4	36
Carbon tetrachloride		None	10	1.33	1.29	3+4	35
1,2-Dichloroethane				0.24	0.13		
1,1,1-Trichloroethane							
Trichloroethylene		None	>10	2.90	0.37	3+4	37
Perchloroethylene							
Monochlorobenzene							
Ketones							
Acetone							
Methyl ethyl ketone							
MIBK							
Cyclohexanone							
N-Methylpyrrolidone							
Ethers							
Diethyl ether							
Diisopropyl ether							
1,4-Dioxane							
Tetrahydrofuran							
Esters							
Methyl acetate							
Ethyl acetate							
Butyl acetate		None	3/5	0.59	0.45	3+4	46
Miscellaneous							
Dimethylformamide							
Dimethyl sulphoxide							
Pyridine							
Acetonitrile							
Water	98	65		4.25	2.08		

[a] The relative volatility between the less volatile component of the binary mixture and the azeotrope.

Recovery notes: furfuraldehyde

Furfuraldehyde (FF) is unstable in conditions of both light and heat. It will polymerize spontaneously at 230 °C and some polymerization is likely to take place at temperatures as low as 60 °C. A stabilizer mixture consisting of N-phenylsubguanidine, N-phenylthiourea and N-phenylnaphthylamine at levels of 0.001–0.1% will prevent polymerization up to 170 °C. It follows, therefore, that care needs to be taken in distilling FF away from heavy residues. This can only be done under vacuum or in a steam distillation.

Other inhibitors against the effect of oxygen are furamide (0.08% w/w), hydroquinone, α-naphthol, pyrogallol and cadmium iodide. Since most of these are less volatile than FF, there is a danger that the latter will be left unprotected if it is distilled, and newly distilled material is likely to need re-stabilizing.

Apart from its tendency to polymerize, FF becomes acidic if stored in contact with air, but can be cleaned up by being passed through an alumina column before use.

The production of FF involves its recovery from a dilute aqueous solution, so much work has been done on its recovery from water. As Figure A.33 shows, its water

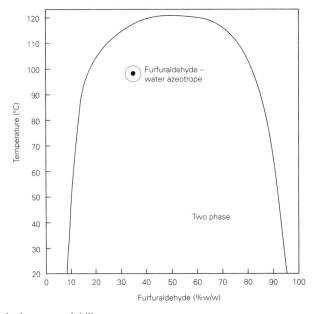

Fig. A.33 Furfuraldehyde–water solubility vs. temperature

azeotrope splits into two phases with an improvement in the recovery of FF-rich phase at low temperature. At 25 °C, the FF-rich phase has a density of about 1.14 and the water-rich phase 1.013, so an adequate density difference exists to make a separation, although it is probably less than a general-purpose plant would be designed for. A conventional two-column system will produce dry FF (Figure A.34).

The stripping of the azeotrope from both water and from dry FF is easy and neither column needs to be operated at high reflux. However, FF can be extracted from water with many solvents, including alcohols, ketones and chlorinated hydrocarbons. Partition

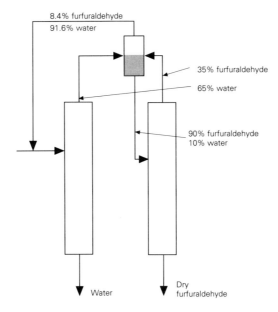

Fig. A.34 Two-column system for drying furfuraldehyde

coefficients for these extraction solvents shows that the last-named are the most attractive (Table A.23).

Table A.23 Partition coefficients of furfuraldehyde between various solvents and water

Solvent	Partition coefficient	
	High concentration[a]	Low concentration[b]
2 Ethylhexanol	8.4	3.9
tert-Amyl alcohol	7.4	4.0
MIBK	13.4	7.1
Dichloromethane (MDC)	14.1	14.4
Trichloroethylene	10.9	6.0
Chloroform	12.0	12.3

[a] Furfuraldehyde in solvent phase ~55% w/w.
[b] Furfuraldehyde in solvent phase ~20% w/w.

This property of FF is useful if modest quantities of FF need to be separated from water on a scale which does not justify a continuous fractionation. Drying can be achieved in a three-stage batch operation:

1 single-stage liquid–liquid extraction;
2 batch distillation of solvent–water azeotrope until batch is dry;
3 flash over of FF as dry distillate.

The only loss of FF is in the first stage, in which the water phase will carry away a small proportion of the FF in the feed.

364 Appendix 1

As an example of this technique, the following is a recovery plan for an 83:17 water–FF mixture.

1. Mix feed and MDC in a ratio of 2:1. This will split to a bottom MDC phase and a top aqueous phase:

	Water	FF	MDC
Feed (kg)	83	17	50
Aqueous (kg)	82.1	1.13	1.1
MDC (kg)	0.9	15.87	48.9

2. Batch distilling wet MDC of FF at a reflux ratio of 1–2:1 with phase separation of the MDC–water azeotrope. The azeotrope has a composition of 98.5:1.5 MDC–water and the solubility of water in MDC is 0.15%. A reflux ratio of 1:1 would therefore remove $48.9 \times 1.5 \times 0.0135 = 0.99$ kg of water by the stage when all the MDC has been stripped off the FF.

3. A side stream of FF from the batch column while the column head is on total reflux, removing any traces of moisture and keeping the last of the MDC. It should be noted that the FF must be inhibited against polymerization since, with MDC at the column head, vacuum distillation is not practicable.

The usual specification of virgin furfuraldehyde allows a moisture content of 0.2%.

WATER

1 Names   Water

2 Physical properties

Molecular weight	18
Empirical formula	H_2O
Boiling point (°C)	100
Freezing point (°C)	0
Specific gravity (20/4 °C)	0.998
Liquid expansion coefficient (per °C)	0.000 207
Surface tension (at 20 °C in dyn/cm)	72.75
Absolute viscosity (at 25 °C in cP)	0.89

3 Fire and health hazards

Flash point (closed cup) (°C)	None
Autoignition temperature (°C)	None
Lower explosive limit (ppm)	None
Upper explosive limit (ppm)	None
IDLH (ppm)	None
Odour threshold (ppm)	None
TWA–TLV (ppm)	None
Saturated vapour concentration (70 °F in ppm)	–
Vapour density (relative to air)	0.63

4 Solvent properties

Hildebrand solubility parameter	–
Dipole moment (D)	1.87
Dielectric constant (20 °C)	79.7
Evaporation time (diethyl ether = 1.0)	–

5 Aqueous effluent characteristics

Solubility in water (at 25 °C in % w/w)	–
Solubility of water in (at 25 °C in % w/w)	–
Biological oxygen demand (w/w)	–
Chemical oxygen demand (w/w)	–
Theoretical oxygen demand (w/w)	–

6 Handling details

Hazchem number	–
Hazard class	–
Vapour pressure (at 21 °C in mmHg)	18.65
Loss per transfer (% of liquid transferred)	0.002

7 Vapour pressure constants (mmHg, log base 10)

Antoine equation	$A = 8.07131$
	$B = 1730.63$
	$C = 233.426$
Cox chart	$A = N/A$
	$B = N/A$

8 Thermal information

Latent heat (cal/mol)	9703
Specific heat (cal/mol/°C)	18
Heat of combustion* (cal/mol)	11 500
Heat of fusion (cal/mol)	1432

*Since the heats of combustion are net, heat needs to be supplied to water in turning it into steam in an incinerator.

Azeotropes and activity coefficients of component X: water

Component Y	Azeotrope °C	Azeotrope % w/w X	α	A_{XY} $=\ln \gamma_X^\infty$	A_{YX} $=\ln \gamma_Y^\infty$	Source[11] Vol.	Source[11] Page
Hydrocarbons							
n-Pentane	35	1			11.6		
n-Hexane	62	6		7.54	13.1		
n-Heptane	79	13	>20[a]		14.5		
Cyclohexane	70	8					
Benzene	69	9		5.42	7.46		
Toluene	84	18	20[a]	8.11	8.13		
Ethylbenzene	92	33					
Xylenes	93	37					
Alcohols							
Methanol		None	5	0.66	0.80	1	48
Ethanol	78	4	4[a]	0.96	1.77	1	181
n-Propanol	87	29		1.16	2.74	1	286
Isopropanol	80	12	5[a]	1.09	2.47	1	330
n-Butanol	93	42		1.14	3.96	1	406
Isobutanol	90	33		1.35	3.71	1	439
sec-Butanol	87	27		1.10	3.56	1	419
Cyclohexanol	98	30	20[a]	1.52	2.76	1	514
Ethylene glycol		None	>20	−0.43	−1.31	2(c)	297
Methyl Cellosolve	99	15					
Ethyl Cellosolve	98	13	5[a]	0.71	1.90	1	450
Butyl Cellosolve	99	21	20[a]	0.47	5.30	1	526
Chlorinated hydrocarbons							
Methylene dichloride	38	1	20[a]	3.89	5.75	1	1
Chloroform	56	3		4.48	6.81		
Carbon tetrachloride	66	4					
1,2-Dichloroethane	72	8		4.01	6.31		
1,1,1-Trichloroethane	65	4		5.43	8.67		
Trichloroethylene	73	6					
Perchloroethylene	88	16					
Monochlorobenzene	90	28					
Ketones							
Acetone		None	10/2	1.35	2.29	1	193
Methyl ethyl ketone	73	11	2[a]	1.83	3.37	1	363
MIBK	88	24		1.85	5.90		
Cyclohexanone	96	55	10[a]	1.76	3.60	1	511
N-Methylpyrrolidone		None		0.17	0.46	1	379
Ethers							
Diethyl ether	34	1	>20[a]	3.53	4.49	1	344
Diisopropyl ether	62	5	10[a]	3.00	1.54	1	525
1,4-Dioxane	88	18		1.35	2.03	1	383
Tetrahydrofuran	64	4	20[a]	1.60	3.44	1	370
Esters							
Methyl acetate	56	5	>10[a]	2.14	3.16	1	264
Ethyl acetate	70	8	10[a]	1.97	3.82	1	393
Butyl acetate	90	27		2.27	6.92	1	515
Miscellaneous							
Dimethylformamide		None	6	0.16	0.82	1	276
Dimethyl sulphoxide		None		−0.83	−1.46	1	119
Pyridine	94	43		0.76	3.43		
Acetonitrile	76	16	10[a]	1.56	2.30	1	78
Furfuraldehyde	98	35		2.08	4.25		

[a]The relative volatility between the less volatile component of the binary mixture and the azeotrope.

Notes on the properties of water

Water has a liquid expansion coefficient very much lower than those of all other solvents. As a result, there is much less thermal convection in tanks containing water than there is in solvent tanks. It is more important, if a sample truly representative of a tank's contents is needed, to blend, the tank contents positively or take samples at frequent levels in the case of water than for other liquids.

Water also has the unusual property of expanding as it cools from 4 °C to its freezing point. If water at 4 °C becomes trapped in a valve and then cools further it is possible that it will develop a pressure high enough to crack the body of the valve. For this reason drain valves that may contain water should not be made of cast iron or other brittle metals.

Frequent accidents take place because liquids, immiscible in water and at a temperature over 100 °C, are pumped into tanks containing a water layer in the bottom. It often takes an appreciable time before the water layer boils but when it does so the rate of steam formation can be very high, causing a violent eruption of steam and the hot tank contents from the tank vents. Liquid droplets can be carried considerable distances causing both a health risk and damage to property. Dipping a tank where water may be present using water-finding paste should be a routine precaution when transferring very hot liquids.

APPENDIX 2

Calculation of vapour–liquid equilibrium

It is possible using the values of γ^∞ from Appendix 1 and from many other sources to calculate the vapour–liquid equilibria (VLE) for binary mixtures. There are a number of equations for doing this but, for use with a pocket calculator, the van Laar equation, although not the most up-to-date, is a good choice. It is accurate enough for engineering purposes and it is easy to visualize the significance of the results as the calculation proceeds.

As an example, the VLE for the binary mixture of n-hexane and ethyl acetate has been calculated using the data in Appendix 1 (Table A.24).

Table A.24 Values of γ_1 and γ_2 using the van Laar equation:

$$\ln \gamma_1 = A_{12}\left(\frac{1}{1+A_{12}x_1/A_{21}x_2}\right)^2$$

			n-Hexane (1)	Ethyl acetate (2)
Cox chart:				
A			6.9386	7.3065
B			1212.1	1358.7
γ^∞			2.62	2.41
$\ln \gamma^\infty$			0.963 $(=A_{12})$	0.880 $(=A_{21})$
Vapour pressure at 70 °C (mmHg)			791.2 $(=P_1)$	599.1 $(=P_2)$
x_1	x_2	$A_{12}x_1/A_{21}x_2$	γ_1	γ_2
1.00	0.00	—	1.000	2.41
0.95	0.05	20.79	1.002	2.23
0.90	0.10	9.85	1.008	2.07
0.80	0.20	4.38	1.03	1.79
0.70	0.30	2.55	1.08	1.57
0.60	0.40	1.64	1.15	1.40
0.50	0.50	1.09	1.25	1.27
0.40	0.60	0.73	1.38	1.17
0.30	0.70	0.47	1.56	1.09
0.20	0.80	0.27	1.82	1.04
0.10	0.90	0.12	2.15	1.01
0.05	0.95	0.06	2.36	1.003
0.00	1.00	0.00	2.62	1.000

Applying the values of vapour pressure, one can calculate the values of x, y and total pressure (Table A.25).

Tabel A.25 VLE for n-hexane–ethyl acetate

x_1	$x_1 \gamma_1 P_1$	$x_2 \gamma_2 P_2$	Total pressure (mmHg)	y_1
0.00	0.00	599.1	599.1	0.00
0.05	93.4	570.8	664.2	0.14
0.10	170.1	544.6	714.7	0.24
0.20	288.0	498.5	786.5	0.37
0.30	370.3	457.1	827.4	0.45
0.40	436.7	420.6	857.3	0.51
0.50	492.9	380.4	873.3	0.56
0.60	545.9	335.5	881.4	0.62
0.70	598.1	282.2	880.3	0.68
0.80	651.9	214.5	866.4	0.75
0.90	717.8	124.0	841.8	0.85
0.95	753.1	66.8	819.9	0.92
1.00	791.2	0.0	791.2	1.00

Plotting the values listed in Table A.25 shows that there is an azeotrope at 0.63 mole fraction. This corresponds reasonably well to the azeotrope at 0.6155 mole fraction at a total pressure of 760 mmHg and a temperature of 65.5 °C reported in the literature (Figure A.35).

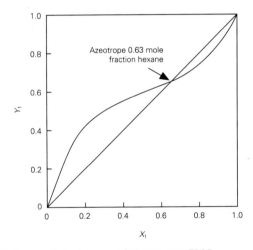

Fig. A.35 Calculated VLE diagram for n-hexane–ethyl acetate at 70 °C

Since there is very much more information in the literature on azeotropes than on activity coefficients, it would be useful to back calculate values of γ from azeotropic data. In theory this can be done.

At an azeotrope the values of x_1 and y_1 are the same. This allows the value of γ_1 and γ_2 to be calculated at an azeotrope since

$$y_1 P = x_1 \gamma_1 P_1$$

where P is the total pressure in the system. Therefore,

$\gamma_1 = P/P_1$

Using the composition, temperature and pressure of the n-hexane–ethyl acetate azeotrope quoted above:

$\gamma_1 = 760/686.7 = 1.316$ (Table A.24: 1.1355)

and

$\gamma_2 = 760/511.1 = 1.487$ (Table A.24: 1.428)

The van Laar equations can be rearranged so that

$$\ln \gamma^\infty = \ln \gamma_1 \left(1 + \frac{x_2 \ln \gamma_2}{x_1 \ln \gamma_1}\right)^2$$

Using $x_1 = 0.6155$ and $x_2 = 0.3845$:

$\gamma_1^\infty = 3.33$ and $\gamma_2^\infty = 2.20$

These results are appreciably different to those obtained from vapour pressure data (2.62 and 2.41, respectively). For practical purposes in solvent recovery this will not make an important difference to the values of α between $x = 0.2$ and 1.00, but it could be appreciable between zero and 0.05 in this case (Table A.26).

Data from the azeotrope would give a higher value for α at the bottom of the column so that the difficulty of stripping the n-hexane out of the ethyl acetate would be underestimated when employing these data.

Tabel A.26 Effect of error in γ on values of α (value of $\alpha^* = 1.30$)

x_1	From VL data		From azeotrope	
	γ_1/γ_2	α	γ_1/γ_2	α
0.20	1.82	2.37	1.81	2.35
0.10	2.15	2.80	2.39	3.11
0.05	2.36	3.07	2.80	3.64
0.01	2.56	3.33	3.21	4.17
0.00	2.62	3.41	3.33	4.33

Hence, although there is a great deal of data available on azeotropes, very small discrepancies can have a considerable effect on the values of relative volatility if operations are to be carried out on dilute solutions and at least three consistent points on the VLE curve should be used to calculate values of γ.

Nonetheless, in the absence of any other information the calculation from azeotropic data is better than nothing provided that a good margin of safety is included in a plant design depending on them.

Figure A.36 shows graphically the effect of activity coefficients on the relative volatility of the system. An ideal Raoult's law mixture would have a relative volatility (α^*) across the composition range of 1.3. This is below the level at which fractional distillation would normally be chosen to separate the components, but not impossibly

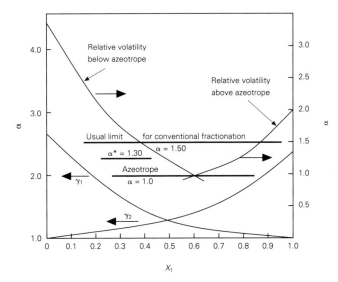

Fig. A.36 n-Hexane–ethyl acetate binary showing the effect of γ_1/γ_2 on relative volatility

low if a very good column were available or if the required purity of the two products was not very high.

The presence of an azeotrope means that straightforward fractionation can never yield two nearly pure products, but it would theoretically be possible to make one pure product and a pure material of the azeotropic composition.

Inspection of Figure A.36 shows that it should be possible to make a nearly pure ethyl acetate ($x_1 = 0$) and a nearly pure azeotrope from a feed of composition between $x_1 = 0$ and 0.60 with a log mean value of α of about 2.0.

If the feed had a composition richer in n-hexane than the azeotrope (i.e. $x_1 > 0.615$), it would be more difficult with a log mean relative volatility of 1.4 to make a good quality n-hexane and a pure azeotrope.

APPENDIX 3

Conversion factors for SI units to those used customarily in solvent recovery

The conversion factors are rounded to three significant figures.

Particular care in using figures from the literature must be paid when units such as gallons and tons are employed, since the American versions have the same titles as but different magnitudes to their Imperial namesakes.

It should be noted that throughout this book mol implies g-moles and kmol implies kg-moles.

Table A.27 Weight: equivalents to 1 kilogram (kg)

6.48×10^6	grains	gr
1×10^6	milligrams	mg
1×10^3	grams	g
35.3	ounces	oz
2.20	pounds	lb
1.10×10^{-3}	short tons	US ton
1.00×10^{-3}	metric tonnes	Te

Table A.28 Pressure: equivalents of 1 kilopascal (kPa)

10^3	pascals	Pa
	newtons per square metre	N/m²
7.50	millimetres of mercury	mmHg
	torr	Torr
4.02	inches of water	in H$_2$O
0.335	feet of water	ft H$_2$O
0.295	inches of mercury	inHg
0.145	pounds per square inch	psi
1.02×10^{-2}	kilograms per square centimetre	kg/cm²
1.00×10^{-2}	bars	bar
0.987×10^{-2}	atmospheres	atm

Table A.29 Energy: equivalents of 1 kilojoule (kJ)

1.00×10^3	joules	J
	newton metres	N m
7.37×10^2	foot pounds (force)	ft-lbf
2.39×10^2	calories	cal
0.102	kilogram-metres	kg-m
9.86	litre-atmospheres	l-atm
1.00	kilojoules	kJ
0.948	British thermal units	BTU
0.527	Centigrade heat units	CHU
0.349	cubic foot-atmospheres	ft^3-atm
0.239	kilocalories	kcal

Table A.30 Industrial energy units: equivalents in MJ

1.015	pound of steam from and at 212 °F	lb.f&a
1.266	tons of refrigeration	
2.684	horsepower hour	hp h
3.600	kilowatt hour	kW h
35.16	boiler horsepower	BHP
105.5	therms	Th

Table A.31 Length: equivalents of 1 metre (m)

1.00×10^{10}	ångströms	A
1.00×10^9	nanometres	nm
1.00×10^6	micrometres, microns	μm
3.94×10^4	{ thou { mils	mil
1.00×10^3	millimetres	mm
39.4	inches	in
3.28	feet	ft
1.00	metres	m

Table A.32 Area: equivalent to 1 square metre (m^2)

1.00×10^4	square centimetre	cm^2
1.56×10^3	square inch	in^2
10.8	square foot	ft^2
1.2	square yard	yd^2
2.54×10^4	acre	acre
1.00×10^5	hectare	ha

Table A.33 Volume: equivalent to 1 cubic metre (cm^3)

1×10^6	cubic centimetre	cm^3, cc
	millimetre	ml
2.11×10^3	US pint	US pt
1.76×10^3	Imperial pint	Imp pt
1.00×10^3	litre	l
264	US gallon	US gal
220	Imperial gallon	Imp gal
35.3	cubic foot	ft^3
6.29	barrel (42 US gal oil)	bbl
1.31	cubic yard	yd^3
6.75×10^{-3}	acre inch	acre in

Table A.34 Customary and SI units

Surface tension	dyn/cm = 1.00 millinewtons/metre (mN/m)
Kinematic viscosity	centistoke (cSt) = square millimetres/second (cm^2/s)
Dynamic viscosity	centipoise (cP) = kilopascal seconds (kPa s)
Specific heat	BTU/lb/°F = 0.239 × joules/gram per degree kelvin (J/g/K)
Density	lb/ft^3 = 0.016 × grams/cubic centimetre (g/cm^3)
Heat transfer	BTU/h/ft^2/°F = 0.0489 × kilojoule/hour/square metre/degree kelvin (kJ/h/m^2/K)
Dipole moment	Debye = 3.33 × 10^{20} coulomb metre (C m)

Bibliography

Reference books and further reading

Since there are few books devoted to solvent recovery, the information that a recoverer needs to carry out his task is scattered through the technical literature, in both books and journals.

It would not be practical to suggest a list of journals that would be worth obtaining. A wide spread of new information on the chemical and physical properties of solvents and on the techniques for separating and dehydrating them is to be found in a great range of publications. Indeed, since the great increase in the number of solvents that took place in the 1950s and 1960s there have been comparatively few new solvents put on the market and therefore few papers on their properties and processing.

At the same time, solvents, which once were specialities carrying technical support from their producers, are now mostly commodity chemicals for which high-grade technical support is seldom available.

The recommended books are for further reading on the various aspects of the subject of solvent recovery or are sources of the information needed to design recovery processes.

See 1–6 for:

Chemical engineering theory and practice relevant to distillation and other solvent treating technologies.

See 7–10 for:

Specific technology on the processing of solvent wastes.

Further physical property information on solvents covered by Appendix 1 and information on solvents not covered there:

See 11–15 for:

Vapour–liquid equilibrium, vapour pressures, activity coefficients and relative volatilities.

See 16–17 for:

Azeotropes.

See 18–23 for:

Solubilities and liquid–liquid equilibria.

See 24–26 for:

Safety and general properties.

Bibliography

1. Rousseau, R.W. (Ed.) (1987) *Handbook of separation process technology*. Wiley, New York. Contents: a good survey of the principles involved in separation and of the processes used industrially, though rather unbalanced in the space devoted to unusual techniques.
2. Schweitzer, P.A. (Ed.) (1979) *Handbook of separation techniques for chemical engineers*. McGraw-Hill, New York. Contents: a survey of the techniques used in separating chemicals with the majority of space being devoted to the sort of processes which might be met in solvent recovery.
3. Robinson, C.S. and Gilliland, E. R. (1950) *Elements of fractional distillation*, 4th edn. McGraw-Hill, New York. Contents: old-fashioned but clear explanation of the principles of fractionation.
4. Van Winkle, M. (1967) *Distillation*. McGraw-Hill, New York. Contents: a good reference book for the whole range of fractionation by distillation.
5. Prigogine, I., and Defay, R. (1954) *Chemical thermodynamics*. Longmans, London. Contents: useful for the understanding of the theory behind activity coefficients and their derivation.
6. Perry, R.H., and Green, D. (Eds) (1984) *Perry's chemical engineers' handbook*, 6th edn. McGraw-Hill, New York. Contents: the most complete collection of information on chemical properties and chemical engineering theory and practice. If only one reference book were available this should be it. Weak on toxicity.
7. Breton, M., Frillici, P., Palmer, S. *et al.* (1988) *Treatment technology for solvent containing wastes*. Noyes Data Corporation, Park Ridge, NJ. Contents: the treatment of effluent water contaminated with solvents is very well and comprehensively treated. The processing of solvents, whether contaminated with water or involatile residues, is less well covered. USA regulations as in place in 1987 are fully described.
8. E.P.A. and I.C.F. Consulting Associates (1990) *Solvent waste reduction*. Noyes Data Corporation, Park Ridge, NJ. Contents: survey of the US requirements on discharge of solvents and the various options, such as reuse, recycling, incineration, cement kiln fuel available. Applicable mostly to the small operator with in-house facilities.
9. Riddick, J.A., and Burger, W.B. (1986) *Organic solvents*, 4th edn. Volume 2 of *Techniques of chemistry* series. Wiley, New York. Contents: a good source of physical data and of laboratory purification methods. Hazards and stabilizers/inhibitors for laboratory work are covered.
10. Freeman (Ed.) (1990) *Incinerating hazardous wastes*. Technomic Publishing. Contents: a survey of the current technology in the field.
11. Gmehling, J., Onken, U., and Arlt, W. *Vapour–liquid equilibrium data collection* (13 volumes, up to 1984). Dechema, Frankfurt/Main. Contents: a very large collection of vapour–liquid data and activity coefficients. Most of the binary mixtures illustrated with VLE diagrams. An invaluable reference work for the solvent recoverer working at the high-tech end of the industry. As a by-product of the vapour–liquid data, there is a good listing of Antoine equation constants.
12. Hirata, M., Ohe, S., and Nagahama, K. (1975) *Computer-aided data book of vapour–liquid equilibria*. Elsevier, Amsterdam. Contents: if the Dechema series is unavailable, this single volume covers a great many common industrial mixtures together with vapour–liquid equilibrium diagrams.
13. Ohe, S. (1976) *Vapour–liquid equilibrium data*. Elsevier, Amsterdam. Contents: more complete than ref. 12, but still a long way short of the Dechema collection.
14. Maczynski, A., and Bilinski, A. (1974–85) *Verified vapour–liquid equilibrium data*, 9 volumes. Polish Academy of Sciences, Warsaw. Contents: a very large number of binary vapour–liquid equilibria but not diagrams, so much less easy to use than the Dechema collection.
15. Dreisbach, R.R. (1952) *Pressure–volume–temperature relationships of organic compounds*, 3rd edn. Handbook Publishers. Contents: Cox chart constants and tables of vapour pressure–temperature data for individual compounds.
16. Weast, R.C. (Ed.) *CRC handbook of chemistry and physics (the rubber handbook)* (editions up to 1975). CRC Press, Cleveland, OH. Contents: apart from a lot of information about very many organic and inorganic chemicals, these editions carry information on azeotropes with the composition of the two phases when they are not miscible.
17. Horsley, L.H. (Ed.) (1973) *Azeotropic data III. Advances in chemistry series*, No. 116.

American Chemical Society, Washington, DC. Contents: easily the most comprehensive collection of azeotropes but needs revision. Nonetheless, essential for anyone engaged in solvent recovery.

18 Wisniak, J., and Tamir, A. (1980) *Liquid–liquid equilibrium and extraction*, 4 volumes. Elsevier, Amsterdam. Contents: a very comprehensive survey (up to its publication date) of technical literature references but no data on the solubility of solids and liquids in liquids.

19 Stephen, H., and Stephen, T. (1963–79). *Solubilities of inorganic and organic compounds*, 4 volumes. Pergamon Press, Oxford. Contents: a very large amount of information collected from the technical literature on binary and ternary mixtures of partly miscible solvents. All produced as tables with no diagrams. A useful but now somewhat dated reference book.

20 Sorensen, J.M., and Arlt, W. (1979) *Liquid–liquid equilibrium data collection*, 3 volumes. Dechema, Frankfurt/Main. Contents: a companion collection to the Dechema vapour–liquid data collection with most ternary systems plotted as diagrams.

21 Hansch, C., and Leo, A. (Eds) (1979) *Substituent constants for correlation analysis in chemistry and biology*. Wiley, New York. Contents: a very complete listing, to the year of publication, of the values of $\log_{10} P$ produced by Pomona College.

22 Horvath, A.L. (1982) *Halogenated hydrocarbons—solubility and miscibility with water*. Dekker, New York. Contents: a comprehensive coverage of water–halogenated hydrocarbon solubility and a thorough theoretical discussion of the mechanism of miscibility.

23 *Solubility data series* (1979 continuing), many volumes. Pergamon Press, Oxford. Contents: a very detailed survey of liquid solubilities at various temperatures, all critically evaluated. The volumes on hydrocarbons and low-boiling alcohols in water have been published.

24 Flick, F.W. (Ed.) (1985) *Industrial solvents handbook*, 3rd edn. Noyes Data Corporation, Park Ridge, NJ. Contents: a collection of the information on physical properties collected from manufacturers' brochures, but nothing on fire and health hazards. Some information on the uses of solvents, solubility of resins and polymers, etc.

25 de Renzo, D.J. (Ed.) (1986) *Solvents safety handbook*. Noyes Data Corporation, Park Ridge, NJ. Contents: for every commonly used solvent this lists the essential physical properties, health data and transport safety information.

26 Sax, N.I. (1984) *Dangerous properties of industrial materials*, 6th edn. Van Nostrand-Reinhold, New York. Contents: the definitive book on toxicity of solvents and other chemicals. Also gives some information on physical properties. There are condensed versions covering a more restricted list of chemicals which may be sufficient for the smaller operator.

Index

Absolute pressure
  condenser vent 75
Absolute viscosity 202–367
Absorption 14–17
Absorption column 16
A.C. *see* Activated carbon
Acetic acid 74–5
Acetic ester 335
Acetone
  activity coefficients 99
  ED entrainers 104
  methanol separation 95–6
  properties/recovery 298–301
  separation 81
  water removal 107
Acetonitrile (ACN) 103, 356–9
ACN *see* Acetonitrile
Activated carbon
  ethyl acetate recovery 28–9
  water recovery 40
Activated carbon adsorption 7, 8–14, 33–5, 40
Activated carbon beds
  construction materials 12
  heating 11–12
  size 10
Activity coefficients
  absorption 15
  extractive distillation 99–103
  liquid-liquid extraction 122–3
  Raoult's Law 35–7
  solvent properties 201–367
  VLE diagrams 198–9
Adronal 256
Adsorption
  activated carbon 7, 8–14
  disposal 134
  effluents 33–5
  water removal 117–19
Agitated thin-film evaporators (ATFE) 69–72, 76
Air
  pollution 7–8, 17–18
  scrubbing 7

  solvent removal 3, 7–21
  stripping 23, 35–7, 39
Air-cooled condensers 46
Airco process 19–21
Alcohols 103, 226–58
Aldehydes
  odour 77
Aliphatic esters
  steam distillation 94
Aliphatic hydrocarbons 18, 186
Alkanes 29–30, 103
Alumina
  water removal 118
Amines 73
Antifreeze 261
Antoine equation 192, 202–367
Aqueous effluents 23–40
Aromatic hydrocarbons
  ED entrainers 103
ATFE *see* Agitated thin-film evaporators
Atmospheric pressure 89–90, 92
Autoignition temperature
  diethyl ether 318
  hazards 143–4
  solvent properties 194, 202–367
Automatic batch steam distillation unit 61
Azeotropes
  ACN 358
  entrainers 111
  pressure distillation 117
  scrubbing liquids 16
  separation method 79
  solvent properties 198, 203–367
Azeotropic distillation
  ethyl acetate 329
  methyl ethyl ketone 304–5
  solvent separation 94–7
  THF 329
  water removal 107–16

Batch distillation 41, 83–8
Batch distillation kettle
  external jacket 44

Batch distillation kettle, *continued*
  internal coils 44
Batch package distillation units 68–9
Batch steam distillation 64
Batch stills
  extractive distillation 101
Benzene
  flammability 191
  properties/recovery 214–16
  scrubbing liquid 15
  solubility/temperature relationship 127
  water recovery 39
  water removal 109
Benzene hydride 211
Benzole 214
Binary mixtures 80, 81–2, 83, 98
Biodegradation 23
Biological disposal 135
Biological oxygen demand (BOD) 194, 202–367
Blockages
  fractionating columns 49
BOD *see* Biological oxygen demand
Boiling points
  ACN azeotropes 358
  azeotropic distillation 114–15
  recovery effects 73–4
  solvent properties 189–90, 202–367
Bund walls 153
Bursting discs 47, 164–5
Butan-2-ol 250
n-Butanol 108, 243–6
sec-Butanol 250–5
Butan-2-one 302
2-Butoxyethanol 268
n-Butyl acetate 340–2
Butyl alcohol 243
Butyl cellosolve 268–71
n-Butyl ethanoate 340

C8 aromatics 223
Calorific value 171–2
Carbinol 226
Carbon adsorption *see* Activated carbon adsorption
Carbon tetrachloride 185, 279–81
Catalytic incineration 133–4
Caustic soda 123, 330
Cement kilns 131–3
CFCs *see* Chlorofluorocarbons
Charging
  previous batches 160
  stills 159–60
Chemical damage 92–3
Chemical oxygen demand (COD) 194, 202–367
Chemisorption 123–5

Chlorinated hydrocarbons 272–97
Chlorinated solvents
  disposal 133, 180–1
  toxicity 146
  vapour pressure reduction 74
  water cleaning 39
Chlorobenzene 295
Chloroethene 285
Chlorofluorocarbons (CFCs) 2, 181, 185, 288
Chloroform 114, 276–8
Cleaning
  fractionating columns 49
  maintenance 161–2
  tanks 162, 164
Coalescer 26
COD *see* Chemical oxygen demand
Coils
  batch distillation kettle 44
  steam heating 44
Colour removal 33
Combustible material 145–6
Combustion
  see also Incineration
  fuels 172
Condensation
  low-temperature 17–19
Condensers 45–8
Consumption
  solvents 186
Contamination 86
Continuous distillation 41, 83–8
Continuously cleaned heating surfaces 69–72
Continuous steam stripping 64
Contract processing 175
Conversion factors 372–4
Cooling towers 46
Corrosion 42, 110, 165
Costs *see* Economic aspects
Cox chart 202–367
Cox equation 83, 89–90, 192
CTET *see* Carbon tetrachloride
Curtain gases 20–1
Cyanomethane 356
Cyclohexane
  azeotropic distillation 114
  properties/recovery 211–13
  residue separation 74
  water removal 109
Cyclohexanol 90–1, 92, 256–8
Cyclohexanone 90–1, 92, 258, 311–12
Cyclohexyl alcohol 256
Cyclohexyl ketone 311

Damage
  see also Corrosion
  chemical 92–3

Dangers *see* Hazards
n-Decane
  ethyl acetate recovery 29
Decanting 25–6, 31
Decomposition odours 77
Degreasing solvents 288, 291
Dehydration methods 105–28
Dephlegmator 87
Design
  installation 140–1
Desorbate treatment 10–11
Dessicants 123–5
Destruction 172–3
Dichloromethane 272
1,2-Dichloroethane 282–4
Dielectric constant 195, 202–367
Diethylene oxide 323
Diethyl ether 204, 316–19
Diisopropylamine 284
Diisopropyl ether 104, 114, 255, 320–2
Dimethylamine 75
Dimethylbenzenes 223, 225
Dimethylformamide (DMF)
  azeotropic distillation 95
  decomposition 75
  dehydration 116
  liquid-liquid extraction 121
  properties/recovery 343–8
  skin adsorption 147
Dimethyl ketone 298
Dimethyl sulphoxide 147, 349–51
Dimso 349
1,4-Dioxane 323–6
DIPE *see* Diisopropyl ether
Dipole moment 195, 202–367
Direct steam injection 43
Discharge limits 7–8, 18, 24
Disentrainer 48
Disposal 129–35
Distillation
  fractional 41–57
  separation methods 79
Distillation unit 38
DMF *see* Dimethylformamide
DMSO *see* Dimethyl sulphoxide
Domestic use 184
Dowtherm E 295, 297
Drain valves 57, 154
Droplet settling speed 25–6
Drums
  filling 151
  handling/emptying 149–50
  storage 148–9
Drying
  diethyl ether 319
Drying solvents 105–28

Dumping 129

Earthing 154
Economics
  disposal 131, 132
  drying solvents 127
  solvent recovery 171–81
  water clean-up 38–40
ED *see* Extractive distillation
EDC *see* 1,2-Dichloroethane
Efficiency
  carbon adsorption 9–10
  fractionating columns 52–3
Effluent
  discharge standards 24
  gas restrictions 131
Electric heating 42
Empirical formulae 202–367
Entrainers
  azeotropic distillation 94–7
  benzene 216
  sec-butanol 252–3
  ethyl acetate 339
  extractive distillation 97–104
  MEK drying 305–6
  n-propanol 237
  water removal 109–16
Environmental damage 1–2
Environmental regulations 186–7
1,4-Epoxybutane 327
Equilibrium temperatures
  air purity standards 17–18
Erosion 42
E.S. *see* Extraction solvents
Esters 103, 331–42
Ethane-1,2-diol 259
Ethane nitrile 356
Ethanol
  azeotropic distillation 114
  chloroform stabilization 278
  drying entrainers 109, 110, 112
  entrainers 110
  FLE 123
  properties/recovery 230–3
  steam distillation 65
  water removal 109, 115
Ether 316
Ethers 316–30
2-Ethoxyethanol 265
Ethyl acetate
  liquid-liquid extraction 121
  properties/recovery 335–9
  recovery 28–9
  vapour pressure reduction 74–5
  VLE 368–71
Ethyl alcohol 230

382 *Index*

Ethylbenzene 93, 220–2, 225
Ethyl cellosolve 265–7
Ethylene dichloride 282
Ethylene glycol 259–61
Ethylene glycol monobutyl ether 268
Ethylene glycol monomethyl ether 262, 265
Ethyl ether 316
Ethyl oxide 316
Evaporation
   heating systems 42–4
   solvent separation 23, 35
   time 195, 202–367
Evaporators
   agitated thin-film 69–72
   temperature 91
Exotherms 46, 59–60, 142
Expansion
   fractional distillation 54
Explosions 142–3
Explosive limits *see* Lower explosive limit;
   Upper explosive limit
External jacket
   batch distillation kettle 44
Extraction solvents 28, 30–1
Extractive distillation
   MEK 307
   solvent separation 97–104
   water removal 116–17

Feed points 51, 86
Feedstock screening/acceptance 158–9
Feedstock tanks 47, 57
Fenske equation 80–1
FF *see* Furfuraldehyde
Fire
   emergency procedure 167–9
   hazards 143–6, 202–367
First aid 166–7
Flammable inventory 141
Flammable limits 191
Flash distillation 48
Flash points 145, 171, 190–1, 202–367
FLE *see* Fractional liquid extraction
Float switch
   still safety 55
Fluxing residue 72–3
Foam formation
   fractionating columns 50
Forced circulation evaporator 68
Forklift trucks 152
Fouling
   fractionating columns 49
   heating surfaces 60–73
Fractional distillation 41–57, 79
Fractional freezing 126
Fractional liquid extraction (FLE) 123

Fractionating columns 48–53, 80
Fractionation
   water removal 107, 114
Freezing
   DMF drying 347
Freezing points 190, 202–367
Freon 113 *see* Trichlorotrifluoroethane
Fuel
   motor 210, 219, 225
   solvents 171–2
Fural 360
2-Furaldehyde 360
Furfural 360
Furfuraldehyde 360–4

Gas blanketing 56, 210, 219, 252
Gas curtains
   Airco process 20–1
Gas-freezing plant 162
Gas-liquid chromatography (GLC) 103
Gas phase
   solvent removal 7–21
Gas-phase adsorption 9
Gilliland correction 86
GLC *see* Gas-liquid chromatography
Glycol 259
Glycol ethers 45, 73, 259–71
Grain alcohol 230
Gravity decanter 111
Gravity separators 25

Handling
   entrainers 116
   solvents 148
Hazard classes 202–367
Hazards 137, 141–7
Hazchem number 202–367
Health hazards 202–367
Heat
   consumption 63–4
   exchangers 43–4
   heat of adsorption 11–12
   heat of combustion 197, 202–367
   heat of fusion 196
   heating surfaces
      continuous cleaning 69–72
      fouling 60–1
   heating systems, evaporation 42–5
   removal, carbon adsorption 13–14
   requirements, azeotropic drying/reflux ratios
      115
Henry's Law 35–7
n-Heptane 208–10
Hexalin 256
Hexamethylene 211

n-Hexane
    ED entrainers 104
    properties/recovery 205–7
    residue separation 74
    scrubbing liquid 15
    VLE 368–71
High-pressure distillation
    THF 329–30
Hildebrand solubility parameter 194–5, 202–367
Hot gas regeneration 11
Hot oil bath 67
Hot oil heating 45
Hot wells 42–3
Hydration 123–5
Hydrocarbons
    chlorinated 272–97
    ED entrainers 103
    fuel 171, 172
    properties/recovery 202–25
2-Hydroxybutane 250

IBA *see* Isobutanol
IDLH *see* Immediate danger to life and health
Ignition sources
    fire hazard 143–5
Immediate danger to life and health (IDLH) 146, 147, 193, 202–367
Impurities
    n-alkanes 30
    solvent recovery 184–5
    water removal 126–7
Incineration
    aqueous effluents 23, 33
    fluxing residue 72
    used solvents 2
    waste disposal 130–1, 133–4
Industrial methylated spirits (IMS) 230, 233
Inert atmosphere 143
Inert gas blanketing *see* Gas blanketing
Inhibitors
    see also Stabilizers
    carbon adsorption 11
    diisopropyl ether 322
    1,4-dioxane 325
    methylene chloride 274
    THF 330
    trichloroethylene 291
In-house recovery 178–80
Inorganic salts 37, 59
Inspections
    routine 164–5
Installation design/layout 140–1
Ion-exchange resins
    water removal 119
IPA *see* Isopropanol

Isoalkanes
    ED entrainers 104
Isobutanol 247–9
Isobutyl alcohol 240
Isopropanol (IPA)
    drying entrainers 113
    properties/recovery 240–2
    steam distillation 65
    water removal 109
Isopropyl alcohol 240
Isopropyl ether 320

Ketones
    carbon adsorption 12
    ED entrainers 103
    properties/recovery 298–315

Laboratory operating procedures 139–40
Land filling 129
Latent heat 11, 115–16, 195, 202–367
Layout
    installation 140–1
LEL *see* Lower explosive limit
Liquid expansion coefficients 190, 202–367
Liquid feed points 51
Liquid head 76
Liquid hold-up
    fractionating columns 53
Liquid-liquid extraction 121–3, 307
Liquid-phase adsorption 9
Liquid side streams 50
Liquid solvent thermal incinerators 129–31
Lithium chloride 125
Los Angeles Rule 66 2, 183
Losses
    solvents 184
Loss per transfer 195, 202–367
Lower explosive limit (LEL) 145, 172, 191, 202–367
Low-temperature condensation 7, 17–19
Luwa evaporator 70
Lyophilic membranes 31

Maintenance 161–5
MCB *see* Monochlorobenzene
MDC *see* Methylene chloride
Mechanical loads
    fractionating columns 53
MEG *see* Ethylene glycol; Monoethylene glycol
MEK *see* Methyl ethyl ketone
Membrane separation 31–2, 120–1
Merchant recoverers 173–5
Metals
    distillation equipment 41–2
Methanol
    acetone separation 95–6

Methanol, *continued*
  extractive distillation 102
  properties/recovery 226–9
  steam distillation 65
  THF separation 96
2-Methoxyethanol 262
Methyl acetate 102, 331–4
Methyl alcohol 226
Methylated spirits 230, 233
Methylbenzene 217
Methyl cellosolve 262–4
Methylchloroform 285
Methyl cyanide 356
Methylcyclopentane 213
Methylene chloride
  ED entrainers 104
  FLE 123
  liquid-liquid extraction 121–2
  methanol recovery 274
  properties/recovery 272–5
  water cleaning 39
Methylene dichloride 65, 96, 272
Methyl ethyl carbinol 250
Methyl ethyl ketone (MEK)
  incineration 134
  properties/recovery 302–7
  relative volatility 83
  salting-out 125–6
  separation 81, 85
Methyl glycol 262
Methyl isobutyl ketone (MIBK)
  properties/recovery 308–10
  relative volatility 83
  separation 85
4-Methylpentan-2-one 308
2-Methyl propan-1-ol 247
N-Methyl-2 pyrrolidone 313–15
MIBK *see* Methyl isobutyl ketone
Miscibility
  see also Solubility
  common solvents 95
  with water 194
Molar volumes
  ED entrainers 98
Molecular sieves
  Rekusorb process 12–13
  THF 330
  water removal 117–19
Molecular weight
  carbon adsorption 10
  solvent properties 202–367
Mole fraction 62–3, 301
Monochlorobenzene 295–7, 307
Monoethylene glycol
  see also Ethylene glycol
  FLE 123
  scrubbing liquid 15
Motor fuel 210, 219, 225

Naphthalenes 103
Nitrogen
  Airco process 19, 20, 21
  molecular sieve regeneration 119
NMP *see* N-Methyl-2 pyrrolidone
Nonane
  liquid-liquid extraction 121

Odour 76–7, 146–7, 192–3, 202–367
Oil prices 186
Operating procedure 137–69
Organic adsorbents
  water removal 119
Overflow line
  still safety 55
Oxygen
  fire hazard 143
Oxygenated solvents
  fuel 171
Oxygen demand 194, 202–367
Ozone
  destruction 185
  generation 183–5
  ozone depletion potential 185, 288

Paint 67, 133
Pallets 152
Paraffins
  ED entrainers 103
Partial condenser 87
Partition coefficients
  ethyl acetate recovery 28
  furfuraldehyde 363
  n-propanol 236
n-Pentane
  entrainer 96
  properties/recovery 202–4
  residue separation 74
  THF drying 108–9
Perchloroethylene 116, 292–4
Perk 292
Permeation membranes 31
Permits 161
Peroxides
  diethyl ether 318–19
  diisopropyl ether 322
  1,4-dioxane 325
  glycol ethers 267, 271
  tetrahydrofuran 329
Pervaporation
  MEK 307
  solvent separation 66
  water removal 31–2, 40, 120–1

Petroleum ethers 204
Phase separation
 solvent separation 96
 waste disposal 135
 water removal 111
Phenylethane 220
Physical properties 189–99, 201–367
Pipelines
 blockage 155
 cracking 204
 expansion 155
 frozen 213, 216
 labeling 154
Plastic tanks 153
Pollution limits 7–8, 17–18, 24
Potassium carbonate 123
Pressure
 absolute 75
 atmospheric 89–90, 92
 steam distillation 62, 63
 testing 164
Pressure distillation
 water removal 117
Pressure drop 49, 75
Pressure-relief devices 47
Process operations 159–61
Product tanks 56–7
Propan-2-ol 240
Propan-2-one 298
n-Propanol 109, 234–9
n-Propyl alcohol 234
Protection 165–6
Pump selection 155
Pyridine
 properties/recovery 352–5
 recovery 181
 salting-out 125–6
 solvent extraction 30
 water removal 109
M-Pyrol 313

Raoults'Law 35–7
Reboiler
 temperature 91
Recovery notes 201–367
Recycling 1
Redistributors 51
Reflux ratio 85–6, 107, 115
Regeneration
 scrubbing liquids 16
Regulations
 environmental 186–7
Rekusorb process 12–13
Relative volatility
 ethanol distillation 232
 extractive distillation 97–8

solvent properties 197
solvent separation 81–3
steam distillation 93–4
vacuum distillation 89–91
VLE diagrams 198–9
Reodorants 77
Residues 160
solvent separation 59–77
Residue tank 55–6
Retrofitting
 fractionating columns 53
Running tanks 56

Safety
 see also Operating procedure
 condensers 46–7
 still design 55
Salting-out 125–6, 306
Salts
 ethanol distillation 232
 inorganic 37, 59
Saturated vapour concentration (SVC) 193,
  202–367
Scraped-surface evaporators 44
Scrubbing liquids 15–16
Sea dumping 129
Sewers
 discharge limits 24
Sextone 311
Shell and tube heat exchangers 43–4
Side streams
 fractionating columns 50
Silica gel
 water removal 118–19
Single-column distillation
 DMF 346
Single-column drying
 sec-butanol 254
Site access 140
SI units 372–4
Smell see Odour
Sodium hydroxide see Caustic soda
Solubility
 n alkanes
 n-butanol 245
 sec-butanol 252
 butyl Cellosolve 270
 furfuraldehyde 362
 Hildebrand parameters 194–5
 measure 27
 MEK 305
 methanol 229
 methyl acetate 333
 MIBK 310
 pyridine 355
 solvent properties 202–367

Solubility, *continued*
  solvents 105–6
  THF 330
Solutes
  solvent removal 3
Solvent extraction 26–31
Specific gravity 30, 190, 202–367
Specific heat 196, 202–367
Splash filling 154
Stability
  entrainers 111
  scrubbing liquids 16
Stabilizers
  see also Inhibitors
  chloroform 278
  1,2-dichloroethane 284
  furfuraldehyde 362
  perchloroethylene 294
  1,1,1-trichloroethane 287
Staff considerations 138
Standing orders 158, 159
Steam distillation 62–6, 93–4
Steam heating 42–4, 47
Steam raising
  waste solvents 133
Steam regeneration 11
Steam stripping 30, 31, 38, 40
Still kettle design 54
Stills
  charging 159–60
  gas-freezing plant 162
  washing 160–1
Stock tanks 56–7
Storage
  bulk 152–5
  fractional distillation 54–7
  solvents 148–58
  tanks 152–3
Stripping column 16–17
Sulphuric ether 316
Sumps
  entry into 163–4
Supercritical fluids 123
Superheating 68
Supersaturation 67–8
Suppliers
  return to 180–1
Surface tension 202–367
Sussmeyer process 66–7
Sussmeyer solvent recovery unit 61
SVC *see* Saturated vapour concentration

TA Luft solvent discharge limit 7, 17–18
Tankers
  loading/unloading 156–8

Tanks
  cleaning 164
  gas-freezing plant 162
  running 56
  storage 55–7, 152–3
  vent inspection 165
Tar 72–3
Temperature
  carbon adsorption 10
  exotherms 59–60
  explosions 142
  low-temperature condensation 17–19
  reboiler 91
Ternary azeotropes
  n-butyl acetate 342
  chloroform 278
Ternary solubility diagrams
  sec-butanol 254
  ethyl acetate 338
  methylene chloride 275
  n-propanol 238
Tetrachloroethene 292
Tetrachloroethylene 292
Tetrahydrofuran
  ED entrainers 104
  methanol separation 96
  properties/recovery 327–30
  water removal 108–9
Tetramethylene oxide 327
Theoretical oxygen demand (ThOD) 194
Thermal expansion coefficient
  n-pentane 204
Thermal incineration 130–1, 133–4
Thermal information 202–367
THF *see* Tetrahydrofuran
Thin-film evaporators 44, 76
ThOD *see* Theoretical oxygen demand
Threshold limit value (TLV) 193, 202–367
TLV *see* Threshold limit value
Toll recovery 175–80
Toluene
  ED entrainers 103
  properties/recovery 217–19
  residue separation 74
  steam distillation 62–3, 93
Toluol 217
Toxic hazards 146–7
Toxicity
  entrainers 110
Tray requirements 82–3
1,1,1-trichloroethane 185, 285–8
Trichloroethylene 62–3, 289–91
Trichloromethane 276
1,1,2-trichlorotrifluoroethane 104, 185
Triclene 289
Trike 189, 289

Turndown
  fractionating columns 52

UEL *see* Upper explosive limit
Ullage 151, 154
Upper explosive limit 145, 191, 202–367
U-tube reboiler 44

Vacuum charging 159
Vacuum distillation 88–93
Valves 47, 154–5
Vapour density 194, 202–367
Vapour distillation 66–7
Vapour feed points 51
Vapour inhalation 146
Vapour-liquid equilibrium (VLE)
  binary mixtures 368–71
  immiscible solvents 64, 65
Vapour-liquid equilibrium (VLE) diagrams
  acetone 300
  n-butanol 253
  sec-butanol 253
  butyl cellosolve 271
  cyclohexanol 258
  1,4-dioxane 325
  DIPE 322
  DMF 345
  ethyl acetate 337
  explanation 198–9
  isobutanol 249, 253
  isopropanol 242
  MEK 305
  methanol 228
  methyl acetate 333, 334
  methylene chloride 275
  n-propanol 239
  relative volatility 82
  THF 329
Vapour pressure
  absolute pressure effects 89
  absorption 15
  extractive distillation 97

Raoult's Law 35–7
reduction 73–6
salt solubility 126
scrubbing liquids 16
solvent properties 202–367
toluene separation 74
Vapour side streams 50
Venting 161
Vents
  emissions 156
  safety 46, 47
  scrubbing 160
Vessels
  design 54–5
  entry into 163
Viscosity 16, 202–367
Volatile organic compounds (VOC) 1, 23, 183

Waste disposal 129–35
Waste water
  solvent removal 23–40
Water
  see also Solubility
  azeotropes 107–8
  carbon adsorption 11
  cooling, sources 45–6
  low-temperature condensation 18–19
  properties 365–7
  purity 24
  relative volatility effects 98
  removal 105–28
  scrubbing liquid 15
  solvent removal 3
  solvent separation 23–40
Wetting
  fractionating columns 52
Wiped-film evaporator 71, 76
Wood alcohol 226

Xylenes 223–5
Xylol 223